Multiaged Silviculture

Multiaged Silviculture

Managing for Complex Forest Stand Structures

Kevin L. O'Hara
Professor of Silviculture, University of California–Berkeley

OXFORD
UNIVERSITY PRESS

Great Clarendon Street, Oxford, OX2 6DP,
United Kingdom

Oxford University Press is a department of the University of Oxford.
It furthers the University's objective of excellence in research, scholarship,
and education by publishing worldwide. Oxford is a registered trade mark of
Oxford University Press in the UK and in certain other countries

© Kevin L. O'Hara 2014

The moral rights of the authors have been asserted

First Edition published in 2014

All rights reserved. No part of this publication may be reproduced, stored in
a retrieval system, or transmitted, in any form or by any means, without the
prior permission in writing of Oxford University Press, or as expressly permitted
by law, by licence or under terms agreed with the appropriate reprographics
rights organization. Enquiries concerning reproduction outside the scope of the
above should be sent to the Rights Department, Oxford University Press, at the
address above

You must not circulate this work in any other form
and you must impose this same condition on any acquirer

Published in the United States of America by Oxford University Press
198 Madison Avenue, New York, NY 10016, United States of America

British Library Cataloguing in Publication Data

Data available

Library of Congress Control Number: 2014936881

ISBN 978–0–19–870307–5

To my wife Jan, who puts up with so much, but is always supportive.

Preface

This book describes the scientific rationale for the silviculture of complex forests. Complex forests include stands with mixed species and multiple age classes. They include diverse structural features such as wildlife habitat or unusual structural elements. Complex multiaged forests range from the simple two-aged structure that might result from a shelterwood with reserve trees to the complex, many-aged forest resulting from traditional selection silviculture. There is science that supports both the need for complex stands and the silviculture to manage them. This silviculture is highly varied and requires a strong foundation in ecological science. However, silviculture is based on more than ecological science: it exists within constraints that limit what is operationally possible and economically feasible.

The intent of this book is to address the need for a strong foundation and justification for designing and implementing multiaged stands. This silviculture can be less regimented and more free-form than the silviculture used to implement forms of multiaged silviculture in the past. The motivation for this book originated from the realization that the preponderance of advanced instructional information for college-level teaching was focused on evenaged silviculture. Although general silvicultural texts present a balanced approach to all types of silvicultural systems, advanced texts have focused more on even-aged silviculture. This is particularly true in the area of regeneration practices, where several individual books are almost exclusively on even-aged forests (Cleary et al. 1978; Lavender et al. 1990; Duryea and Dougherty 1991; Hobbs et al. 1992). There are also several books featuring plantation forest management (Savill and Evans 1986; Evans and Turnbull 2004; West 2006).

This book can serve as a supplemental text in introductory silvicultural classes. Advanced silviculture courses can use this book to address complex stand structures, or along with texts on even-aged silviculture. It is also intended to serve as a reference book on advanced silvicultural concepts for managers and scientists.

The book begins with an introductory chapter and a chapter on the history and origin of multiaged silvicultural systems (Chapter 2). Chapters 3 through 5 present the dynamics of multiaged stands, beginning with disturbance dynamics, development of multiaged stands, and gap dynamics and ending with a description of the management of multiaged stands. Chapter 6 introduces basic multiaged management systems. Stocking control is presented in Chapter 7 as the core area for designing and maintaining multiaged stand structures. Basic concepts related to regenerating and tending of multiaged stands are presented in Chapters 8 and 9.

The next section of the book covers a variety of specific issues related to managing multiaged forests. Transformations, or treatment regimes that move stands from even-aged to multiaged structures are presented in Chapter 10. Chapter 11 discusses using multiaged silviculture to achieve objectives such as timber, biodiversity, or carbon storage. Concepts related to estimation of site productivity, growth and yield, optimization of stand structure, and others are presented in Chapter 12. Chapter 13 discusses the relative volume growth and economic productivity of multiaged stands. Silviculture affects the genetic structure of forests, and multiaged stands have been criticized for potentially damaging genetic structure. These issues are addressed in Chapter 14. Wind, insects, and pathogens, as well as treatment options related to these threats, are discussed in Chapter 15. Social drivers affecting both the management and acceptance of multiaged structures are presented in Chapter 16. Finally, an epilogue provides a synthesis of the overriding issues affecting the perceptions of multiaged silviculture and the role of multiaged silviculture as we move forward in an era of uncertainty over climate, invasives, and the role of forests in societies undergoing complex changes.

I'm indebted to the many people who made suggestions and offered encouragement to complete this book. Helpful chapter and section reviews were provided by John-Pascal Berrill, Kathleen Cahill, Rolf Gersonde, John Helms, Steve Jack, Dryw Jones, Christopher Keyes, Susan Kocher, Erkki Lähde, Olavi Laiho, David Larsen, Kathryn McGown, Hiromi Mizunaga, Linda Nagel, Lakshmi Narayan, Benjamin Ramage, Tudor Stancioiu, Kristen Waring, and Robert York. Frieder Schurr provided the group selection diagram from Blodgett Forest in Chapter 6. There were also many anonymous reviewers who provided helpful suggestions. Although I have attempted to develop a book that is correct and complete, there are unavoidably errors. I hope that readers will notify me of mistakes and make suggestions for improvement.

Kevin L. O'Hara
Professor of Silviculture
University of California–Berkeley

Contents

1 Introduction — 1
  1.1 Multiaged silviculture — 1
  1.2 Terminology — 2
  1.3 Moving forestry forward — 3

2 History of multiaged silviculture — 5
  2.1 Introduction — 5
  2.2 Origins of multiaged approaches — 5
  2.3 Cyclic patterns — 6
  2.4 Recent trends — 6
  2.5 Synthesis — 10

3 Disturbance dynamics of multiaged stands — 11
  3.1 Introduction — 11
  3.2 Disturbances and age structures — 11
  3.3 Disturbance effects and silviculture — 12
  3.4 Disturbance regimes — 12
  3.5 Disturbance types — 16
  3.6 Synthesis — 21

4 Dynamics of multiaged stands — 22
  4.1 Introduction — 22
  4.2 Dynamics of light-limited systems — 22
  4.3 Dynamics of moisture-limited systems — 27
  4.4 Tree and stand architecture — 29
  4.5 Growth patterns — 32
  4.6 Mortality — 33
  4.7 Synthesis — 35

5 Dynamics of forest gap and group openings — 37
  5.1 Introduction — 37
  5.2 Gap dynamics — 38
  5.3 Gap studies—gap variability and implications for management — 44
  5.4 Gap replacement models — 47
  5.5 Synthesis — 47

6 Multiaged management systems — 48

- 6.1 Introduction — 48
- 6.2 Multiaged systems — 50
- 6.3 Synthesis — 57

7 Multiaged stocking control — 59

- 7.1 Introduction — 59
- 7.2 Stocking control concepts — 59
- 7.3 Multiaged stocking control — 68
- 7.4 Synthesis — 81

8 Regenerating multiaged stands — 84

- 8.1 Introduction — 84
- 8.2 Natural regeneration — 84
- 8.3 Artificial regeneration — 93
- 8.4 Cultural treatments — 95
- 8.5 Synthesis — 97

9 Tending multiaged stands — 98

- 9.1 Introduction — 98
- 9.2 Thinning treatments — 98
- 9.3 Release treatments and vegetation control — 102
- 9.4 Pruning treatments and wood quality — 102
- 9.5 Fertilization and forest nutrition — 104
- 9.6 Prescribed burning — 105
- 9.7 Synthesis — 106

10 Transformations to multiaged stand structures — 107

- 10.1 Introduction — 107
- 10.2 The silviculture of transformation — 108
- 10.3 Sustainability and transformation — 111
- 10.4 Constraints on transformation — 114
- 10.5 Synthesis — 116

11 Managing multiaged stands for diverse objectives — 117

- 11.1 Introduction — 117
- 11.2 Timber—quantity and quality — 117
- 11.3 Biodiversity and wildlife habitat — 118
- 11.4 Climate change — 120
- 11.5 Carbon — 120
- 11.6 Restoration — 121
- 11.7 Sustainable forest management — 123
- 11.8 Other ecosystem services — 124
- 11.9 Synthesis — 124

12 Growth projection in multiaged stands · **125**

12.1 Introduction · 125
12.2 Site quality assessment · 125
12.3 Growth projection · 128
12.4 Growth projection tools · 129
12.5 Optimization tools · 132
12.6 Synthesis · 133

13 Volume and economic production of multiaged stands · **136**

13.1 Introduction · 136
13.2 Stand-level volume production relationships · 136
13.3 Volume production of even-aged and multiaged stands · 139
13.4 Economic productivity · 142
13.5 Synthesis · 143

14 Genetics and multiaged silviculture · **145**

14.1 Introduction · 145
14.2 Genetic conservation and silviculture · 145
14.3 Reproductive processes · 147
14.4 Genetics and natural disturbances · 150
14.5 Genetics and multiaged silviculture · 151
14.6 Synthesis · 156

15 Multiaged structures and stand health · **158**

15.1 Introduction · 158
15.2 Wind and other storm damage · 158
15.3 Insects, pathogens, and invasives · 160
15.4 Fire · 164
15.5 A dynamic climate · 166
15.6 Management options · 166
15.7 Synthesis · 167

16 Social justifications for multiaged silviculture · **169**

16.1 Introduction · 169
16.2 Aesthetics · 169
16.3 Naturalness · 171
16.4 Social acceptability of multiaged silviculture · 174
16.5 Naturalness and the "easy way out" · 174
16.6 Synthesis · 175

Epilogue · **177**

List of Species · 180
References · 182
Index · 211

CHAPTER 1

Introduction

1.1 Multiaged silviculture

Developing and maintaining forests with complex stand structures has become a worldwide priority. Changing social values are motivating this change by demanding multifunctional forests that are also more natural. This interest is driven by the recognition that complex forest structures provide a variety of ecosystem values and services. Perhaps, unlike any other area of forestry, the expanding interest in multiaged systems has corresponded with an international movement to promote and develop alternative approaches to forest management. This is largely driven by a backlash against even-aged, and particularly plantation forestry, because these types of management are generally assumed to simplify structures and reduce diversity. There is also a strong perception that multiaged or uneven-aged systems are more natural forms of forest management. The emergence of certification systems also has promoted greater stand complexity in forestry.

The central thesis of this book is that **multiaged stands**, or stands with two or more age classes (Helms 1998), offer a great diversity of management options for contemporary forestry and for achieving different forms of stand-level complexity. Complex stands may include multiple age classes or mixed-species stands. They may include snags or forest floor biomass in the form of coarse woody debris. **Stand structures**—or the horizontal and vertical distribution of components in a stand including the heights, diameters, crown layers, shrubs, herbaceous plants, standing and down wood (Helms 1998)—may vary from two-aged/two-strata stands to stands where many canopy strata and species are intermingled in both horizontal and vertical dimensions. Multiaged stands encompass relatively simple two-aged stands to very complex stand structures formed by traditional uneven-aged silviculture. They can range from intensively managed stands to stands where management interventions are light and infrequent. Management objectives for this suite of options can include enhancing biodiversity, timber production, carbon storage, representing natural processes, or enhancing aesthetics. Likewise, the tools to achieve these objectives can vary from very intensive regeneration operations to simply letting natural processes determine trajectories related to stand development, regeneration, or mortality. The "toolbox" for multiaged silviculture is generally the same as for even-aged silviculture; however, it is how these tools are used that results in the fundamental differences in the stand structures produced by these two approaches. This highly diverse and flexible set of tools provides great latitude in how forests are managed. Forests may be managed to provide only individual ecosystem services such as timber or water resources. Alternatively, they may be managed as multiple benefit forests that provide many ecosystem services and in which efforts are also made to maintain ecosystem processes. In either case, the broad rubric of multiaged silviculture offers many alternatives to meet objectives.

Even-aged silviculture represents a standard in forestry that has been refined by centuries of scientific and operational progress. There are excellent texts focused on even-aged and plantation silviculture (Savill and Evans 1986; Evans and Turnbull 2004; West 2006) and others that focus on regeneration of primarily even-aged stands (Cleary et al. 1978; Lavender et al. 1990; Duryea and Dougherty 1991; Hobbs et al. 1992). Because even-aged silviculture is much more developed and extensively practiced, multiaged silviculture has traditionally viewed even-aged silviculture as a reference condition from which it is compared (e.g., see Assmann 1970). As different forms of management, both even-aged and multiaged silviculture are used on our forest landscapes and possess a range of potential advantages and disadvantages. In this book, various forms of even-aged stands are often used as a reference condition with differences discussed and even highlighted. All of these management options represent potential treatment alternatives, or methods of forest stewardship, on our forested landscapes. Most importantly, in isolation from even-aged silviculture, the science and practice of multiaged silviculture is much weaker.

The range of silvicultural systems has traditionally been viewed as a dichotomy between even-aged and multiaged systems. This is consistent with recent and traditional classifications of silvicultural systems (Matthews 1989; Smith et al. 1997; Nyland 2002) that separate even-aged from uneven-aged systems. However, there are many options for silvicultural systems that use common tools and procedures. Rather than viewing them as alternatives, they actually exist on a continuum of stand management strategies (Bradshaw 1992; O'Hara et al. 1994). Whereas it is useful to have a classification of these systems, this classification should neither define nor constrain the organization of treatments used to manage forests or their outcomes. The traditional classification may have implied a greater suite of options associated with even-aged systems; however, the current reality is that there are many options to create and maintain multiaged stand structures as well. By including all stand structures with two or more age classes, the term multiaged is inclusive of a great range of silvicultural options.

The proposed alternatives to even-aged forestry originate from a variety of locations and have taken on a great diversity in names (Table 1.1). These alternative approaches vary from being descriptive to strongly based scientific appeals for new philosophies and foundations for stand management. However, what is lacking from all of these is a detailed and scientifically founded synthesis on the silviculture of creating and maintaining complex forest structures.

Silviculture is traditionally a stand-level art and science. It provides the tools to modify stands by directing stands on pathways or trajectories to different structures. This book presents the science of multiaged silviculture at the stand level. Stands form the building blocks for larger landscapes, and are therefore critical elements of meeting large-scale objectives (O'Hara and Nagel 2013). However, this book is intentionally limited to stand-level management with only minor references to meeting objectives at larger scales.

This book is written on the foundation that there are many commonalities worldwide between forest types related to stand dynamics. This was point advanced in Oliver and Larson's seminal book *Forest Stand Dynamics* (1990) and further developed in Oliver (1992). Wadsworth (1992) also detailed how tropical silviculture was derived from temperate silviculture and made the point that there are overriding similarities in silviculture regardless of species or ecosystem. The processes of regeneration, competition for light and moisture, density-related mortality, and many others are common to all forests in various degrees. The concepts presented in this book are therefore meant to apply to boreal, temperate, and tropical forests, to wet and dry forests, and to forests of shade-tolerant or shade-intolerant species.

1.2 Terminology

The term "multiaged" was chosen as the focus for this book because, by including two-aged stands,

Table 1.1 Alternative silvicultural terms for systems that create multiaged stands.

Name	Source
Selection (group and single tree)	Generic (see Matthews 1989)
Plenterwald	Generic (see Schütz 2001a)
Polycyclic	Generic
Green-tree retention	Franklin 1989, North et al. 1996
Multicohort	Oliver and Larson 1990
Multiaged	O'Hara 1996
Variable retention	Mitchell and Beese 2002, Beese et al. 2003
Dauerwald	Möller 1922, Huss 1990, Helliwell 1997, Schabel and Palmer 1999

it is more comprehensive and inclusive than the more traditional term "uneven-aged." Multiaged encompasses much of contemporary silviculture that leaves significant retention or reserve trees after harvest treatments. Traditional forms of uneven-aged silviculture are also included. These include single tree selection and group selection systems, which leave patterns of residual vegetation that may range from homogeneous to heterogeneous or patchy. Also included are emerging systems such as variable retention and other less formalized systems that attempt to emulate natural disturbances with live and dead tree retention. This suite of systems could also be termed "irregular forestry" (i.e., Susse et al. 2011), because they promote variation in spatial patterns, tree sizes, and age structures (Schütz 2002a, 2002b). **Even-aged** systems—also called rotation forest management, age class forests, or monocyclic systems—manage a single age class of trees over a rotation. Forest plantations are usually even aged and almost always established through planting, or less commonly, artificial seeding (Helms 1998). **Planted forests**, or stands with a component of trees from artificial regeneration, can be either even aged or multiaged.

The assumption upon which this book is based is that the broad array of stand structures with multiple age classes have more similarities than differences regardless of whether they have two or many age classes. It also assumes that the differences between multiaged and even-aged stands are the primary determinant of management strategies. Regeneration is often a limiting process in silviculture, and the regeneration process is more complicated under a stratum of existing trees or in openings where surrounding trees may influence regeneration. Hence the difference in regenerating new age classes of trees also separates multiaged structures from systems that regenerate trees under conditions approaching full sunlight.

Multiaged stands often include mixtures of tree species. Attaining greater tree or plant species diversity is often a goal of attaining complex forests. Because it is impossible to separate multiaged silviculture from mixed species compositions in practice, this book not only focuses on multiaged stands but also discusses species mixtures because of their essential contributions to stand complexity.

Other terminology follows *The Dictionary of Forestry* (Helms 1998) as the standard English language source for forestry terms. However, because the surge in interest in multiaged forms of silviculture is a global movement, there are a great many terms for various forms of multiaged silvicultural that have come into common usage in recent years (Table 1.1). These include traditional single tree and group selection, and many others that are descriptive or attempt to differentiate themselves from more traditional approaches. These newer approaches are not specifically discussed in this book because they are seen as variations on theme, or different combinations of treatments that result in relatively complex stand structures. Hence, multiaged silviculture can be viewed as a series of low-intensity treatments using the terminology of Duncker et al. (2012).

The calls for new and more natural forest management have also generated a suite of new names that describe these strategies. Since many of these movements call for greater use of multiaged systems, there is much overlap in the names for the movements and those of the management strategies (Table 1.2). As a result, the names for the general management strategies and the new systems have become confusing, and it can be difficult to separate the philosophies of the new strategies from the systems used to manage individual stands.

1.3 Moving forestry forward

Forestry is facing unprecedented challenges due to a variety of societal and environmental changes. Society is placing increasing demands on forests as places to live, recreate, or simply to find solace, as well as a traditional source for wood products. A prerequisite for many of these uses is esthetic quality, and there are also increasing demands for forests that are perceived as natural. The inherently long-term nature of managing forests is also complicated by anticipated changes in global climate. As a result, there is a need to develop forests that are resistant to disturbance and resilient to changes in climate, and this need requires that management anticipates long-term environmental changes and their effects on forests. Multiaged silviculture is well suited to address these challenges (O'Hara

Table 1.2 Labels used to describe alternative forms of forest management (adapted from O'Hara 1998, 2001; Pommerening and Murphy 2004; Helms and Porter 2009).

Name	Source
New forestry	Franklin 1989
New perspectives	Kessler et al. 1992
Ecosystem management	Salwasser 1994
Sustainable forestry	Maser 1994
Restoration forestry	Pilarski 1994
Excellent forestry	Robinson 1994
Close-to-nature forestry	Mlinšek 1996
Nature-oriented silviculture	Lähde, Laiho, and Norokorpi 1999
Diversity-oriented silviculture	Lähde, Laiho, and Norokorpi 1999
Ecoforestry	Drengston and Taylor 1997
Continuous cover forestry	Yorke 1992, Garfitt 1995, Mason et al. 1999, Helliwell 2013
Holistic forestry	O'Keefe 1990, Koch and Skovsgaard 1999
Ecological forestry	Seymour and Hunter 1999
Ecological silviculture	Benecke 1996
Close-to-nature silviculture	Kenk and Guehne 2001
Common sense forestry	Morsbach 2002
Back to nature	Gamborg and Larsen 2003
Positive impact forestry	McEvoy 2004
Near-natural forestry	Larsen 2005
Nature-based forestry/forest management	Diaci 2006, Larsen and Nielsen 2007
Irregular forestry	Susse et al. 2011
Pro Silva	Pro Silva 2012
Retention forestry	Gustafsson et al. 2012, Lindenmayer et al. 2012

and Ramage 2013). It recognizes the need and value of complexity and the importance of management approaches that provide flexibility for individual stand-level solutions (Messier et al. 2013). It also provides options for management where overlapping generations of trees within a single stand can provide options for gradual change and resiliency to changes in social demands or the environment.

Along with the surging interest in multiaged management strategies has been an increase in research into the ecology and management of complex forests. This research has revealed many things. Recent findings correct many historical misconceptions about silviculture of complex forests systems, such as claims that complex systems are less productive, dysgenic, or result in stand health problems. There is also a growing consensus that the management of multiaged stands is a sound strategy for a world that is placing greater demands on forests while adapting to a changing climate. Most important is the growing realization that making silviculture effective involves adjusting desired stand structural goals. Rather than a finite group of possible stand structures, there are many possibilities that vary with local environments and societal demands. Multiaged stands therefore encompass a vast array of possible structures and management options. Multiaged stand management is also more than simply leaving structural elements: it involves balancing tradeoffs between different ecosystem services, and managing light levels, competition, and regeneration to provide for long-term sustainability.

This book updates the science of multiaged silviculture by synthesizing available research and describing a broader application for multiaged systems. Multiaged silviculture encompasses a vast array of many structures and options for management. In the past, it has been viewed more narrowly, as though there were a single optimal solution that would be universally best suited to a wide range of conditions and objectives. This book indicates there are many pathways to sustainability and alternative ways to meet management objectives. It specifically addresses issues that have hindered the use of multiaged systems in the past, and those issues that will allow multiaged systems to be a major part of modern forestry.

CHAPTER 2

History of multiaged silviculture

2.1 Introduction

The history of any field of science or management is important for understanding its current state. This is true for silviculture, which has a long and interesting history and is central to forestry. Rather than providing a complete review of silvicultural history, this chapter summarizes important historical trends related to multiaged silviculture, with several specific examples. It focuses more on how history has shaped the current state of multiaged silviculture rather than attempting to be a comprehensive historical review. More complete histories of silviculture can be found in Fernow (1911), Troup (1928), Makkonen (1975), Puettmann et al. (2009), and reviews of multiaged silvicultural history in Mustain (1978), Schütz (1994), O'Hara (2002), Pommerening and Murphy (2004), and Schütz et al. (2012). Later chapters in this book explore some of the issues that have historically influenced multiaged silviculture, such as relative productivity, effects on genetics, and stand health implications.

2.2 Origins of multiaged approaches

One might imagine that the first timber harvests in any region were the removals of individual trees. Cutting, moving, and processing individual trees was likely exceedingly labor intensive, thereby necessitating careful outlays of labor and judicious selection of which trees to cut. Although these initial harvests probably paid little attention to regeneration or sustainability, they were a form of selection harvest within the broad rubric of multiaged silviculture. Multiaged silviculture—in any form or under any of its many names—did not originate in one place, nor was it only one thing. Multiaged silviculture does not have an inventor, or a place and date of origin. Instead it was a logical, but highly varied, way to efficiently manage a forest when the technology was in its infancy.

Published accounts of multiaged silviculture date to the Middle Ages in the Mediterranean (Thirgood 1981) and at least to a similar period in Japan (Totman 1989). Other references to early silviculture date back several thousand years (Fernow 1911; Troup 1928; Heske 1938) and encompass many ancient civilizations. Indeed, as forestry is, in part, the outcome of shortages, the independent development of forestry in so many different places is not surprising. Shortages or scarcity, in the case of forests, may refer to many forest-based resources such as timber, forage, places for hiding, food sources, or protection of water resources. Whereas the origins of forestry date back millennia, the development of formal silviculture and methods of regulation, at least as we know them, largely originated in recent centuries. The contemporary silviculture alternatives that lead to relatively complex stand structures are derived from these early practices. Regenerating the forest probably became a concern as timber became more scarce, marking the beginning of sustainable approaches to multiaged silviculture. Mustain (1978) noted that silviculture is often thought to begin when the need for regeneration becomes paramount. But the act of removing trees and affecting stand structures is part of the practice of silviculture.

Much of contemporary silviculture originated in central Europe (Troup 1928; Pommerening and Murphy 2004; Puettmann et al. 2009; Bončina 2011). Although regeneration methods in silviculture can be viewed as a continuum of options (O'Hara et al. 1994), the way this continuum has been segmented

Multiaged Silviculture. Kevin L. O'Hara.
© Kevin L. O'Hara 2014. Published 2014 by Oxford University Press.

into individual regeneration methods is largely based on developments in central Europe. Multiaged silviculture also has strong roots in central Europe. Schütz et al. (2012) described the Plenter system as the "archetype" for single tree removal systems throughout the world. The Plenter system originated as a means of single tree selection. Largely discredited at times, the system was illegal in some countries. The root of the term Plenter is found in "plunder" because of the early interpretation of the method. However, it was well suited for use in the central European forests, which contain silver fir, European beech, and Norway spruce, because of the shade tolerance of these species, the ease of obtaining regeneration, and the inherent productivity of forests consisting primarily of species with excurrent growth forms (Puettmann et al. 2009; Schütz et al. 2012). Even in central Europe, the Plenter system has not been widely applied due to a variety of social pressures (Schütz et al. 2012). Indeed, the implementation of multiaged systems throughout Europe has never been common in terms of the total area applied, and it has experienced cycles of favor and disfavor (Pommerening and Murphy 2004).

Contemporary silviculture also has roots in other parts of the world where local species and environments drive the development of silvicultural practices. Silviculture has a tradition of being highly variable based on regional differences in environment and species. Methods based by local practitioner experience often are designed to meet the limitations of local environments as well as the requirements of species. However, these systems are often based on practitioner experience and may be poorly documented in the scientific literature.

The influence of the Plenter system from central Europe as the "archetype" (Schütz et al. 2012) for the development of systems around the world cannot be understated. H. A. Meyer, an immigrant from Switzerland, is generally credited with bringing the Plenter system to North America (O'Hara 2002). Along with the selection concepts of the Plenter system, Meyer (1943, 1952; Meyer and Stevenson 1943) was a strong proponent of reverse-J or negative exponential diameter distributions as stocking targets for uneven-aged stands. Meyer was also the first to describe these distributions as "balanced." Fujimori's (2001) Japan-oriented silviculture text also cites the central European origins of single tree selection.

2.3 Cyclic patterns

Similar patterns of cyclic changes in the use of silvicultural systems are apparent worldwide. Over the last century, multiaged stands were in common use; then, a strong shift to even-aged systems occurred in the mid-twentieth century. The pendulum has begun to swing back in recent decades, and now there is a renewed interest in management for complex stand structures and emulating natural processes. The catalysts for recent cycles have largely been poor applications of multiaged silviculture or poor interpretations of science that are not exclusive to either even-aged or multiaged silviculture (Box 2.1). These trends gained inertia from political actions that often worked to ban certain practices. For example, the cycles in Europe associated with the Plenter system were largely due to poor interpretations of the science of this emerging system (Pommerening and Murphy 2004; Puettmann et al. 2009; Schütz et al. 2012). In Scandinavia, the "green lie" began with poor interpretations about multiaged systems (Box 2.2). In the United States, several controversies over clearcut systems in the 1960s and 1970s were catalysts for a resurgence of more complex systems. These cases in the United States, most visible in western Montana and West Virginia (Burk 1970; Fairfax and Achterman 1977; Dana and Fairfax 1980; Swanson 2011), were examples of foresters' overzealous efforts to manipulate environments to implement even-aged systems (Figure 2.3). The resulting backlash from the environmental community was enormous (e.g., Wood 1971; Fritz 1989; Devall 1993).

2.4 Recent trends

Silvicultural science has made great advances in recent decades. These advances include new developments in silvicultural technology and a greater ecological understanding of stand and landscape dynamics. For example, the technology of plantation management has provided for highly productive and predictable stand growth. Research has also demonstrated the importance of maintaining

Box 2.1 The "selective" cutting controversy in coast Douglas-fir

One of the most influential events affecting the implementation and acceptance of multiaged silviculture in the United States was the controversy related to "selective" cutting in coast Douglas-fir. This controversy is an example of the potential problems with silviculture terminology, and how a perceived rigidity in treatment design and implementation can lead to poor treatment application. Early in the twentieth century, forestry in the Douglas-fir region of coastal Oregon and Washington was overcoming problems with wildfire control and developing the technology to reforest cutover lands. The development of tractor-based log removals and truck transport in the 1920s and 1930s led to proposals to "selectively" harvest these productive timberlands. Prior to this time, harvesting had been primarily railroad-based, necessitating the nearly complete removal of all trees. However, an influential report by Kirkland and Brandstrom (1936) urged the implementation of forms of what they termed "selective" silviculture as part of an effort to achieve sustainability. The proposed management system was thought to be more economically efficient because only mature or declining trees would be removed and more sustainable because, with a complete system of roads, the lands could be harvested on a sustained yield basis. The system was also thought to produce high value timber and other forest values.

Beginning in the mid-1930s, "selective" harvests were carried out on federal ownerships. A series of study plots were established and monitored in old Douglas-fir forests that received cuts that removed, on average, approximately 37% of the merchantable volume (Curtis 1998). Most of these plots were in older, decadent stands (up to 600 years of age). Subsequent analysis of these plots indicated the treatments resulted in excessive windthrow and encouraged regeneration of shade-tolerant species such as western hemlock instead of Douglas-fir (Munger 1950; Isaac 1956; Figure 2.1). The system was largely abandoned in the region in favor of the wide-spread adoption of even-aged systems using dispersed clearcuts.

More details on these events are available from Issac (1956, 1967), Steen (1990), Curtis (1998), and O'Hara (2002). Kirkland and Brandstrom (1936) actually proposed

Figure 2.1 Western hemlock regeneration beneath a thinned Douglas-fir stand in western Washington, US.

Box 2.1 Continued

a more flexible system than was implemented in the region. They recognized the sunlight requirements of coast Douglas-fir and, in addition to the light selection treatments they were credited with, advocated treatments ranging from light salvage treatments to "clearcutting of small groups" of up to about 4 ha. The failure of these light treatments was, in part, a combination of poor terminology and the poor economic conditions of the time.

The term "selective" means many things and has largely been replaced with more precise terminology (Zingg 1999, O'Hara 2002). There were likely many reasons why these light treatments were more commonly implemented instead of the more flexible systems with larger openings advocated by Kirkland and Brandstrom. Perhaps it was greed, poor terminology, or poor economic conditions. The abrupt shift to even-aged systems is strong evidence of the cyclic nature of how forestry is practiced. In this case, the prominence of the Douglas-fir region within the federal land holdings of the US government caused these forms of silviculture to be discredited not just in coast Douglas-fir, but throughout the United States. It has only been in recent decades that multiaged practices have once again become more common.

Box 2.2 The "green lie" in Scandinavia

In boreal forests in Scandinavia, multiaged approaches were relatively common until the publication of a strongly negative article in 1916 (Barth 1916). This article by an influential Norwegian forester described the ruin of forests in Scandinavia due to the common practice of "selective" cutting (Lie et al. 2012). Barth cited a pattern of harvest exceeding growth that was attributed to the preferential cutting of larger trees. Following Barth's article, the selective cutting era of Scandinavian forestry became known as the "green lie" and was replaced by even-aged systems.

Lie et al. (2012) examined records of stand structure and harvesting for a large property in southeast Norway to document patterns of cutting from the period encompassing the late 1800s to the time of Barth's article. Their study used a variety of stand records, stand reconstruction, and tax records of larger spatial scales to describe stand structures and harvest patterns. They compared their results to current stand-level stocking and large-scale patterns of both unmanaged landscapes and landscapes managed with even-aged approaches. Stocking in the stands of the "green lie" era were understocked in comparison to the unmanaged landscape but exceeded the stocking in the landscapes managed with even-aged systems. A common short-coming of multiaged approaches is their potential to homogenize stand structures and create small management units (O'Hara 1998; Kerr 1999; Shifley et al. 2006; Franklin and Johnson 2011). However, Lie et al. (2012) found diverse stand structures and larger stand sizes than with the current even-aged approaches.

Finland went through a similar period when multiaged systems were essentially banned in the 1940s to favor even-aged systems. Pukkala et al. (2012) attributed this change to

Figure 2.2 Multiaged research stand in southern Finland.

the demands of the pulp and paper industry for small wood. Multiaged stands are still present in Finland (Figure 2.2)

HISTORY OF MULTIAGED SILVICULTURE 9

> **Box 2.2** *Continued*
>
> and national inventory data allows long-term comparisons of growth of even-aged and multiaged forests. Analyses of these data indicated the multiaged stands (uneven-sized) were more productive (Lähde et al. 1994a, b; Laiho et al. 2011). Separate analyses indicated multiaged options in Finland were also more productive and economical than comparable even-aged options (Chapter 12; Pukkala et al. 2010; Tahvonen et al. 2010).
>
> These cycles of silviculture in the boreal forests of Fennoscandia have probably been more extreme than elsewhere in the world. These cycles have resulted in wide swings in how science was interpreted and demonstrate the potential hazards in the misinterpretation of science. They also demonstrate the human tendency to shift dramatically from one extreme to another. In this case, the people of Fennoscandia are closely tied to their forests, and their intentions were always sound. However, with such an important resource and the long time periods needed to grow a forest and make such substantive changes in stand structure, these cases demonstrate the potential harm in such wide-ranging shifts in silviculture policy.

functioning ecosystems at variable spatial scales. At the stand level, this research has provided insights into the complex interactions between disturbances, stand growth, and stand structural development. Greater use of complex stand structures has emerged from this science because of the recognition that these structures support many ecosystem processes and many values not provided by simpler structures.

There is also a recognition that silvicultural practices should vary by management objectives and therefore by land ownership. Several authors have captured this idea by describing land allocations where some forest lands might be designated for

Figure 2.3 Clearcut and terraced hillside from western Montana, US.

preservation, others for intensive management, and others for a multiple benefit focus (Salwasser 1992; Seymour and Hunter 1992). These land allocations have demonstrated that a variety of silvicultural options are highly desirable on diverse forest landscapes that include many different stand structures (O'Hara et al. 1994). Although the pendulum has recently shifted toward a greater inclusion of more complex stand structures on forest landscapes, there is also a recognition that even-aged structures also provide important ecosystem and social values (Drever et al. 2006; Puettmann et al. 2009).

2.5 Synthesis

Multiaged silviculture has many origins. Any action where humans manipulated the forest to meet their need for wood or other resources was a form of silviculture. Their limited technology to remove trees would force decisions to carefully choose the trees to fell based on needs and the difficult transportation problems of moving the raw material to its place of need. The removals of individual trees may have been among the first silvicultural interventions and continues today as one of the most important.

There is no question that the individual tree selection of multiaged silviculture has been used opportunistically to remove more valuable trees without regard for future stand structure or stand health. This represents only one explanation for the rejection of these practices by the forestry profession or the political restrictions on these processes. The rejection of these practices has also been based on poor science and changing social values. However, even-aged silviculture has also been used opportunistically and has resulted in legislative restrictions on practices. Neither even-aged nor multiaged silviculture has license for the practice of sound silviculture without the benefit of sound professional judgment.

History reveals cyclic patterns of highs and lows of interest in both even-aged and multiaged silviculture. David M. Smith (1972) once said "no silvicultural procedure is so universally applicable that it deserves to be viewed as anything approaching standard operating procedure. The history of silviculture in this country [United States] is long enough to reveal that there has been too much tendency for methods of cutting to vacillate between extremes that are partly fads and partly reactions to problems of a temporary nature." These cycles are the result of complex patterns related to social or political values as well as silvicultural science. The cycles reveal that human opinion often drifts from one extreme to another like a pendulum. This pendulum process obscures the reality that both extremes—and all the options in the middle—offer viable alternatives for forest management that can and should be part of our managed forest landscapes. Either extreme may be a logical way to manage forests: the key is finding sustainable ways to do so.

CHAPTER 3

Disturbance dynamics of multiaged stands

3.1 Introduction

Silviculture has become recognized as a form of applied ecology (Smith et al. 1997) that exists within an understanding of stand dynamics that is applied to direct stands on certain developmental trajectories. Understanding disturbance dynamics is central to silviculture because many, but not all, silvicultural activities attempt to emulate natural disturbances. Natural disturbance dynamics are particularly important for guiding the design of complex stand structures and particularly, multiaged structures (O'Hara and Ramage 2013).

The dynamics that separate even-aged from multiaged stands are the differences in the sequence of disturbance, and the understory dynamics that form unique conditions for regeneration to develop. Disturbances that form natural multiaged stands consist of a series of low or mixed-severity disturbances that initiate new age classes or cohorts without destroying existing age classes. Multiaged management emulates these low-severity disturbances with treatments that encourage the regeneration of new age classes without completely removing existing trees.

This chapter provides a brief overview of disturbances with a focus on how disturbances may result in multiaged stands. Understanding disturbance dynamics is important for managing multiaged stands and disturbance emulation. The concept of disturbance regimes is presented, and several of the more important types of disturbances that result in multiaged stands are discussed. The intent is to provide an overview of how disturbances may result in multiaged stands and the many sources of variability that can result in multiaged structures. Other important references on disturbances and forests include Sousa (1984), Pickett and White (1985), Agee (1993), Oliver and Larson (1996), Frelich (2002), Franklin et al. (2002), Perera et al. (2004), and Turner (2010).

3.2 Disturbances and age structures

A central dynamic in the development of any forest is the effect of disturbances. This is particularly true in multiaged stands where disturbances are the primary determinant of age-class formation and the multiaged character of these stands. A forest **disturbance** is defined as a discrete event that disrupts ecosystem, community, or population structure and changes the availability of resources or the physical environment (Helms 1998). Runkle (1985) defined disturbance as a force that kills at least one canopy tree. Oliver and Larson (1996) described it as an event or series of events that changes the availability of **growing space** or an intangible measure of resource availability.

Disturbances are often classified as either natural or human caused. Natural disturbances are caused by a wide variety of climatic, geologic, biotic, and abiotic factors. They occur at a range of frequencies and severities, they may occur in an instant or over long periods of time, and they may occur over variable spatial patterns. Natural disturbances interact to affect stand structure and age-class formation in many ways, thus providing for great variation in stand structure. Human-caused, anthropogenic, or artificial disturbances include silvicultural interventions as well as other forest perturbations that

Multiaged Silviculture. Kevin L. O'Hara.
© Kevin L. O'Hara 2014. Published 2014 by Oxford University Press.

may not have a forest management objective. Both natural and artificial disturbances can stimulate the development of a new age class or cohort of trees. These disturbances can also affect species composition, tree vigor, and available growing space without affecting regeneration. For example, an ice storm may cause a large, temporary reduction in photosynthetic capacity and tree vigor but may not stimulate regeneration of a new cohort. Likewise, an insect defoliation may reduce the vigor of some species, making them less competitive, without stimulating the regeneration of a new cohort.

Silviculture emulates natural disturbances by attempting to mimic the same stimuli that favor certain species and the development of certain stand structures. For example, some aspects of even-aged regeneration methods resemble stand-replacement disturbances such as severe fire. Others such as the overstory removal in a shelterwood treatment may resemble large wind disturbances. Other attempts to favor certain species may be more subtle, such as the timing of harvesting to coincide with a species' seed fall or a shelterwood preparatory cut to encourage development of advance regeneration. **Advance regeneration** is a term that refers to the seedlings or saplings that develop in the understory beneath an intact canopy (Helms 1998). Antos et al. (2000) described advance regeneration as a "tree seedling bank" because it represents a reservoir of potential post-disturbance trees. Although disturbance-based silviculture has been encouraged on a formal basis only recently (Hunter 1993; Attiwell 1994; Haila et al. 1994; Seymour et al. 2002; Perera et al. 2004; Bergeron et al. 2006; Franklin et al. 2007; North and Keeton 2008; Fenton et al. 2009; Long 2009; Geldenhuys 2010; O'Hara and Ramage 2013), the conceptual basis for emulating natural processes is as old as silviculture (Hawley 1921).

3.3 Disturbance effects and silviculture

Disturbances have effects beyond creating new age classes or cohorts. Different disturbances favor different tree species and thus affect species composition. Insects and pathogens are often host specific and serve as regulators of species composition. Species also vary in their resistance to fire, wind, and other disturbances. Disturbances have other effects that are relatively subtle. Lugo (2008) described the "invisible effects" of hurricanes in tropical and temperate forests . These effects, which are generally overlooked, include the effects of small-scale redistributions of growing space that increase stand and forest complexity, serve to rejuvenate forests, and affect evolutionary development through species selection.

Disturbances that kill trees result in standing dead trees (**snags**) or downed logs (**coarse wood debris**). Trees may also be damaged but not killed, resulting in trees with cavities, reiterated stems (as described in Ishii and Ford 2001), and other unique features that make them important habitat features for a variety of invertebrate and vertebrate species. These features, also referred to as **biological legacies**, are important consequences of natural disturbance regimes that are often lost when strand structures are homogenized by various management strategies, including multiaged silviculture. The maintenance and creation of these features is a critical part of maintaining diversity and functioning ecosystems.

Although there are logical reasons to emulate disturbances with silviculture, there are operational limits that constrain our ability to fully integrate a disturbance-based silviculture. Our knowledge of disturbance dynamics is also based on historical disturbance regimes that have already changed and are likely to change further with climate change and ongoing introductions of nonnative pests and pathogens. Sound approaches to disturbance emulation in changing ecosystems will require compromises between the emulation of historical disturbance regimes and the operational constraints of contemporary forest management (Box 3.1).

3.4 Disturbance regimes

The term **disturbance regime** is used to characterize disturbances and their effects on a forest ecosystem. Another definition is the aggregate behavior over long time frames and large areas (Spies and Turner 1999). Runkle (1985) described the disturbance regime as the pattern of death of dominant individuals. Several compilations of these characteristics or descriptors have been assembled. For example, White and Pickett (1985) published a list that has since been refined by many others (Agee

Box 3.1 "Multicohort management" in Canada's southeastern boreal forests

Several disturbance types combine to affect boreal forests in eastern Canada. Fire is the primary natural disturbance type, but various insects and wind are also important factors. Stand-replacing fires frequently result in the establishment of even-aged stands dominated by shade-intolerant broadleaved trees, particularly on mesic sites. The process of stand development, in the intervals between fires, is generally characterized by a gradual transition from broadleaved to mixed stands, and eventually to stands dominated by conifers. This change in species composition is accompanied by a transition from even-aged to multiaged structures (Bergeron and Harvey 1997). With a dispersion of these fires across the boreal landscape, this transition from the initial broadleaved to conifer stands results in a suite of different stand structures and compositions. Current forest management uses almost exclusively even-aged systems to establish conifer species on rotations that are shorter than the mean fire return interval, which has ranged historically from approximately 100 to 350 years (Cyr et al. 2009). The result is a large-scale rejuvenation and homogenization of these landscapes to relatively young stands, including plantations, of conifer species that historically dominated late in stand development at much older stand ages.

A "multicohort management" strategy aimed at providing a diversity of structures at landscape scales has been proposed for these forests (see Malcolm and Harvey 2013). However, the strategy is not to establish multiaged stands across the landscape but rather to use three different but overlapping management approaches to maintain three types of stand structure that are characteristic of broad stand development stages (Harvey et al. 2002, Figure 3.1). One approach is similar to current even-aged management systems but where shade-intolerant hardwoods and other post-fire stand types are regenerated naturally, if possible, and managed on even-aged rotations (Bergeron and Harvey 1997). Some of these stands are treated by partial harvesting, the second approach, to develop multiaged stands characteristic of later "successional" stand structures and compositions. Finally, the third approach is a lighter harvest that forms small gaps aimed at regenerating shade-tolerant species and maintaining stands with more old-growth attributes.

This overall approach would not emulate the historical development patterns of these stands before active forest management. Instead, it would utilize silvicultural treatments to provide many of the stand structures found on those historical landscapes in shorter time periods. It represents a compromise between a full disturbance emulation model and the current timber production model using even-aged silviculture. It demonstrates the limitations of disturbance emulation and the potential of creative management approaches.

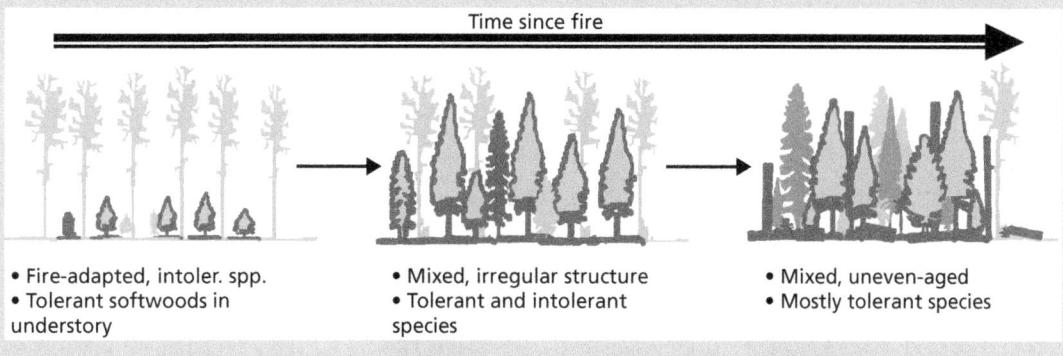

Figure 3.1 Three stand structure/management approaches showing their general position on a stand development trajectory since fire and variation in species composition.

1993; Turner et al. 1998; Frelich 2002; Suffling and Perera 2004). The disturbance regime concept has primarily been applied to natural disturbances, but similar concepts and some of the same terms are also applied to managed forests and silvicultural treatments. Components of the disturbance regime generally include: the type of disturbance and its duration, severity, frequency, size, seasonality or timing, and spatial pattern. Disturbance types include fire, wind damage, ice storms, insects and pathogens, flooding, herbivory, other animal activities, and geologic events such as volcanoes, mass movement, erosion, and others.

The **duration** of the disturbance event refers to the period of time the disturbance is stressing trees. For a severe wind storm or an intense fire, the duration may be very short, lasting only minutes or hours. For other disturbances, such as flooding, the duration may last days or weeks; and for some disturbances, such as drought or the effects of some pathogens and insects, the duration may be long term or even continuous. Many root diseases may be continuously present and slowly weaken or kill trees over long periods such as years or decades. Foster et al. (1998) described five different types of large, infrequent disturbances with durations ranging from minutes to months. Duration interacts with intensity to affect severity for a disturbance such as fire. A fire of a given intensity that maintains that intensity for a longer period will result in more mortality. For example, a rapidly moving fire may have a high intensity but kill very few trees. However, it should be noted that long-term ongoing perturbations that do not produce a discrete pulse of mortality (e.g., climate change, decades-long droughts) are often considered **stressors** as opposed to disturbances.

Severity refers to the proportion of aboveground vegetation damaged or removed, as well as the degree of damage to the forest floor and soil (Frelich and Reich 1999; Keeley 2009). A more general definition is simply the effect of the disturbance on the ecosystem (DeBano et al. 1998). Severity is often assumed to be equivalent to disturbance **magnitude** (Runkle 1985), although sometimes they are treated separately (see Edmonds et al. 2000). Severity is a description of the effects of the disturbance on an organism, population, community, or ecosystem. It represents the amount of heat released by a fire, and therefore it is also dependent on fuel conditions and fire behavior. Characterizing disturbance severity often focuses primarily on the trees. However, understory plants and the soil surface are important attributes that affect ecosystem response to the disturbance. Roberts (2007) developed a three-dimensional graphical model for presenting disturbance severity based on canopy, understory plant, and forest floor/surface disturbance. The model can also be applied to characterize the effects of silvicultural activities (Figure 3.2). Hence the model provides a visual array of how disturbances might differ in their effects on subsequent stand development.

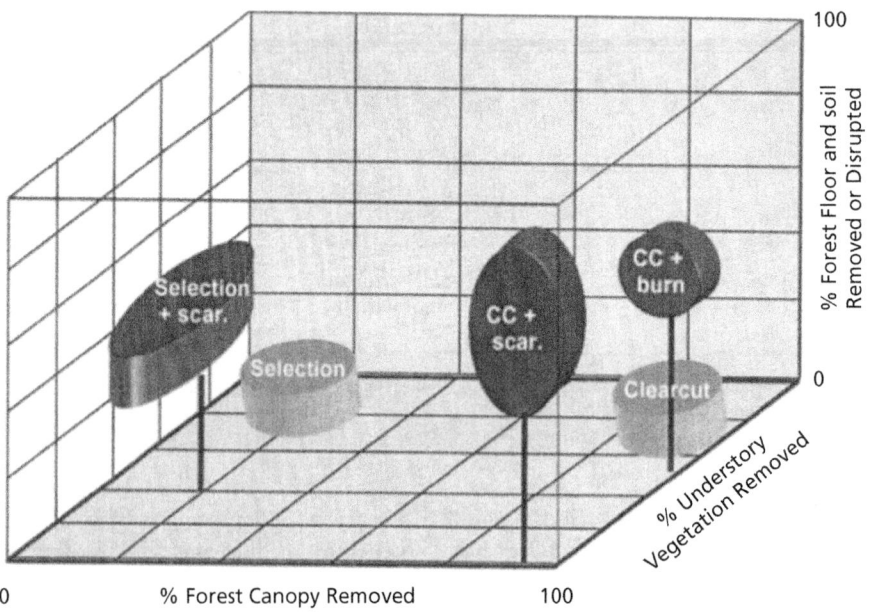

Figure 3.2 Three-dimensional model characterizing disturbance severity. CC = clearcut, Scar. = scarification of soil surface (from Roberts 2007; reproduced with permission from Elsevier).

A similar term to severity and magnitude is **intensity,** which has been used to describe the physical energy released by the disturbance process. For most disturbance types, intensity and severity will be highly correlated. The energy that causes windthrow and tree damage, for example, will be highly correlated. Fire intensity is often analogous to flame length and is measured in the heat energy released per unit of fireline. Since an intense fire may have short duration, there is a weaker correlation between fire severity and intensity than exists between severity and intensity for other disturbance types. For example, van Mantgem et al. (2013) found that a warmer climate may result in greater tree mortality with constant fire intensity demonstrating interactions with other variables.

Frequency of disturbance is usually expressed as the average number of disturbances per time period at a given point on a landscape. For example, a forest might have a frequency of five insect outbreaks per 100 years. Frequency affects which species are competitive and is instrumental in forming subsequent stand structures. Very frequent disturbances will favor species that can reach sexual maturity in the interval between disturbances, or species that can survive the disturbances. This is important for multiaged stand dynamics because systems that experience frequent but minor disturbances—where some trees survive and others regenerate due to the disturbance—form multiaged stands. The disturbance **return interval** is simply the inverse of the decimal form of frequency (1/frequency), assuming this frequency is expressed as an annual probability of occurrence: five disturbances per 100 years equals 0.05, and the return interval would be 20 (e.g., 1/0.05). The return interval is therefore analogous to the cutting cycle of the managed multiaged stands discussed in Chapter 7. All of these metrics provide historical averages of frequency or return interval of disturbance but do not express the variation that exists. Fires or windstorms or other periodic disturbance types occur over variable intervals that may encompass large ranges. Cutting cycles that attempt to emulate these disturbances can also be highly variable in length.

Rotation period is the length of time for an area equivalent to the whole forest or landscape of interest to be disturbed. This is analogous to the **fire cycle** for a particular forest. Rotation period integrates disturbance frequency with the size of a disturbance event. For example, if windthrow affects 100 ha of a 30,000 ha area per year or 0.33%, the rotation period would be 333.3 years. The rotation period is unique to the disturbance type: hence an ecosystem would have one rotation period for windthrow and a different one for flooding. For a managed forest, the **rotation** is the length of time between regeneration of an even-aged stand and its final harvest. For a regulated forest of even-aged stands, the rotation period is equal to the rotation. Whereas an idealized managed forest of even-aged stand would be completely treated in the rotation period, the randomization of disturbances in the unmanaged forest would result in some areas being disturbed more, and some less, frequently than others.

Seasonality or **timing** of disturbance events can affect subsequent stand development and stand structure. Many of these seasonal patterns are related to phenology. Trees and shrubs may be more susceptible to defoliation after a spring foliar flush, or surface roots may be more susceptible to fire damage when soil moisture is low. Disturbances that affect the seedbed may favor species that seed post-disturbance or consume seed that falls pre-disturbance. The timing of disturbances can therefore affect both the survival of existing trees or individual species, and whether species are able to reproduce. As a contributor to multiple age classes and stand complexity, the timing of disturbance events is a relatively subtle but important factor affecting stand structure.

The **size** or **extent, spatial pattern,** and **shape** of the disturbed area interact to affect stand structure and are other mechanisms for the formation of multiaged stands. A patchy high-severity disturbance can form a multiaged stand structure in a manner similar to the way in which frequent low-severity disturbances can form multiaged stands. The former might be analogous to traditional group selection approaches, or other approaches where small areas are periodically affected by disturbance. In these cases, different areas are affected at periodic intervals. Frequent low-severity disturbances over contiguous areas are more analogous to traditional single tree selection.

Variation in disturbance characteristics also contributes to the disturbance regime. It is common to label disturbance regimes based on descriptors that may represent average characteristics: for example, a stand-replacement fire regime or a snow avalanche zone. But variation in these disturbance characteristics over time and space contribute to variability in resultant stand structures, and may even determine the presence of certain structural features or species. The stand-replacement fire regime includes variation in fire severity, and snow avalanches occur on highly irregular intervals. An area with a dominant disturbance regime will therefore consist of ranges of characteristics that are as much a part of the disturbance regime as the average frequency or severity. Prediction of disturbance regimes and their effects is therefore more difficult because of this variation. However, natural vegetation patterns and structures are the result of these disturbance regimes and their variation; the management of complex stand structures should attempt to integrate these features into management regimes.

Disturbance synergism or **disturbance interactions** may also enhance effects on a forest. Examples of this include fire predisposing trees to insect attack, flooding making trees more susceptible to windthrow, or stem decay leading to stem breakage. The net effect of these interactions is a greater total severity of disturbance effects. Trees that might have survived a single disturbance may be killed by the interaction. Frelich (2002) used a "cumulative disturbance severity" to integrate the effects of multiple disturbance events. Similarly, the variable spatial patterns of multiple disturbances can lead to heterogeneous spatial patterning. These effects can greatly enhance stand structure complexity.

3.5 Disturbance types

A variety of disturbance types affect forests and result in multiaged stands. Several overviews of forest disturbances have already been published (Pickett and White 1985; Oliver and Larson 1996; Frelich 2002). Rather than summarizing the range of disturbance types that affect forests, this section will focus on how disturbances types affect the development of multiaged stands. Beyond the common disturbances discussed below, many other disturbances affect forests, and these may also contribute to the development of multiaged stands.

3.5.1 Wind

Strong winds can affect trees by breaking stems, uprooting trees, removing foliage and branches, or even modifying tree form. The source of these winds can be from large-scale frontal systems, hurricanes or cyclones, localized microbursts, thunderstorm activity, or the persistent winds that affect high elevation or coastal forests. They can also be augmented by localized or large-scale topography. A common forest disturbance is the loss of trees by either uprooting or stem breakage, collectively referred to as **windthrow** (Hale et al. 2012; Schindler et al. 2012; Mitchell 2013). Windthrow increases growing space availability and can expose mineral soil (Figure 3.3; Putz 1983). This releases understory trees and may stimulate the regeneration of new trees. Many windthrow disturbances stimulate the formation of new age classes, thus forming multiaged stands. The windthrow of individual trees may also affect species composition, as some species will be more resistant to wind, more likely to exist as advance regeneration, or more competitive on the limited mineral substrates of disturbed sites. In New England, Foster (1988) reported that more shade-tolerant species were in lower canopy positions and had greater survival following hurricane-force winds than less tolerant species that were more dominant. In southeast Alaska, Sitka spruce regenerated on windthrow mounds more often than western hemlock (Deal et al. 1991). In addition, greater soil disturbance following windthrow led to greater relative proportions of spruce in these stands.

Wind disturbs forests at a variety of scales, ranging from small perturbations to massive events such as hurricanes (Everham and Brokaw 1996). Wind affects tree crowns by breaking branches and physically removing foliage. As crowns sway under light wind conditions, they abrade the crowns of surrounding trees and thus cause "crown shyness" where crown contact declines with increasing tree height (Figure 3.4). With heavy winds, branch breakage and foliage loss can be substantial. Herbert et al. (1999) reported leaf area index reductions of up to 59% in a montane rain forest in Hawaii, US, following a hurricane. These reductions in canopy cover lead to increases in light penetration and increased growing space availability. The

Figure 3.3 Root mounds from windthrow in upland hardwood forest in Tennessee, US.

Figure 3.4 Crown shyness displayed in Malaysian broadleaved forest.

crown abrasion from endemic wind contributes to understory reinitiation, as greater light penetration gradually allows new regeneration to become established. The more severe winds can lead to greater increases in growing space. The result is a gradient of understory effects ranging from gradual increases in light or moisture availability to more sudden changes.

3.5.2 Fire

Fire is a dominant disturbance type in many forests in tropical, temperate, and boreal regions. The potential for fire severity and frequency to interact provide much opportunity to affect stand structure and species composition. Many forests experience low-severity or mixed-severity fire regimes where many trees may survive. In some cases, these low intensity fires occur at moderate frequencies and may stimulate the development of a new age class of trees. This pattern was followed by many mixed conifer forests of the inland Pacific Northwest. For example, forest-swestern larch, ponderosa pine, interior Douglas-fir, and grand fir would all experience periodic fire, but regeneration of all the species could occur with each fire (Barrett et al. 1991; Agee 1993). In other cases, fire may occur at high frequencies, and successful establishment of new age classes may occur only occasionally. Ponderosa pine forests of the southwest historically experienced fire on intervals of less than 10 years, but regeneration was sporadic (Fulé et al. 1997). In southeast Australia, alpine ash is maintained by a frequent fire regime where large trees are resistant but small trees have little resistance and need to achieve a fire-resistant size between fires to survive (Bowman and Kirkpatrick 1986).

Fire also affects species composition by favoring more resistant species. Resistance comes from several mechanisms. Some species are able to survive fire because of fire-resistant bark, deep root systems, or reduced branch retention in the lower bole. Many pines and other conifers such as coast redwood, giant sequoia, and species in the cypress family have bark that is fire resistant or insulates the stem from heat damage. Some eucalypts also have fire-resistant bark. Most larch species have low branch retention, and many pines have deep root systems.

Another common response mechanism is the ability to regenerate quickly after fire through dormant, fire-resistant seeds stored in surface layers or in crowns, or sprouting from stumps, lignotubers, or root suckers. Sprouts from established root systems typically grow quickly, providing an initial competitive advantage over many seedling species (Figure 3.5). Hence these regeneration mechanisms often determine which species will dominate after a fire. Similarly, the sprouting ability of some species extends up the tree bole, enabling rapid refoliation following crown scorch through the formation of epicormic, or water, sprouts (Figure 3.6). These epicormic sprouts originate from dormant or adventitious buds, as do stump sprouts, and are stimulated by heat or light (Kozlowski and Pallardy 1997).

The effect of low- and moderate-severity fire on forest structure is to enhance the development of multiple age classes and promote fire-resistant species. Multiaged stands are often formed because

Figure 3.5 Sprouting black cherry stump in New York, US.

Figure 3.6 Epicormic sprouts on the stem of a recently burned coast redwood in California, US.

some trees survive while others are killed by fire. Fire may reduce species diversity by selecting species that are resistant, or increase diversity by reducing the dominance of species that are highly competitive only in the absence of fire. As an example of the former, the effect of fire suppression on many forest ecosystems in western North America has been to increase diversity by increasing the relative abundance of species with low fire resistance.

3.5.3 Insects

Many types of damage caused by insects can stimulate the initiation of new age classes of trees. Some insect damage can result in stand-replacement disturbance, particularly in single-species stands attacked by bark beetles. Many other forms of insect disturbance are minor and affect only tree vigor, certain species, or reproductive organs. For example, defoliations reduce tree vigor, and most insects focus on a single or small group of species. Insects that feed on seeds and other reproductive tissues affect the regeneration of those species as well as the species composition of the ensuing stand. Any of these minor disturbances have the potential to affect the formation of new age classes of trees, thereby maintaining a multiaged stand structure.

In some cases, the frequency of insect attacks may also approach a regular pattern. Bark beetles often attack trees of certain size. The mountain pine beetle generally attacks trees greater than approximately 20 cm in diameter (Fettig et al. 2007). These attacks are often fatal to affected trees and can kill individual trees, groups, or entire stands and thereby cause a variety of spatial patterns to form. Since smaller trees are rarely attacked, this creates a frequency pattern where attacks occur over the minimum time period required to grow trees of the minimum size. In northern Minnesota, US, Reinikainen et al. (2012) reported that the combined effects of spruce budworm and tent caterpillar resulted in a multiaged structure in mixed forests.

Another example of an insect that causes cyclic effects on forest stands is the eastern spruce budworm. This defoliator will often kill balsam fir and several spruce species in spruce–fir forests in northeastern North America. Cyclic patterns where major outbreaks occur on 30–40 year intervals are linked to maturity of balsam fir and other factors (Sprugel 1976; Royama 1984; Porter et al. 2004; Royama et al. 2005). These cycles favor the formation of multiaged stands because many overstory trees are killed during epidemic levels of budworm activity while other overstory trees and understory trees survive.

3.5.4 Pathogens

Forest pathogens include all of the fungal diseases and heart rots that infect trees, as well as plant parasites such as mistletoes. Many pathogens can form multiaged stands by causing mortality that encourages new age-class development or by affecting certain species or size classes. Root diseases commonly have significant effects on forest structure. Many root diseases, such as laminated root rot or armillaria root disease, affect only certain species and create root disease mortality centers that may form canopy gaps. However, the persistence of these disease organisms will result in long-term species conversions. Other pathogens, such as stem cankers, foliar diseases, or rusts, generally have relatively minor effects on age structure but may nonetheless affect species composition. Mistletoes can kill trees but often only reduce vigor and cause deformities. Their method of seed dispersal is aided by a multiaged canopy where seeds are likely to infect younger trees.

Nonnative pathogens often represent a more extreme disturbance event because they may virtually eliminate highly susceptible species. Examples include chestnut blight and various *Phytophthora* species, which are currently affecting forests in a variety of regions. The effect of these invasives is similar to that of endemic pathogens in that mortality is often limited to individual species. However, due to the absence of historical exposure, susceptible tree species may have little or no evolved resistance, and thus effects may be especially rapid and comprehensive.

3.5.5 Ice storms

Ice storms (also known as ice accumulations or glazing events) are weather events where rain freezes on objects near the ground, including trees. These events result in breakage of tree branches, whole tree breakage, deformity of residual trees, and loss of leaf area. These events occur with some regularity in eastern North America and are a major cause of forest damage (Irland 2000). Effects on forests are highly variable depending on the often confounding interactions of species composition, stand structure, and the characteristics of the storm event. Most studies on ice storms are post-disturbance reconstructions that are limited in their occurrence and may be highly spatially variable.

Species vary in their ability to shed ice, sustain bending, or recover from ice loading events. For example, in northern hardwoods, Weeks et al. (2009) found ice storms increased the proportion of American beech because of the persistence of this species in the understory. Increased understory light levels are also a common result of ice storms (Beaudet et al. 2007) resulting in pulses of regeneration. Trees also vary by size in their resistance to damage. Small trees can often recover from bending displaying a resilience not present in larger trees. Understory trees in complex forest structures may also be protected from damage by overstory trees (Bragg et al. 2003). Kenderes et al. (2007) found that increasing structural heterogeneity increased resistance to ice and wind damage to forests in Hungary. Hence the effect of ice storms on multiaged stands may be to release understory trees and promote the multiaged character of these stands, but there may also be more resistance to this effect in more complex structures.

3.5.6 Other types of forest disturbance

There are many other disturbance types that can form or maintain multiaged stands, or contribute to stand complexity. These disturbances may only affect a single species and may alter either age or species structure. For example, mammalian herbivory will often express strong species preferences, which can modify species composition. However, species-specific herbivory does not normally result in new age classes or cohorts of trees, and mammalian herbivory may actually prevent the development of new age classes. Removal of small trees often favors other existing trees that can occupy the available growing space. When mature trees are killed by disturbance agents, this typically results in a regeneration event. When only some mature trees are killed, the resulting regeneration will lead to a multiaged stand. Animals will also damage trees, resulting in the death of mature trees and changes in age structure. Bears will strip the bark from trees to eat the cambium and thereby either kill the tree entirely or kill the top and thus stimulate the formation of new sprouts. Volcanic eruptions can also have species selectivity. The ash deposition from the 1980 eruption of Mt St Helens in Washington, US was more likely to kill Pacific silver fir than the other species present (Segura et al. 1994). There are many other types of disturbances that contribute to stand complexity, and different types of disturbance can combine to have other unique effects on structure.

3.6 Synthesis

Disturbances provide a number of critical processes in stand development. They recycle nutrients, renew forests, and control species composition and stand structure. They contribute to stand structural diversity and species diversity. What is often viewed as damage is actually the renewal of a forest or the creation of important stand structural elements such as snags or coarse woody debris. Disturbances are best described with whole regimes that capture the variability in characteristics such as frequency and severity. The types of disturbance are also variable, with many different types contributing to the formation and maintenance of multiaged stands.

A wide variety of disturbances produce multiaged stands, and the variation in disturbance regimes and time since disturbance produces an endless variety of multiaged forms. With a multitude of different multiaged stand structures formed by natural disturbances, the lesson for silviculture is that managed structures can also be highly variable even within a single forest type. They can be variable in structure and—like the natural disturbance regimes they may emulate—highly variable in the management regimes used to create them. If disturbance emulation is a silvicultural objective or guding principle, natural disturbance regimes demonstrate that there are a great many ways to manage multiaged stands.

Disturbance emulation is the foundation for a limited assortment of silvicultural treatments. Whereas all silviculturists should have a strong understanding of the basis of treatment as well as stand or broad-scale responses in relation to disturbances, there are constraints on what can be emulated and what should be emulated. There is also a need for caution in using historical disturbance patterns as models for future treatment at any scale. The current emphasis on describing silviculture as an applied ecology is well supported. However, the ecology that is applied in the future, particularly as related to disturbance, will likely be less grounded in historical patterns and more based on compromise between societal constraints and the limitations of a changing climate.

CHAPTER 4

Dynamics of multiaged stands

4.1 Introduction

Understanding stand dynamics is prerequisite for silviculture and particularly for the silviculture of complex stands. The simultaneous growth and development of multiple age classes of trees complicates our interpretations of within-stand growth and competition. This is particularly true in the understory, where small trees are important future overstory trees but may exist in variable states of competition for light or moisture. Some of these understory trees must be able to persist and grow to maintain vigor and form and to advance their position in the stand. Growth and persistence of understory trees in multiaged stands is therefore a process of obtaining adequate resources while maintaining an overstory. This chapter discusses the dynamics of overstory/understory relations through the limitations of light and moisture in complex stands. Shade tolerance is presented as a dominant factor that limits options in multiaged stands, and tree architecture is presented as a diagnostic variable. This information is used to define tree growth patterns in multiaged stands as trees move through the canopy and respond to increases in growing space. More detail on stand dynamics, and particularly the dynamics leading to the formation of complex stand structures, is available in Oliver and Larson (1996). Gap dynamics are presented in Chapter 5.

4.2 Dynamics of light-limited systems

4.2.1 Light regimes

The reduced light in the lower strata of multiaged stands has a dominant influence on growth and form of regeneration. Because multiaged stands typically consist of multiple canopy strata, upper strata will receive plentiful sunlight whereas lower strata will be in much poorer light environments. The term **light regime** is used to describe the dynamics of light quality, duration, and intensity on a daily, seasonal, or annual basis. A common means of describing the light regime in understory canopy strata is the ratio of below-canopy light to above-canopy light. This is often expressed as the percent of above-canopy light (PACL) or percent of photosynthetically active radiation (PAR). The PACL typically varies inversely with the light extinction characteristics and number of the canopy layers overhead. Thus, a single-stratum overhead might have a steep PACL gradient as shown in Figure 4.1. As stand structure increases in complexity, the light regime would also become more complex (Figure 4.1). PACL is a useful tool for expressing the relative light intensity under a forest canopy; however, it does not provide much indication of the quality of the light that is received. Light quality can vary with type of foliage overhead, the amount of diffuse and direct light, and the quality of the light.

Most vascular plants increase their growth with increasing light intensity until their light needs are met. An asymptotic or saturation curvilinear relationship is often used to describe these relations (Figure 4.2; Coates and Burton 1999; Stancioiu and O'Hara 2006c; O'Hara et al. 2007b). The implication of this form is that plant growth increases with light intensity until some maximum. Above this maximum, growth is generally unaffected by increases in light intensity. The shape of this light-response relationship would vary with the species in both the understory and overstory, and their shade tolerance.

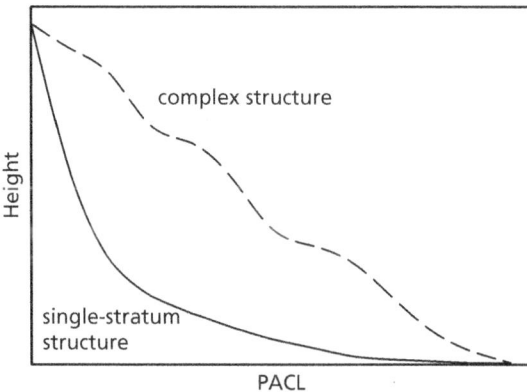

Figure 4.1 Percent above-canopy light (PACL) as a function of height in a simple single-stratum structure with a mature overstory (solid line) and in a complex structure with multiple canopy strata (dashed line).

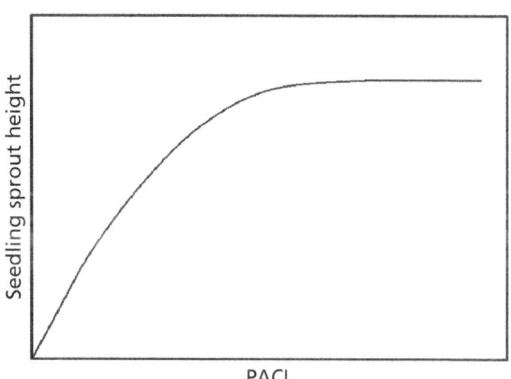

Figure 4.2 Seedling or sprout height with increasing percent above-canopy light (PACL).

A key component of multiaged silviculture is finding the appropriate stocking relationship that provides for a healthy overstory and adequate growth of the understory. This is often viewed as a simple trade-off that is somewhat representative of the more complex trade-offs in stocking control of stands with multiple canopy strata. For a two-strata stand, Figure 4.3 demonstrates that an expectation for rapid understory growth can only be achieved with a corresponding decrease in the overstory stocking or growing space occupancy. Understory growth that is comparable to growth of trees growing in even-aged stands will only be possible at low overstory stocking levels. Acker et al. (1998) and Zenner et al. (1998) describe the growth trade-offs for two-strata stands in the Pacific Northwest, US. The slower growth of understory trees has been cited as a reason multiaged stands are inherently less productive than even-aged stands (Assmann 1970), although such reasoning ignores the potential benefits of low overstory stocking on overstory tree growth.

In some forests, clumpy patterns of overstory trees may be necessary to provide adequate light to the understory. Gersonde et al. (2004) used a

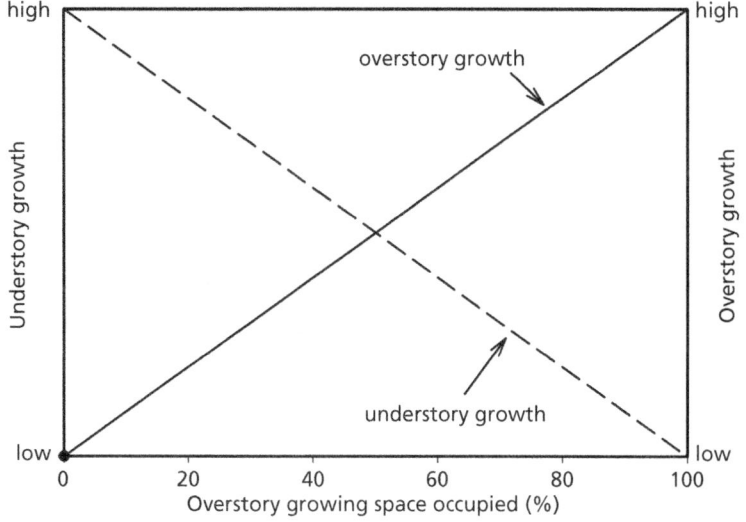

Figure 4.3 Trade-offs between age classes or canopy strata in a simple two-aged or two-strata stand. An allocation of growing space to one age class reduces the potential of the other and represents a trade-off between providing enough space for the older age class to grow and ensuring that there is enough space for the younger age class to persist (adapted from O'Hara 1998).

light model to characterize the light regimes in multiaged conifer forests in California, US, and found irregular canopies allowed more light to intermediate canopy strata. Battaglia et al. (2002) described the enhanced light regime in stands of aggregated patterns of longleaf pine overstory trees. Although their treatments left similar amounts of residual basal area, the light regimes were very different because of the variable patterns of residual trees (Figure 4.4). The effect of these treatments is to have different parts of the stand in different light regimes or different positions in the diagram shown in Figure 4.3. Some areas may be to the left, where understory light is more available, whereas others are further to the right, where the overstory density is higher. Whereas aggregated spatial patterns may create favorable light regimes in some situations, in other cases, other variables may be important. For example, Peck et al. (2012) found no clear patterns with three pine species in Minnesota, US. Other factors, such as latitude, or variations in shade tolerance among species, may override the advantages of having gaps overhead. Loftis (1990a, b) concluded lower and mid-story shade was more important in affecting oak regeneration in the southern Appalachians, US.

Sunflecks are brief periods of irradiance under relatively closed forest canopy conditions (Chazdon and Pearcy 1991). They are an important part of the understory light regime in many forests because they supplement diffuse light energy, which may be more generally present. In some forest types, sunflecks may account for up to 90% of the total PAR, but usually they are considerably less (Lieffers et al. 1999). Sunflecks are also highly dynamic in that they may last only seconds and can vary daily and seasonally in both light quality and duration (Baldocchi and Collineau 1994). The contrasts in light intensity from the sunfleck area to the general forest floor can also be quite variable, and all sunflect radiation may not be used (Wayne and Bazzaz 1993). As a result, trees vary in their ability to react to sunflecks due to differences in patterns of stomatal closure and other adaptations—such as in tree architecture—to light regime.

Light quality is another important variable affecting understory dynamics in multiaged stands. At one level, light quality might refer to the contrasts between direct beam light, diffuse light through an overhead canopy or **high shade**, or the combination of diffuse and direct light near the forest canopy or **low shade** (Chapman 1945; Oliver and Larson 1996;

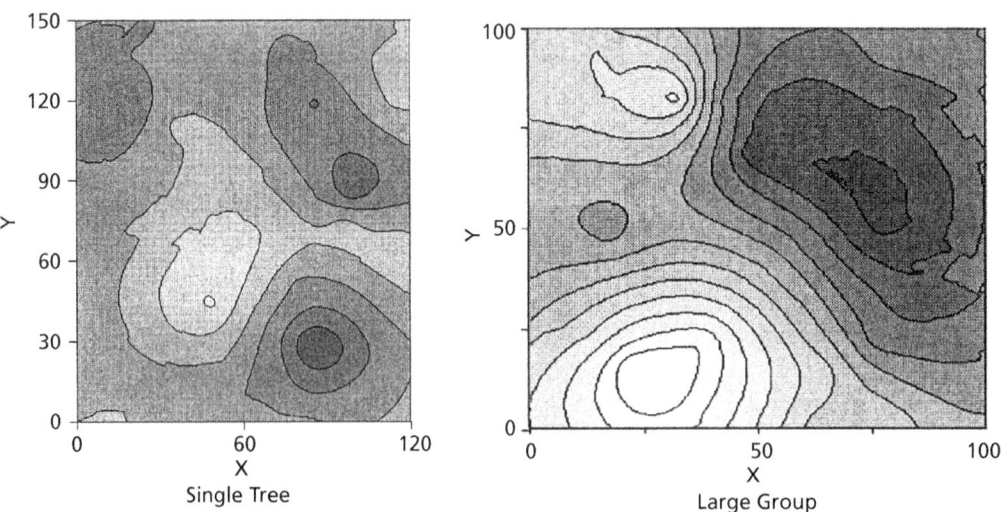

Figure 4.4 Spatial patterns for light in longleaf pine stands in Georgia, US. The figure on the left shows the result of a single tree selection treatment and on the right, that of a large group treatment. Dark patterns represent areas in poorer light regimes and light areas, the best light regimes (from Battaglia et al. 2002. The effect of spatially variable overstory on the understory light environment of an open-canopied longleaf pine forest. Can. J. For. Res., 32, 1984–1991, © *Canadian Science Publishing or its licensors*).

Leiffers et al. 1999). Light quality may also refer to the proportion of the light spectrum available to plants. The structure of the forest can lead to a variety of different light qualities depending on whether the light passes through a forest canopy or through large or small gaps (Endler 1993). The forest canopy can therefore affect the range of the light spectrum that is transmitted and thus benefit certain trees over others. For example, Lieffers et al. (1999) showed how the light quality—as represented by the ratio of red to far-red wavelengths—varied with light intensity to give conifers a general advantage at some light intensities but not at others. Clearly, there is more to light regimes than simply the light intensity, and these light regime dynamics cannot be represented with simple stocking or stand-density measures that are used to approximate overstory light interception. Instead, the light environment for the understory of a multiaged stand is highly complex, and, in the absence of detailed data about light quantity and quality, the performance of trees on the site may be the best indication of the present light regime.

4.2.2 Shade tolerance

The relative shade tolerance of trees is often a limiting factor for multiaged silviculture, because it determines the structural arrangement of trees in mixed-species stands. Physiologically, shade tolerance is the result of relative light compensation points for different species. The light compensation point is the light level where respiration and photosynthesis are equal. Below this light level, the plant is unable to produce enough carbon to sustain itself, but above this level, it produces a surplus. Species vary in their compensation points and their ability to tolerate low-light regimes. Species also vary in their leaf structure in relation to shade tolerance and light regime (Abrams and Kubiske 1990). Trees may therefore have both physiological and morphological responses to shade (Canham 1989). Ranking species' shade tolerances is often used to show the relative ability of different species to endure shade as as would be encountered in a multiaged system (Malcolm et al. 2001). The primary ranking for many North American species can be found in a table produced by Daniel et al. (1979) that places species in five shade tolerance categories (Table 4.1).

Table 4.1 Shade tolerance classes for selected North America species (from Daniel et al. 1979).

Tolerance class	Species
Very tolerant	American beech
	Balsam fir
	Eastern hemlock
	Sugar maple
	Western hemlock
	Western redcedar
Tolerant	Coast redwood
	Grand fir
	Red maple
	Spruce species
	Tanoak
	White fir
	Yellow-cedar
Intermediate	Coast Douglas-fir
	Eastern white pine
	Radiata pine
	Red oaks
	Sugar pine
	Western white pine
	White ash
	White oak
	Yellow birch
Intolerant	Black cherry
	Eastern redcedar
	Loblolly pine
	Lodgepole pine
	Noble fir
	Ponderosa pine
	Shortleaf pine
	Yellow poplar
Very intolerant	Aspen species
	Black locust
	Cottonwood species
	Jack pine
	Larch species
	Longleaf pine

A shade tolerance ranking can be conceptualized for multiaged stands by determining which species can endure the low light that penetrates the foliage of another species in an overstory position. If a species endures this light level, that species is more tolerant than the overstory species. Hence, the ranking represents a guide for how species might be vertically arranged in multiaged structures. Species with low tolerance to shade could be above more tolerant species but not below (Figure 4.5). These types of structures demonstrate the limitations that differences in shade tolerance place on mixed stands. A mixture with a more shade tolerant species dominating the understory is likely to shift in species composition to that more tolerant species. In a 50-year study on eastern North American hardwoods, Schuler (2004) reported that multiaged treatments encouraged less desirable, shade-tolerant species. Similar patterns are common in other forest types. This is one of the major impediments to the use of multiaged systems in managing mixtures of shade-intolerant and shade-tolerant species.

Whereas shade tolerance rankings generally focus on the ability of a tree to survive or endure the shade of another, growth rates under various light levels are also an important and more complex consideration. Understory trees generally have asymptotic growth relationships with increasing light (Figure 4.2). These light-response relationships vary with species. Figure 4.6 shows the probability of mortality for several Himalayan species, and Figure 4.7 shows growth patterns for three species in eastern Europe. In both cases, different patterns were evident by species. However, these relationships generally correspond to shade tolerance rankings, with more tolerant species showing greater growth at low light levels. But the crossing of these relationships indicates stand tolerance is more complicated than a simple ranking.

Other factors may affect a species response to its light regime. Site quality was shown to be a factor with western redcedar and coast Douglas-fir in British Columbia, Canada (Drever and Lertzman 2001). For young broadleaf seedlings in eastern North America, allocation to carbohydrate reserves varies

Figure 4.5 Relative shade tolerance of Norway spruce and Scots pine in southern Finland. The Norway spruce crown in the center extends to the ground, whereas the Scots pine crowns are all quite high.

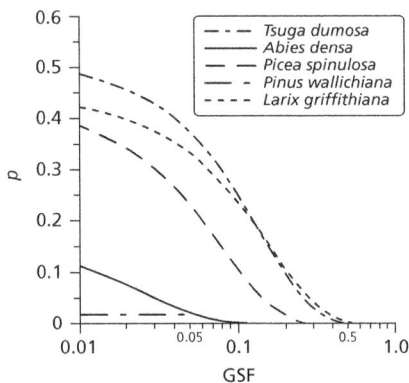

Figure 4.6 Probability of mortality (p) over a two-year period as a function of light availability (GSF) for five Himalayan species. The highest risk of mortality was for the shade-intolerant *Larix* and *Pinus* species (from Gratzer et al. 2004, Interspecific variation in the response of growth, crown morphology, and survivorship to light of six tree species in the conifer belt of the Bhutan Himalayas. Can. J. For. Res. 34, 5, 1093–1107, © *Canadian Science Publishing or its licensors*).

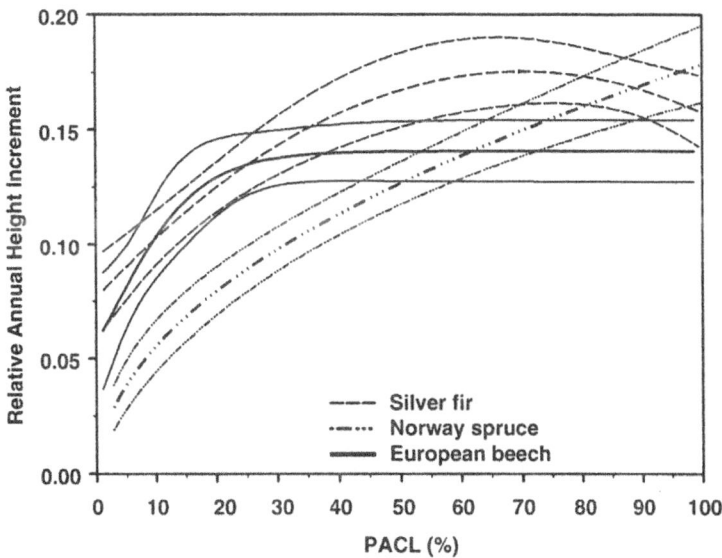

Figure 4.7 Relative annual height growth for silver fir, Norway spruce, and European beech in Romania as a function of percent above-canopy light (PACL). Patterns are shown for each species with standard error bands. At the lowest light intensities, the more shade-tolerant silver fir grows the most and Norway spruce the least. But at higher light intensities, the relative increment patterns vary (from Stancioiu and O'Hara 2006c).

with light levels (Canham et al. 1999), and this allocation, in turn, affects survival and growth (Webb 1981; Kobe and Coates 1997). Likewise, Kneeshaw et al. (2006) found changes in shade tolerance ranking with increasing size in small boreal trees in eastern Canada. Ranking shade tolerance is therefore a simplification of relatively complex relationships describing relative tree or plant growth in a range of light environments.

Despite its complexity, relative shade tolerance is still a guiding concept in multiaged silviculture. Species vary in their ability to endure low light levels. Some will be more competitive in complex structures than others, and managers need to be aware of these relationships and how they might vary with site quality, tree age, or tree size.

4.3 Dynamics of moisture-limited systems

The limiting factor for some multiaged forests is soil moisture. The dynamics and recruitment of new trees and competition for growing space is considerably different when soil moisture is limiting instead of light. At one extreme, you have complex stands, where moisture limitations may affect only some of the species. At the other, you have open stands, where leaf area index (LAI) and stocking may be low and the plant communities are quite open. In the former, stands may appear closed, but available soil moisture may limit complete canopy closure. In these situations, light may limit the development of some species, and moisture may limit others within a single stand. This is demonstrated in Figure 4.8, which shows how oaks in the eastern US become more competitive on dryer sites regardless of the light regime. In contrast, on moist sites, they are most competitive under only a narrow range of light regimes. The phenomenon depicted in Figure 4.8 is representative of the strong interaction between light and soil moisture that exists for many species and affects competition between trees and other vegetation. It may be a seasonal limitation or a limitation that occurs during only a few years. Likewise, limitations of nutrients or other soil conditions, such as depth or pH, may favor one species over another.

On much dryer sites, where stands may be open (Figure 4.9), the limiting factor is primarily soil moisture or some other below-ground limitation. It is difficult to generalize about these sites, which may be called **woodlands** or **savannah woodlands** (Helms 1998) or have local names, such as chaparral in California, US, dehesas in Spain, or maquis around the Mediterranean. They may have open spaces between trees in a form of **open stem exclusion** (O'Hara et al. 1996). Trees may be relatively short and small in these structures, and below-ground competition may be intense. Age

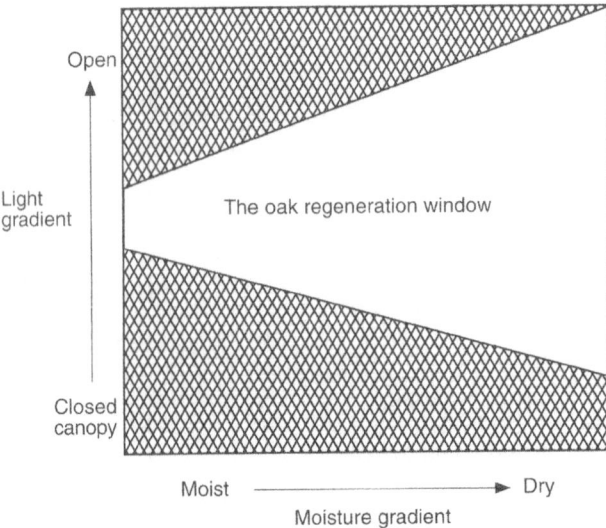

Figure 4.8 Relationship between moisture and light for oaks in the eastern US (from Johnson et al. 2009; reproduced with permission from CABI).

Figure 4.9 Dryland juniper—pine woodland in California, US.

structures may be diverse due to highly variable disturbance histories and episodic regeneration periods correlating with climatic events or disturbances (Tyler et al. 2006; Shinneman and Baker 2009). A variety of adaptation strategies favor different types of plants on these moisture-limited sites. These strategies may be related to rooting depth, resistance to desiccation, or regeneration properties that allow plants to establish quickly and exclude other plants. Unlike many light-limited systems, where regeneration may be common or even continuous, on the more extreme of these moisture-limited sites, a convergence of high seed production with the creation of growing space by disturbance may be necessary for regeneration.

Many climate-change scenarios predict warmer, dryer climates, which would lead to an increase in these moisture-limited forest or woodland types. In some cases, their present existence may be due in part to anthropogenic activities (Martín Vicente and Fernández Alés 2006; Blondel 2006). Although not traditional multiaged forests managed for timber production, these moisture-limited forests may be an expanding proportion of the world's forests. The management of these types will be based, in part, on their age structure and use many of the concepts related to the dynamics of more light-limited systems.

4.4 Tree and stand architecture

The architecture of individual tree crowns is strongly affected by the light regime where they develop. For small trees, tree architecture can be diagnostic as to potential changes in light environment that may affect tree vigor and form. Understory tree growth rates vary with light regime in relatively predictable ways. Crown form also varies in relatively predictable ways (Oliver and Larson 1996). Conifers with strong **epinastic control**—the ability of the terminal bud to control the angle and length of lateral branches—tend to form conical or **excurrent** crown shapes in full sunlight but transition to flat, "umbrella-shaped" crowns with greater shade (Figure 4.10). The difference is generally a shortening of the internode distance and a lengthening of the branches in conifer trees in greater shade. Species with weaker epinastic control or **decurrent** growth forms, such as many broadleaved species, will also vary in crown form in different light regimes, but the differences typically are not as pronounced as those in trees with strong epinastic control (Figure 4.11).

Figure 4.10 Flat-topped grand fir in Washington State, US, showing the umbrella form which is characteristic of trees with strong epinastic control in shaded environments.

Figure 4.11 Decurrent growth pattern of European beech in a multiaged stand in Slovenia.

The maintenance of a central leader results in a tree with a single bole, but the tendency in poor light regimes is for young trees to form flat tops with multiple leaders. For trees with both weak and strong epinasty, control grows stronger as the light regime improves. There is also an interaction with shade tolerance, as species with greater tolerance have greater variation in crown form or crown plasticity in different light regimes (Givnish 1988; Chen et al. 1996; Williams et al. 1999). An intolerant pine has little variation in form in different light regimes, but a tolerant fir would display a great range in form.

These morphological characteristics related to light regime have been quantified in a number of studies for small trees. For example, trees in low-light regimes have larger crown width-to-length ratios than trees in better light regimes (Messier et al. 1999; Ruel et al. 2000). The terminal-to-lateral ratio compares the lengths of vertical and lateral extensions and is greater in trees in better light regimes (Canham 1988a; Messier et al. 1999; Stancioiu and O'Hara 2006b). Leaf morphology may also vary (Abrams and Kubiske 1990). **Specific leaf area**, or the ratio of leaf surface area to leaf mass, is greater in trees grown in poorer light regimes (Williams et al. 1999; Grassi and Bagnaresi 2001). Conifer needle morphology also varies with greater stomatal density in trees from better light regimes (Youngblood and Ferguson 2003).

A reduction in the quality of light regime with time may result in a tree changing from a conical form or a form dominated by a central leader to a flat-topped form or one with multiple leaders, with the shape dependant upon the epinastic control of the species and the degree of change. Changes in crown form are indicative of deteriorating light regimes as a tree becomes more suppressed. Likewise, improvement in the light regime results in trends toward more conical or excurrent forms, or crowns with a more dominant leader. For conifers, improvement in the light regime often leads to a lengthening of the leader relative to branch length and the development of a more conical crown form (Tucker et al. 1987). Figure 4.12 shows this for hiba, a Japanese cypress species, as it changes from a flat-topped, suppressed form to a conical form. For broadleaved species with weak epinastic control, trees often do not return to a single-bole form and instead may become forked trees with a strongly decurrent growth form (Stancioiu and O'Hara 2006b).

Management of multiaged stands can therefore affect light regimes and the form of understory trees. Maintaining adequate light will prevent the extreme suppression that leads to top dieback or mortality, or to the stem forking that may make a tree less desirable for timber. Although many studies have successfully studied light regimes in relation to small-tree morphology (e.g., Mitchell and Arnott 1995; Chen et al. 1996; Sprugel et al. 1996; Messier et al. 1999; Williams et al. 1999; Grassi and Bagnaresi 2001; Stancioiu and O'Hara 2006b) or for growth responses of trees to management (e.g., Tucker et al. 1987; Lieffers et al. 1993; Ruel et al. 2000), information about light thresholds is lacking for most species. The relationships between small-tree crown form and light are probably similar to

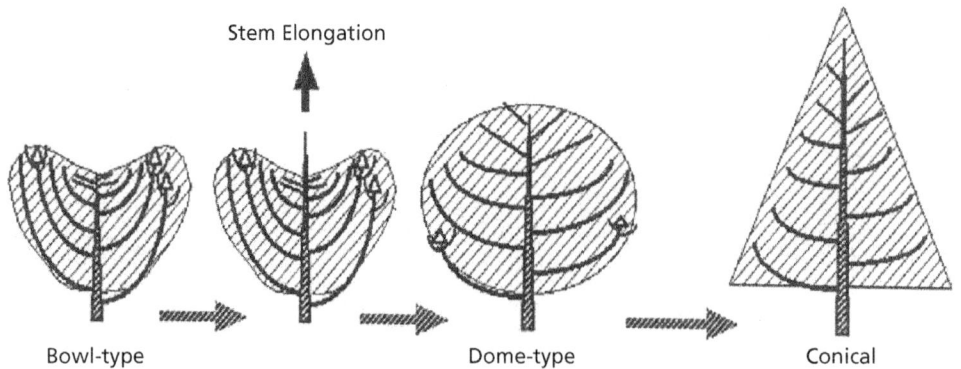

Figure 4.12 Crown development of hiba in Japan after release showing elongation of the leader and a complete change from a flat-topped to a conical crown form (from Hitsuma et al. 2006; reproduced with permission from Springer Japan and the Japanese Forest Society).

those for small-tree growth and light. For example, Stancioiu and O'Hara (2006b, 2006c) reported that light levels above approximately 30% or 40% were necessary to maintain growth and crown form for Norway spruce, European beech, and silver fir.

The crown architecture of large trees in multiaged stands is less affected by suppression and is similar to trees in even-aged stands. Although the architecture of these large crowns is much less studied than for understory trees, there is likely a gradual transition from trees with understory features to upper canopy trees. Tree leaf area increases dramatically with increasing dominance in multiaged stands (O'Hara 1996). Additionally, specific leaf area decreases with increasing dominance, as larger trees have less leaf surface area per unit of leaf mass (O'Hara and Nagel 2006). However, O'Hara and Nagel found the range in specific leaf area was less in multiaged stands than in comparable even-aged ponderosa pine stands. This suggests some subtle differences in crown architecture in larger trees between multiaged and even-aged stands that have not been fully explored.

For stands, **leaf area index** (LAI) is a measure of the ratio of leaf surface area to ground area covered by the canopy. An LAI of 3.0 indicates an average of three layers of foliage over an area. LAI provides a measure of the light resources used by a stand and therefore provides a measure of occupied growing space (O'Hara 1988). Following a stand-replacement disturbance, LAI increases gradually and, barring further disturbance, eventually reaches a maximum for a site. LAI is a common driver in ecosystem process models (Running and Coughlan 1988; Landsberg and Waring 1997; Battaglia et al. 2004) and has been used to describe growth efficiencies of individual trees and age classes in multiaged stands (O'Hara 1996). LAI is not generally comparable between different species: a unit of LAI for a shade-intolerant species will have different light reception and production relationships than a unit of LAI for a shade-tolerant species.

The distribution of biomass in any stand structure is highly variable due to species, age, and site characteristics. Figure 4.13 displays the vertical

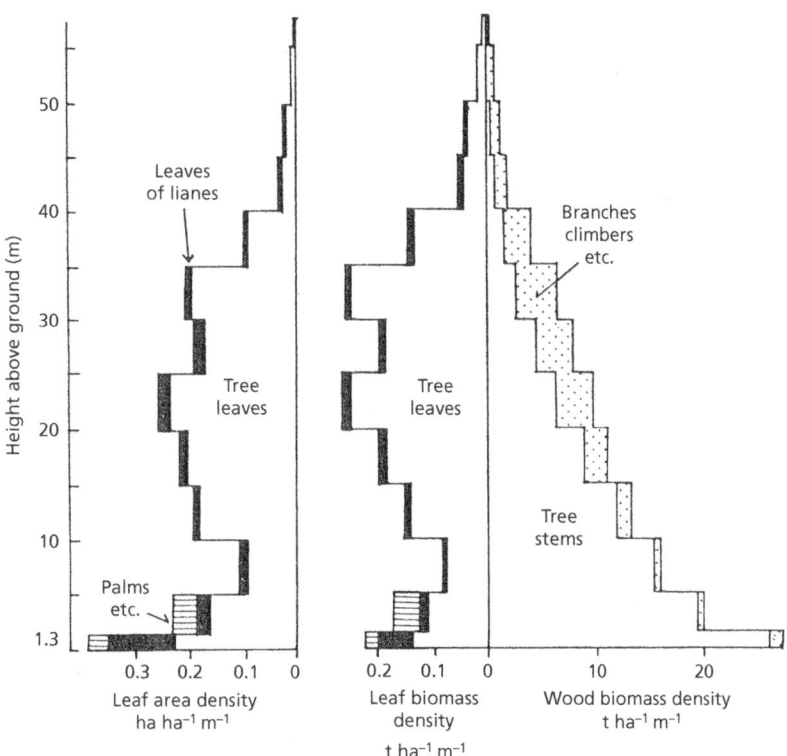

Figure 4.13 Vertical biomass profiles for tropical forest in Malaysia (from Kira 1978; reproduced with permission from Cambridge University Press).

distribution of biomass in a tropical forest in Malaysia. In this forest, which reaches nearly 60 m above ground, most foliage and branch biomass is at the mid-height of the main canopy, and there is very little in the top 15 m. Stem biomass is primarily near the ground. The presence of foliage biomass in the lowest 5 m suggests there are regenerating trees, besides the palms and lianas, but this diagram does not reveal which species are present. Stand-level architecture is difficult to interpret with height profile diagrams. However, the distribution and density of foliage does provide information on the light environment in the lower canopy.

4.5 Growth patterns

Tree growth patterns in multiaged stands are variable from one age class to another and are also influenced by variations in the canopy around them. Oliver and Larson (1996) described the idealized development of a stand containing four cohorts or age classes as a series of increasing height growth curves with advancement in the structure (Figure 4.14). Trees in lower canopy positions struggle to develop in poor light environments or with limited growing space (Figure 4.10), whereas trees in upper canopy positions are vigorous in an environment with relatively little competition. Over succeeding cutting cycles, each age class advances in rank and, with these idealized relationships, assumes a growth rate like that of the next older age class until it is harvested.

These within-cutting cycle growth curves are sigmoid in shape, based on the assumption that their growth is greatest early in the cutting cycle when growing space is most available but decline later in the cutting cycle as available growing space declines.

Actual growth relationships are, of course, more complex. Normal variations in cutting cycle length, unequal volume removals, variations in age class densities, or different species composition would lead to variable growth patterns. What is clear is that trees left too long in suppression or under extreme states of suppression may not respond to increased growing space. The term **release** has two related meanings in silviculture. It commonly refers to treatments that free young trees from competition (Helms 1998), but it also is used to describe the response effect of the subject trees. Both definitions are pertinent to discussions of growth relationships in multiaged silviculture. Each cutting cycle treatment essentially represents a *release* of remaining age classes, and the remaining trees should show a *release* as they respond to the improved competitive environment. This latter type of release is critical to many types of multiaged silvicultural applications, from shelterwood removals to more complex systems.

Many studies have attempted to predict release from tree characteristics. These are useful approaches because tree crown architecture is strongly affected by suppression, and the degree of suppression corresponds to the subsequent response or release to improved conditions. Generally these studies have

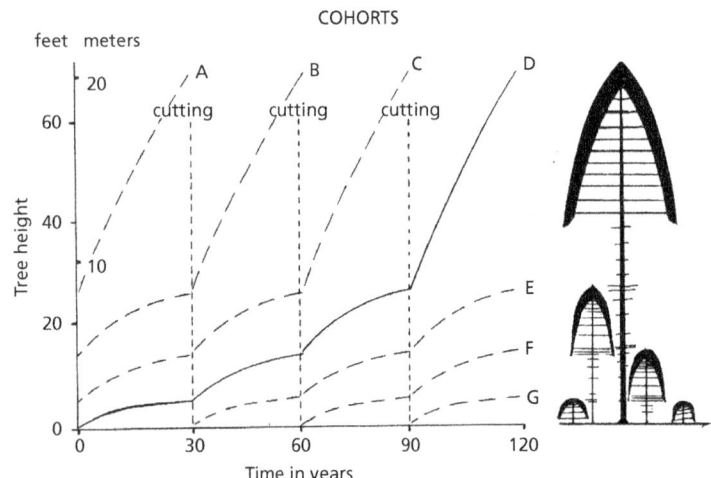

Figure 4.14 Idealized growth patterns for a multiaged forest with four age classes and repeated 30-year cutting cycles (Oliver and Larson 1996; reproduced with permission from John Wiley & Sons).

focused on advance regeneration that may have potential for release into a future overstory position. These trees can be completely released to become part of an overstory after a shelterwood removal, or part of a developing stratum in a multiaged stand.

For evergreen conifers, studies have found that prerelease growth rates correspond well to postrelease growth in many forest types (Ferguson and Adams 1980; Seidel 1985; Helms and Standiford 1985; Tesch and Korpela 1993; Metslaid et al. 2007). Trees with larger crowns (greater internode lengths) or narrower crowns generally have greater vigor and respond better to release. A released tree suddenly exposed to a better light regime may have to convert foliage to adapt to the new light conditions (Tucker and Emmingham 1977; Tucker et al. 1987), but a tree in a better prerelease light environment will have less foliage to convert (Oliver and Larson 1996). Deciduous trees respond more rapidly because foliage conversion is more rapid. Larger trees may actually exhibit slower growth after release than smaller trees, as larger trees have a greater respiration sink that requires a large part of photosynthetic production of the released tree (Figure 4.15; Oliver 1976; Ferguson and Adams 1980; Kneeshaw et al. 2002). This observation agrees with the report by Messier et al. (1999) that there is a maximum sustainable height for boreal species in Canada, where larger trees are less able to respond to release. The best trees for release in either overstory removal operations or in more complex multiaged stands are trees with crown characteristics indicating a vigorous state. Smaller trees also are more likely to respond to release than larger trees.

In contrast to evergreen conifers, broadleaved understory trees have more variable growth habits. These trees not only have a different architecture than conifers do but also have different life strategies for response to disturbance and release (Bond and Midgley 2001). The most significant factor is that many of these broadleaved species can reproduce through stump or seedling sprouts. Postrelease growth rates are affected by prerelease vigor, with trees with high vigor responding more rapidly than trees with low vigor. When suppression results in poor form, release often results in retention of the poor form characteristics of the released tree. In these cases, sprouting trees may be cut at the base so that they resprout, forming a new stem or stems with better form (Johnson et al. 2009).

4.6 Mortality

Tree mortality is an important process for any stand, including complex stand structures such as multiaged and mixed-species forests (Franklin et al. 1987). However, tree mortality patterns are poorly understood in any stand structure. Most research has focused on even-aged stands, but some information is available on complex stands. A primary focus of many studies is the change in the density of these stands over time. For even-aged forests, these density trends are well documented as self-thinning curves (Figure 4.16). Similar limitations on size

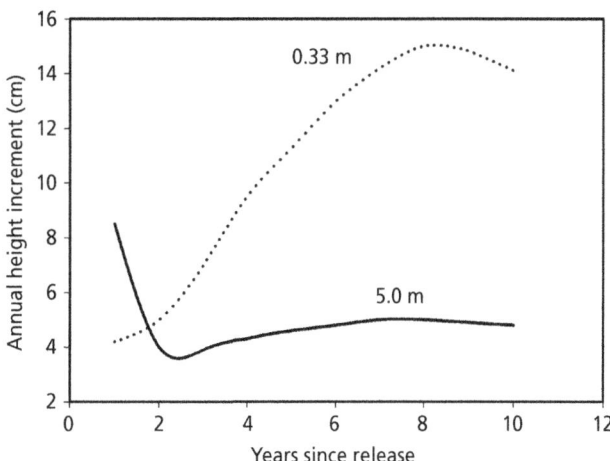

Figure 4.15 Height increment for grand fir trees following release in north Idaho, US. Trees that were 5.0 m in height at the time of release (solid line) grew rapidly the first year and then slowed. Trees that were 0.33 m in height (dotted line) grew slowly at first but increased their growth rapidly over the 10 years following release (adapted from Ferguson and Adams 1980).

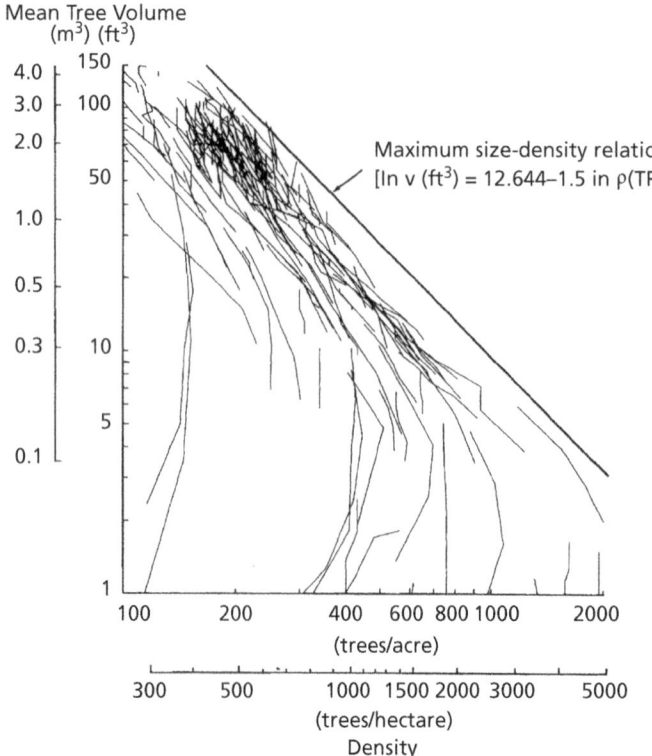

Figure 4.16 Self-thinning relationship for coast Douglas-fir in the Pacific Northwest, US. Lines represent repeated measurements of permanent plots that generally move upward and to the left on the diagram and are limited by the maximum size-density relationship. Vertical movement on the diagram represents an increase in average tree size. Movement to the left represents mortality, and movement to the right is ingrowth (from Drew and Flewelling 1979; reproduced with permission from the Society of American Foresters).

and density must surely exist for complex strands. However, whereas these self-thinning relationships indicate that there is a limiting condition for average size and density, they do not give insight into the dynamics of mortality.

For even-aged stands, the process of **differentiation**, or separation of trees into crown classes (O'Hara and Oliver 1999), is the first step in the suppression and mortality of small trees. Vigorous trees become dominant, and less vigorous trees fall into subordinate positions. For all trees but the dominants, there is a general decline in status, even though they may be increasing in size, and the number of dominant trees declines with time. In young, even-aged stands experiencing stem exclusion, mortality is concentrated in smaller trees as a result of competition and various forms of mechanical damage (see Lutz and Halpern 2006).

In older and more complex stands, mortality patterns become more complex. Diameter frequency distributions are often used to describe the structure of complex stands, particularly multiaged stands.

These distributions are discussed in more detail in Chapter 7. A diameter distribution with a negative exponential shape is a common form used to describe multiaged stands and as a target for management. The assumption imbedded in the use of these models is that trees in small classes eventually replace trees in large classes and that, consequently, the decline in the number of trees with increasing diameters is indicative of a high rate of mortality. But these negative exponential diameter distributions may be poor representations of mortality in natural stands. In northern hardwoods in North America, Goff and West (1975) concluded mortality was high in small trees, low in the main canopy, and high again for the largest trees. Mortality in the understory was high because of the large numbers of trees and the poor light conditions. The mortality in large trees was due to senescence and the reduced vigor in the oldest trees.

Because of the abundance of understory trees in some forest types—either to due to ease of regeneration or intentional management—most mortality in multiaged stands occurs in the understory (Harcombe 1987). Many young seedlings never become fully established, succumbing to an assortment of biotic or abiotic factors such as desiccation, herbivory, insufficient solar radiation, etc. (Swaine and Hall 1983; Kitajima and Augspurger 1989; Collet and Moguedec 2007; Keyes et al. 2009). In addition, mortality rates may be lower for some species because of their shade tolerance or preferential grazing by herbivores. For example, Petritan et al. (2007) found that in Germany, the more shade-tolerant European beech had a higher survival rate than the less shade-tolerant maple and ash did. In Slovenia, Klopcic and Bončina (2011) attributed a decline in the numbers of silver fir to ungulate browsing. Similarly, in tropical north Queensland, Australia, seedlings protected from browsing had a much higher rate of survival than unprotected seedlings did (Osunkoya et al. 1992). Understory trees in boreal, temperate, or tropical forests experience similar barriers to development, and mortality rates may be very high.

As trees develop, they experience an evolving set of stresses that lead to mortality. For saplings, shade tolerance is a key variable in light-limited systems. For example, in coastal British Columbia, Canada, mortality was directly related to shade tolerance, with the shade-intolerant paper birch experiencing a rate of mortality that was more than 10 times that of the more tolerant western redcedar (Kobe and Coates 1997). In addition, for a group of boreal species in eastern Canada, Kneeshaw et al. (2006) concluded that size was also a factor affecting sampling mortality because of the greater respiration costs of larger trees. However, on sites with moisture limitations, shade and drought tolerances may be conflicting factors and may lead to circumstances in which less shade-tolerant species are more likely to survive (Caspersen and Kobe 2001).

Managed multiaged stands will typically be managed at considerably less than maximum density through their cutting cycles (O'Hara and Valappil 1999). At lower densities, well below the maximum, competition-related mortality should be minimal. But older trees may die from pathogens, wind, or general senescence. Moreover, management operations can contribute to mortality. Trees can be damaged by harvesting operations or as a response to sudden changes in environmental conditions.

When Casperson (2006) measured background mortality in managed and unmanaged northern hardwood stands in eastern North America, they found that although the rate of postharvest mortality was relatively minor, it was highest in smaller trees and thus could lead to unexpected changes in stand structure and development. In a study of North American boreal forests, postharvest mortality was high in the first year after harvest but tapered over the next 10 years to background levels (Thorpe et al. 2008). Windthrow, a major cause of postharvest tree mortality, is greater following more severe treatments that result in greater exposure to residual trees (Gardiner et al. 2005). Mortality will occur even in stands managed well below densities where competition-related mortality usually occurs. This noncompetition mortality is due to logging damage to roots or stems, greater exposure of trees to wind, or simply background mortality factors that are independent of harvest activities.

4.7 Synthesis

The dynamics of multiaged and other complex stand structures govern many of the management options for these stands. These dynamics indicate the importance of recruiting and maintaining regeneration in an understory canopy position. In some situations, light limitation may be a factor, whereas moisture or nutrient limitations may be important in others. Moreover, in any situation, understory development is controlled, at least in part, by overstory competition, which may itself limit light or moisture. Shade tolerance is a dominant factor limiting options to design stand structures; for example, it would be counterproductive to design a stand with shade-tolerant species in the overstory and intolerant species in the understory. Similarly, mortality is also a function of these overstory/understory trade-offs but can be reduced by maintaining total stand density at less than maximum levels.

Growth patterns of multiaged stands may follow the idealized pattern shown in Figure 4.14, or, more likely, follow countless other variations. For example, subtle changes in overstory density will affect understory growth. Likewise, variations in understory density will affect understory growth when the overstory is held constant. Species vary in their ability to respond to increased growing space and in their ability to endure low levels of resource availability.

Although the multiaged stand dynamics presented in this chapter highlight the limitations of managing multiaged stands, they also demonstrate the many options available for managing these stands. Managers have the flexibility to slow or accelerate the growth of understory cohorts, thereby concentrating growth in either the understory or the overstory. They can also modify overstory density to affect species composition and control mortality. Understanding the dynamics of complex stand structures is not only important for determining the limitations of managing these structures but also for discovering the possibilities.

CHAPTER 5

Dynamics of forest gap and group openings

5.1 Introduction

Gaps in the forest canopy represent opportunities for forest regeneration and are an important feature of complex stands. A common theme in contemporary silviculture is the emulation of natural disturbances with silvicultural treatments (e.g., Attiwill 1994; Fries et al. 1997; Seymour et al. 2002; Perera et al. 2004; Drever et al. 2006; Long 2009, Geldenhuys 2010). Systems which emulate gap formation and development provide opportunities to develop both complex structures and the processes that typically occur in stands where gaps are a normal part of the disturbance regime. Understanding gap dynamics is therefore a central part of understanding how complex multiaged stands are renewed and develop over time.

Gaps are usually visualized as small openings or clearings in a mature forest (Figure 5.1). They can vary in shape and size, and from minor structural anomalies in a canopy to complete clearings. **Gaps** are defined as spaces in forest stands due to individual tree or group mortality (Helms 1998), although the term was originally used and is often limited to single canopy tree openings (Watt 1947). Gaps may be formed by disturbances such as wind, root diseases, insects, large tree falls, branch fall, and many others. When these gaps are created through forest management, they are called **groups**, and occasionally coupes, or patches. **Coupes** may be small or large forest openings or clearings. For large openings, the terms clearing or clearcut may be used when the disturbance is large and the result of harvesting.

Patches have a broader definition as homogeneous stand structures that may be small or large in scale, are distinct from surrounding structures, and may be either cleared areas, mature forest structures, or some other structure (Helms 1998). White and Pickett (1985) described patches as having no constraints on size. Bugmann's (2001) review of gap models described a stand as an abstract composite of many patches, each with a homogeneous structure.

Whereas there is a confusing and overlapping terminology for these canopy openings, gaps are often single tree or small multiple tree openings. Patches can assume many sizes and be components of stands or consist of multiple stands (O'Hara and Nagel 2013). The homogeneity within a patch implies they cannot be multiaged, but patches may form multiaged stands.

Forest openings have unique stand structures that change over time and interact with surrounding stand structures. Hence the development of each gap is unique and mostly independent of nonadjacent gaps. They are an important component of multiaged stand dynamics in both unmanaged and managed stands. For silviculture, openings are a means of regenerating a forest, and specifically a multiaged forest. The most appropriate definition of gaps for multiaged silviculture are areas small enough so that the majority of the gap area is influenced by the surrounding edges; gaps below this threshold are selection systems, and above, they are clearcuts (Daniel et al. 1979; Bradshaw 1992; Smith et al. 1997). Gap diameters are also defined as less than the surrounding canopy height. **Group selection** (see Chapter 6) is a regeneration method that relies on larger gaps or group openings to regenerate shade-intolerant species or create stand

Multiaged Silviculture. Kevin L. O'Hara.
© Kevin L. O'Hara 2014. Published 2014 by Oxford University Press.

Figure 5.1 A small 400 m² experimental gap in a hinoki forest in Honshu, Japan.

structures with coarse-scale heterogeneity. The focus in this chapter is on how canopy gaps respond post-disturbance and, to a lesser extent, on the processes of gap formation. Hence, gaps and groups arising from both natural disturbances and silvicultural treatments are discussed together.

5.2 Gap dynamics

5.2.1 Gap size

Gaps are often assumed to be the result of the death of a single tree. However, gap formation is highly variable. Trees vary in crown size, height, and architecture, creating variability even when only a single tree falls. Tree stems can be broken or they can be uprooted, and they may damage other trees when they fall (Yamamoto 1996). Other natural disturbances, such as root disease, bark beetle mortality, wind events, and harvest treatments, can result in multitree gap formation. Gap formation from large branch fall, without tree mortality, is also possible (Yamamoto 2000). In tropical forests in Gabon, 10% of the gap area was from branches and other tree parts (Brokaw 1985b). **Lianas**, or woody canopy vines, can bind trees together, increasing multiple tree falls (Putz 1984). Throughout a given forested area, natural gaps form a wide range of sizes. However, usually a skewed distribution of gaps results in more small than large gaps but with a greater proportion of total gap area in large gaps (Brokaw 1985b; Liu and Hytteborn 1991; Coates and Burton 1997). Within a species, tall forests apparently form larger gaps than shorter forests (Clebsch and Busing 1989; Spies et al. 1990; Yamamoto 2000). Different-sized gaps in the same forest type may also serve different ecological functions (Lorimer 1989). Additionally, disturbance events that form gaps, such as a mixed-severity fire that burns through the understory and results in some gaps where trees are killed, can also affect the rest of the stand. Gaps can therefore be quite variable in size, shape, and in their frequency across a stand.

Because of these many variables, there is no standard definition of gap size. They may be defined by the relative height of the surrounding canopy or the amount of direct sunlight they receive (York et al. 2003). Gap areas are usually measured between driplines of the surrounding trees, but an **extended gap** includes the area between tree stems (Runkle 1982). Definition of a gap area is difficult and can vary by a factor of 2.0 depending on how the gap is measured (Brokaw 1985a). Gaps also have a vertical dimension: Brokaw (1982) described gaps as holes in the forest canopy down to a 2 m height above ground. In many gap studies, a large gap is defined for convenience as greater than 1000 m² (Schliemann and Bockheim 2011). In New Zealand beech forests, Stewart et al. (1991) considered a gap to be anything greater than 20 m². In studies of gaps or groups created through silvicultural interventions, sizes up to 1 ha are common. Bradshaw (1992) speculated that—using the traditional silvicultural definition that a gap is anything influenced by edges (Daniel et al. 1979)—in a forest with very tall trees, a gap could be 10 ha in size. This range in gap/group size, by itself, creates a lot of variability in post-disturbance response.

5.2.2 Gap development

Gaps change in many ways after they are formed. A gap formed by the felling of a single tree often leaves an opening in the canopy previously occupied by the crown and disturbed areas where the tree and crown fell (Figure 5.2). For tropical forests, this has been called a *chablis*, which recognizes the heterogeneity in the areas affected by the gap formation (Oldeman 1990). The gap may have considerable soil disturbance if the tree was uprooted, which would result in the presence of bare mineral soil in the gap, an elevated tree stem, and the damage and debris associated with the crown impact (Figure 5.3). Some stems are only broken instead of uprooted, thereby exposing no mineral soil. In other cases, the fall of one tree can knock over another tree or predispose others to windfall. The size and architecture of the fallen tree also affects the gap environment. Each disturbance may therefore form a heterogeneous within-gap environment as well as heterogeneity between gaps.

After gap formation, the growing space available from the disturbance is reoccupied. This may involve new seedlings, vegetative regeneration, canopy expansion from around the gap, and encroachment of roots and plants from the ground level. Gap edges are therefore critical areas where many changes occur. An edge is affected by the environments on either side. For example, the light passing through the gap opening may penetrate the stands on the side of the gap away from the sun (Canham et al. 1990; Figure 5.1). Likewise, on the shaded side of the gap, there may be no direct light due to the shade from the surrounding stand (Oliver and Larson 1996). Studies have also documented increased growth of trees surrounding gaps (York and Battles 2008), and regeneration around openings (Golser and Hasenauer 1997). Edges are complex environments and interact with gaps in a number of ways.

Canopy expansion into the gap from surrounding trees can be rapid in species with a high level of crown plasticity, such as many broadleaved trees in temperate or tropical forests. Small gaps may be completely closed, whereas larger gaps may only experience an expansion around their edges. Alternatively, many disturbances may incrementally expand the gap area, creating areas within a single gap that are different ages since gap establishment. For example, wind may gradually remove trees once they become exposed. Root disease or some insects may attack trees in gaps and also attack

Figure 5.2 Gap in mixed beech forest in New Zealand. This gap was created with a selection treatment and shows a variable canopy pattern.

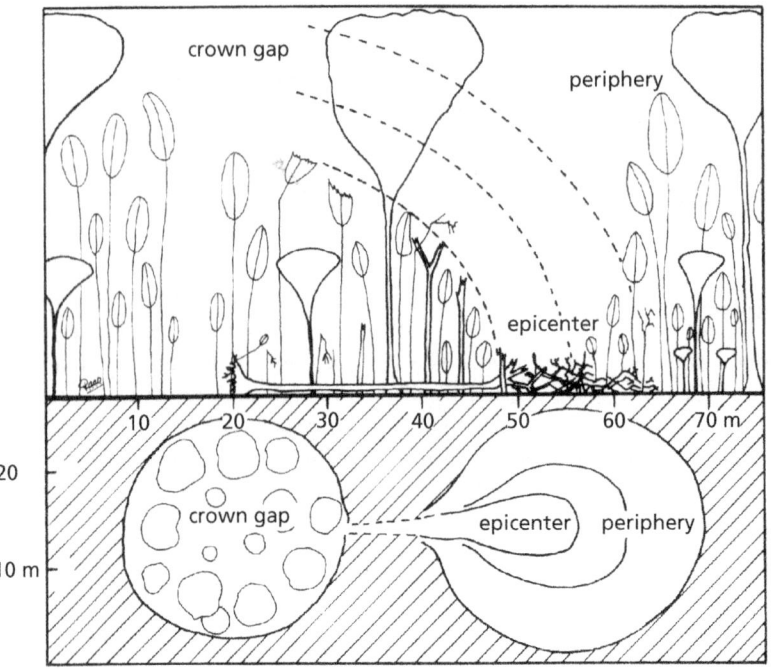

Figure 5.3 Diagram of chablis or gap, showing the gap in the canopy as well as the damage from the fallen crown (from Oldeman 1978; reproduced with permission from Cambridge University Press).

them incrementally, thereby expanding the gap. Conversely, some gaps may be so small that any regeneration that does occur becomes suppressed and excluded from the expanding canopy above. In these cases, a gap may be a temporary phenomenon that acts more like a thinning than a regeneration event.

The emerging gap flora rises upward as it develops. It typically rises faster in the gap center where resources are most available (Figure 5.4). Gray and

Figure 5.4 Group selection opening in a ponderosa pine stand in Montana, US. Note the increase in tree size with distance from the edge (left).

Spies (1996) observed faster growth of three Pacific Northwest species in larger gaps and in gap centers. Trees, shrubs, and herbaceous plants will originate from seeds, advance regeneration, sprouts, layering, or root sprouts/suckers. **Self-thinning**, or competition-induced mortality, will reduce tree and shrub density as the gap vegetation moves upward. The development will resemble even-aged stem exclusion (Oliver and Larson 1996), with more competitive individuals dominating and suppressing those that are less competitive. McDonald and Reynolds (1999) note increasing growth rates for conifers in gaps through 28 years in California's mixed conifer forests. The edges will encroach, providing additional competition to new plants in the gap, and alter inter-tree competition within the gap. Eventually, without further disturbance, a few trees will enter the upper canopy, and the former gap will, more or less, blend in with the surrounding forest. This process is referred to as **gap phase** reproduction (Brokaw 1985b) or **silvigenesis** (Oldeman 1978), and it varies widely across different forest types due to a great many environmental factors.

In theory, a stand or forest may replace itself over extended time periods with a revolving series or **shifting mosaic** steady state of gaps (Watt 1947; Whitmore 1975; Bormann and Likens 1979). This shifting mosaic forest may consist of many gaps at different stages of development, such that the entire forest would periodically be replaced and at any given time is all aged. This theory often assumes a consistency of autogenic disturbances, such as tree falls from senescent trees, over time. These systems are largely theoretical in their exclusion of external factors and simplification of forest development and replacement in forests affected by minor disturbances. In reality, disturbances vary in type and in their regimes (see Chapter 3), and the attempts to classify processes as autogenic or allogenic often results in missing the joint effects of both types (Box 5.1; Oliver and Larson 1996; Kimmins 1997).

Small-scale disturbance regimes do produce multiaged stands and can serve as models for disturbance emulation. However, disturbance effects are never so idealistically homogeneous over space and time as to indicate the narrow range of options represented by a shifting mosaic process and the resultant all-aged forest. Instead, patterns of gap

---

Box 5.1 "Succession," disturbance, and forest management

Succession is an old ecological concept describing the sequence of plant community changes toward a **climax** or self-perpetuating community. The term "succession" can be traced back centuries (Kimmins 1997), but the self-perpetuating climax community is generally attributed to Clements (1916). This model describes a deterministic succession of plant communities that moves toward a climax community through a "relay floristics" pattern (Egler 1954). It implies a relatively stable process where the directional succession of plant communities is normal and perturbations, or disturbances, to this process are the exception (Pickett and White 1985; Oliver and Larson 1996). For forestry, the succession to the climax concept has been the basis for site classification in some regions (Daubenmire 1952, 1976).

The implications of the climax model for forest management were relatively simple: vegetation change was predictable, even orderly, and ecosystems existed in a steady-state equilibrium where disturbance and recovery are balanced (Bormann and Likens 1979; Sprugel 1991). Also, climax forests were the natural community on a site, and these communities were therefore desirable (Sprugel 1991; Oliver and Larson 1996). For multiaged silviculture, the self-perpetuating climax model implied these communities had all-aged stand structures. This promoted single tree selection approaches under a related assumption that natural stand replacement was a shifting mosaic of small gaps. Even high grading (see Chapter 7), where high-value overstory trees were preferentially cut, could be justified because it simulated the shifting mosaic pattern.

Our current understanding of stand development processes reveals forests are much more dynamic than previously assumed (Oliver and Larson 1996). Ecosystems are neither deterministic nor orderly. Instead they are chaotic. Disturbances are relatively common and are a dominant force in directing change in any forested ecosystem. In some regions, site classification continues to be based on climax theory (Cook 1996; O'Hara et al. 1996), and vegetation change or stand development is often referred to as "succession." Nevertheless, there is a growing recognition that the chaotic nature of forests limits the applicability of the traditional succession model. It implies that there are many developmental pathways that vary with disturbance, climate, or even the unpredictability of seed production. Christensen (1988) wrote that land managers

continued

> **Box 5.1** *Continued*
>
> should recognize that "untidiness may be an integral part of maintenance of many ecosystems." Rather than being an obstacle to management, this untidiness justifies varied and flexible approaches for managers. It provides an opportunity to meet objectives in different ways that are not constrained by outdated notions of a "stable Eden" (Marris 2011). It is an empowerment for foresters to use their creativity to meet objectives.

dynamics indicate great ranges in natural processes and great ranges in opportunities for disturbance emulation.

5.2.3 Gap light regimes

As gaps in the canopy, the transmission of light is a dominant factor in vegetation response. Consequently, much of the research work associated with gaps has focused on light or on vegetation responses in light-limited environments. The light regime within a gap is dependent on the size, shape, orientation, latitude, slope, and aspect of the gap, as well as the height difference of the surrounding forest. The light regime also changes diurnally and seasonally. A series of simulated 400 m² gaps are shown in Figure 5.5, where shape and orientation interact with time of day to have different effects on light and shade in the gaps. In these northern-hemisphere gaps, the best average exposure is found in a variety of shapes and orientations, but differences are relatively small. In tropical forests at lower latitudes, much of the gap area may receive full direct sunlight (Canham et al. 1990). Small gaps at higher latitudes may receive no direct sunlight if the angle of sunlight does not approach overhead. Large gaps or small clearings, particularly in a north–south direction (Figure 5.5) may receive considerable direct light. The height of the surrounding trees can have an important effect on the gap light regime. In an old forest in the Pacific Northwest of North America with a 70 m canopy height, a small gap received no direct sunlight, whereas similar gaps in other forest types with shorter canopy heights did receive direct light (Canham et al. 1990).

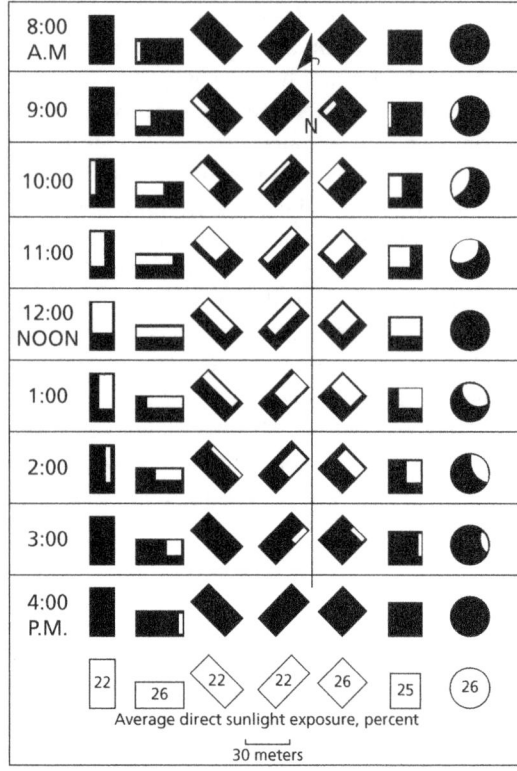

Figure 5.5 Changes in light environment with 400 m² gaps of various shapes on June 7 at 44° N latitude (from Marquis 1965).

In other forest types, gaps of 1000 m² generally received less than four hours of direct sunlight. Light regimes in gaps can be amplified or tempered by slope and aspect. A gap on a slope facing the sun has a similar light regime as a larger gap, whereas a gap on a slope facing away from the sun may get no direct light (Figure 5.6). An additional element of gap light regimes is whether the canopy trees are deciduous or evergreen, which may result in dramatic annual variation in light quality.

In an effort to quantify gap light regimes, the gap light index GLI was developed as a function of diffuse and direct light to quantify the transmission of light over a growing season (Canham 1988b). The GLI requires some direct measure of solar radiation, such as through light sensors or hemispherical photography, to estimate the light regime. A refinement was the gap light index GAPLI, which uses average weather, gap characteristics, latitude, and longitude, but not light measures (Dai 1996). In either

DYNAMICS OF FOREST GAP AND GROUP OPENINGS 43

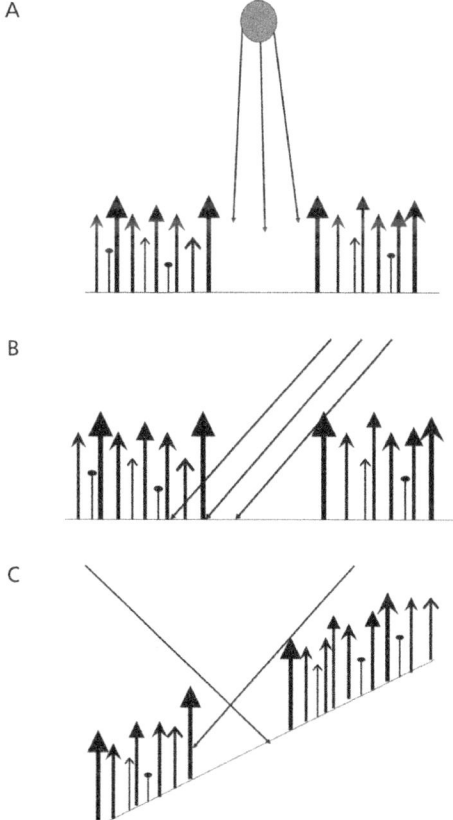

Figure 5.6 Diagrams of three gaps at different latitudes and on different slopes. Subfigure "A" shows a low latitude gap with light penetrating the entire gap. "B" shows the gap at a high latitude, where only part of the gap receives direct sunlight. "C" is a gap on a slope with arrows showing two possible angles for sunlight. If the slope faces the light, the gap receives direct light. If the slope faces away from the light, the same-sized gap may receive no direct light.

case, these indices demonstrate the complexity of understanding light regimes (see Section 4.2) and the need for simple quantitative measures to describe these them.

5.2.4 Gap below-ground competition

Most gap studies have focused on the above-ground competition factors because the major changes in gap formation are in the canopy and these are much easier to measure. However, below-ground competition is also affected by the above-ground formation of the gap, including the light regime.

For example, an increase in solar radiation and the continued presence of the root system from neighboring trees can negate any potential increase in soil moisture levels from a within-gap reduction in the trees or other vegetation. Gap size may exacerbate these trends. For example, in larger gaps, more soil radiation would raise air temperatures, possibly leading to faster litter decomposition as compared to smaller gaps. Comparisons of different gap sizes in both mesic silver fir forests and dry stone pine forests in Italy indicated larger gaps have higher soil temperatures, lower soil moisture, and less organic matter (Muscolo et al. 2007; 2010). In these studies, smaller gaps had greater organic matter, higher carbon-nitrogen ratios, and a greater availability of nitrogen and phosphorus. Not all studies have shown similar results. On a tropical rainforest site, nitrogen and phosphorus were not significantly different in relation to gap size (Vitousek and Denslow 1986). In European beech forests, larger gaps resulted in increased soil temperatures, but gaps generally had few effects on nutrients or mineralization (Bauhus 1996; Bauhus et al. 2004; Ritter 2005). Although gap formation may generally change the microclimate of the gap area by increasing soil temperatures and soil moisture, the effects on nutrient availability and litter decomposition may be more dependent on other variables such as gap size, climate, and other species present. Climate becomes an important factor because decomposition may be greater when moisture and temperature regimes coincide, regardless of gap size. These relatively subtle changes in below-ground and ground-layer processes may therefore be highly important in determining gap development patterns.

5.2.5 Gap partitioning

Species may be more competitive in different parts of a gap because of the edge-to-interior gradients that occur (i.e., **edge effect**) coupled with the highly variable microenvironments within gaps. The occupation of different environments or niches is described as **gap** or **niche partitioning**, where resources are partitioned among species or species groups (Ricklefs 1977). For example, a gap study in eastern North America found three species occupying different environments within gaps, with striped maple having the greatest variability in growth and photosynthesis, sugar maple with the least, and red maple with an intermediate overall response (Sipe and Bazzaz 1994, 1995). Herbaceous plants can also compete with trees to different degrees depending on gap size or location within gaps. In Slovenia, the shade-tolerant wild garlic was more competitive in small gaps, pushing the European beech to gap edges (Diaci et al. 2012). Denslow (1980) concluded that species diversity in tropical forests was partially due to variations in gap size and gap partitioning within gaps to favor the establishment of many species.

The presence of a species in a particular niche or gap microenvironment may be the result of successful regeneration or successful growth. Coates and Burton (1997) referred to the **regeneration niche**, where regeneration occurs because of desirable microsite conditions. The subsequent growth of established trees, or the **growth niche**, indicates these trees are competitive in their environment. It is possible that a species may regenerate simply because seed landed on a suitable substrate (Hofgaard 1993). If this suitable *regeneration* substrate is unsuitable for competitive *growth*, the species will not persist. Hence, long-term competitiveness is multifaceted and requires a sequence of adequate conditions or niches.

Regeneration mechanisms may also be important in affecting species composition in the gap. Advance regeneration that predates the gap-initiation disturbance and other forms of vegetative regeneration is fixed in location. Root sprouts may emerge from adjacent trees after a gap is formed. These individuals have some level of established root system that provides rapid initial growth in the new gap. However, these forms of vegetative regeneration may not be competitive in all cases due, for example, to the light regime. If they are successful, their growth and development would be an example of a growth niche, since their original position in the gap was primarily by chance. These forms of regeneration may also represent competitive advantages in terms of postregeneration or postrelease growth (Bond and Midgley 2001).

The minor disturbance patterns that result in repeated gap formation across a stand or forest are an important contributor to diversity. This diversity is partially the result of the regeneration and growth advantages associated with certain microsites in gaps. Niche partitioning is therefore a means of explaining not only the competitive advantages of species in different parts of gaps but also the high species diversity often associated with small-scale disturbance patterns. The importance of random events or chance in affecting species composition is also important (Brokaw and Busing 2000). In either case, multiaged silviculture that uses artificial gaps to create variable environmental conditions is a key to enhancing stand-level species diversity.

5.3 Gap studies—gap variability and implications for management

Most ecological studies of gaps have focused on natural or small artificial gaps created to simulate natural disturbance effects. Openings are also created for study in managed systems as regeneration treatments. In these management systems, the openings are harvest treatments where a new age class of trees is regenerated. Both types of studies provide opportunities to examine how a range of variables interact to affect post-disturbance development of the canopy opening. The managed systems vary considerably in different regions and use two different forms of selection silviculture (see Chapter 6). At one extreme, individual trees may be removed to create small forest openings; and at the other, groups of trees are removed, resulting in larger openings. These larger openings may range up to approximately 1.0 ha or larger. Although these openings in managed systems are often larger than natural gaps, the processes that affect light, soil moisture, and other ecological processes are similar, and these large openings provide valuable experimental insights into gap effects. This section presents a series of gap and group studies that show interactions between variables and the complexity of studying vegetation responses in forest openings.

Gap responses are complicated by many variables. Latitude is assumed to be important because of its effect on sun angle and shading in gaps (Figure 5.6). For example, in small windthrow gaps that were no larger than 360 m^2 in boreal forests in Sweden (59.88° N), regeneration of the relatively shade-tolerant Norway spruce was favored

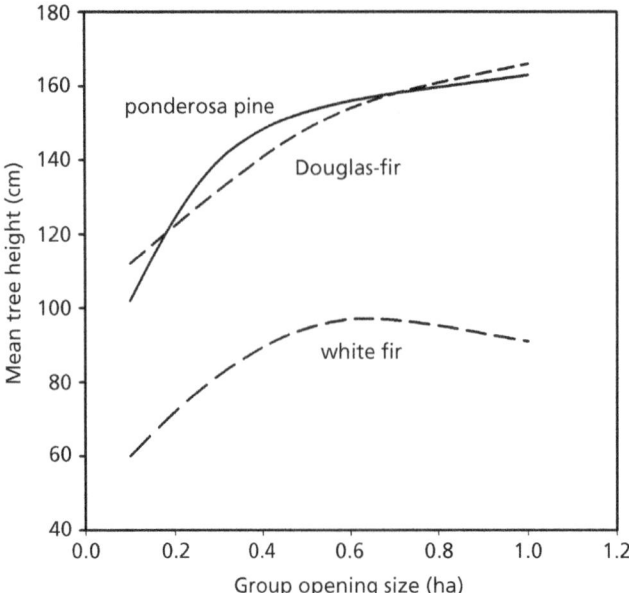

Figure 5.7 Height development with increasing opening size for three species in Sierra Nevada mixed conifer forests in California. Ponderosa pine and coast Douglas-fir both reach their greatest growing in the 1.0 ha openings, but white fir grows slightly better in 0.6 ha openings (adapted from York et al. 2004).

1.0 ha (Figure 5.7). Partitioning was also evident, with giant sequoia growing fastest in group centers and generally being the most site-sensitive species. White fir was the only species that was smaller in the largest group size than in the 0.6 ha group (Figure 5.7).

Canopy gaps are often visualized as uncommon holes in the continuous forest canopy that gradually refill to restore the continuous canopy. Forests affected by tree falls and other gap-forming disturbances are much more complex. For example, Hanson and Lorimer (2007) found that wind-disturbed northern hardwood stands in North America received greater light on the forest floor than comparable managed stands. Development of a stand following gap formation is also complex. Under an idealized model, forests that experience gap disturbance regimes would be undergoing continuous gap formation, with regrowth at different stages in different gaps. This has been referred to as turnover rate, the mean time between gap-creation events at any point in the forest, or the gap area created per unit of forest area per unit of time (Brokaw 1982). Brokaw (1985a) reported turnover rates from 60 years to an extreme of 416 years, implying, at the lower levels, continuous states of gap replacement and development. In several forest types in Japan, old forests experienced repeated gap formation at rates as high as approximately 0.2%/year (Torimaru et al. 2012). These repeated gaps were also often formed adjacent to previous gaps, indicating complex patterns of long-term gap formation. This area of long-term spatial and temporal dynamics of gap formation is largely unexplored but represents a major area of variation in gap responses. For example, disturbances affecting areas at different stages of gap recovery, adjacent gaps and their effects on light regimes, or the availability of regeneration propagules when adjacent areas have different compositions or are at different stages of development are all sources of dynamism in gap development.

This overview of gap/group studies is not meant to be comprehensive but rather to show the great variation in factors affecting gap development. Brokaw and Busing (2000) noted the lack of patterns in species establishment in gap studies and attributed it to chance. However, it could also be interpreted as the complexity resulting from so many variables interacting to affect gap or group responses after disturbance. There is also a need to separate processes that affect establishment from those that affect subsequent growth, or what Coates and Burton (1997) described as the regeneration and growth niches. The competitive advantages that allow species to regenerate and become established on certain substrates may not result in advantages, as species subsequently begin to compete for scarce resources. There is also clearly an element of chance in these processes. For example, the happenstance of a gap forming where advance regeneration is already present or the random seedfall on an ideal microsite for germination and growth are chance events. These factors are stochastic and are largely beyond our prediction capabilities.

5.4 Gap replacement models

As forest openings are complicated by a myriad of variables that make experimental approaches difficult, there have been a number of models developed to simulate gap processes. These models aid our understanding of gap and larger-scale processes by simulating process of stand development of many individual forest patches. There are many of these gap models, and there are, of course, differences between models. Generally, modeled patches are assumed to be homogeneous and exist at different stages of development. Processes of regeneration and mortality are also simulated. Gap models often include ecophysiological parameters that allow them to simulate effects of climate change. These models generally ignore interactions between patches, so they generally do not simulate edge effects between gaps and surrounding forest. In this sense, they are not gap models but instead are patch models that simulate the development of many small, homogeneous groups of trees. More information is available from Shugart (1984), Bugmann (2001), and Schliemann and Bockheim (2011). Modeling approaches for multiaged stands are presented in Chapter 13.

5.5 Synthesis

The dynamics of forest openings, whether they are called gaps, patches, or groups, are complex by any measure. There are a few patterns that emerge as relative constants in post-disturbance gap dynamics: 1) the importance of edge effects in affecting light regimes and affecting regeneration processes, such as the spread of seeds or the expansion of edge plants into the opening; 2) an improved light environment with increasing gap size, as, in many gaps, there are gradients in resource availability that result in gap partitioning to form areas where some species are more competitive than others (e.g., it is common to see greater tree growth in the center or the sunlit side of gaps); and 3) the gap environment is dynamic in nature on both a diurnal and long-term basis. Despite these commonalities, there are seemingly more ways that gaps are dynamic in the sense that consistent patterns are not evident. For example, patterns are not always clear relative to soil nutrients, soil moisture, or the existence of gap partitioning in relation to shade tolerance. As with many ecological relationships, many variables interact to affect gap responses. We can neither characterize all of these variables that affect gap response nor generalize about response relationships across all forest types. This was evident in the extensive comparison table of gap study results which was compiled by Coates and Burton (1997) and where results from many gap studies in boreal, temperate, and tropical forests were efficiently summarized in one large table. It is possible that future, more in-depth study of gap processes will only reinforce the notion that generalizations are risky.

Since it is difficult to generalize about gap responses, we also cannot assume that emulating gap processes is as simple as removing individual trees and watching the forest regrow. Gaps may experience multiple disturbances, or disturbances may expand gaps over time, creating other confounding variables. Disturbance emulation of gap-based disturbance regimes should therefore be as variable as these natural disturbance regimes. In other words, this may warrant gaps of many sizes, gaps that expand over time, and gaps that are recreated on variable intervals. This indicates that a variable silviculture is needed to emulate highly variable processes and that silviculture needs to be flexible to avoid a homogenization of treatments.

CHAPTER 6

Multiaged management systems

6.1 Introduction

Silvicultural systems are a critical means of organizing the treatments in multiaged stands or for other complex stands. The multiaged **silviculture system**—or the planned series of treatments or operations for tending, harvesting, and re-establishing a stand (Helms 1998)—provides the means to maintain a structure over time, or modify it if needed. Unlike even-aged systems, established multiaged systems can be assumed to have no beginning and no end. The severity of periodic treatments and the intervals between treatments, or **cutting cycles**, can be established in the silvicultural system. Adjustments to the silvicultural system can be made to meet different objectives, reduce effects of an insect or pathogen, or adjust for effects of climate change. The silvicultural system is therefore a plan.

The normal convention in recent and past silvicultural textbooks regarding multiaged systems is to classify them as either single tree selection or group selection (Daniel et al. 1979; Matthews 1989; Smith et al. 1997; Nyland 2002; Tappeineir et al. 2007). **Single tree selection** is the removal of individual trees or small groups of trees of different sizes more or less uniformly in the stand (Helms 1998). **Group selection** systems periodically remove trees in groups and then regenerate groups of trees as small, even-aged patches. Together these groups or patches form multiaged stands. Both traditional selection systems may include treatments to tend the stand, such as thinning or pruning younger trees. A point of confusion is where groups become so large that they themselves become stands and should be called clearcuts. The same unit may be perceived as a stand by one person and a subunit of a stand by another because these individuals have different perceptions based on their objectives, training, or outlooks. The distinction between a group selection and a clearcut is generally based on the ecological effects of the unit's size. If there is sufficient edge effect from surrounding trees, then it constitutes a group; otherwise, a stand. The distinction on what constitutes a group might also be based on the intent of the manager: if the unit is regarded as a self-contained, sustained-yield unit that provides periodic and sustained volume production, then it might be classified as a stand (Smith et al. 1997; O'Hara and Nagel 2013).

The distinction between single tree and group selection is also based on the opening's microenvironment, which may affect productivity (Webster and Lorimer 2002). If the entire opening experiences edge effects, then it would typically be called a single tree selection system. Forest stands may also be managed to have other forms of spatial irregularity. This is particularly true for multiaged stands, which may have variable patterns of reserve or retention trees, or heterogeneous patterns of groups or patches.

The group and single tree selection systems are traditional silvicultural systems in forestry dating back centuries (Matthews 1989). However, the distinctions between them are essentially artificial. Edge effects are highly variable within a single stand due to variable heights of trees, aspect, gap orientation, and other factors (see Chapter 5). An alternative way to view multiaged systems is on a continuum from large openings on one extreme and very small openings on the other. Instead of being just two multiaged systems, there are many. This is analogous to a continuum of all types of silvicultural systems, shown in Figure 6.1. This continuum may also extend to other complex structures, such as those used in agroforestry (Box 6.1).

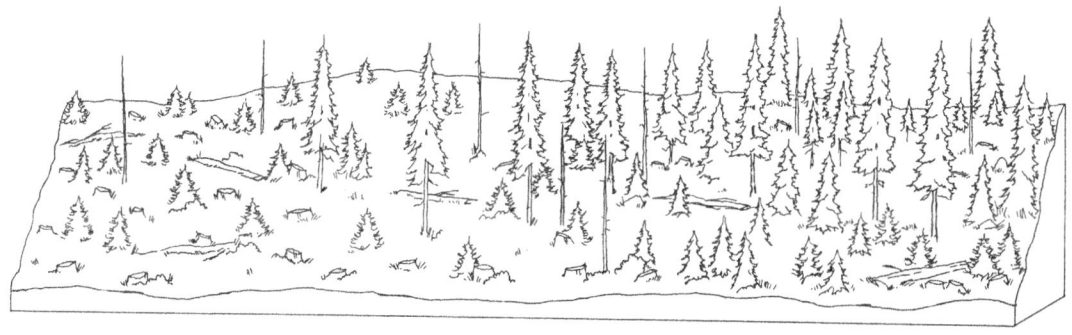

Figure 6.1 Silvicultural systems can be viewed as a gradient of options from complete removal of all trees (left) to removal of only some trees (right); (from O'Hara et al. 1994).

There are other silvicultural systems that result in multiaged stands. Retaining overstory or reserve trees in classical shelterwood (i.e., shelterwood with reserves; Nyland 2010) or seed-tree treatments will form two-aged stands. Coppice-with-standards is a system that maintains multiaged stands by managing multiple cycles of coppice intermixed with seed-origin trees. Likewise, variable retention systems that leave reserve trees for wildlife or aesthetics result in multiaged stands, as does any system that leaves reserve trees. Multiaged stands, and the systems that maintain them, therefore offer

---

Box 6.1 Multiaged silviculture and agroforestry

Agroforestry is defined as "intentional management of trees with agricultural crops" (Helms 1998). The management of, for example, two-strata agroforestry systems therefore involves trade-offs between an overstory and an understory, much like managing a two-strata forest. Hence agroforestry and multiaged silviculture have some similarities in both structure and ecosystem function. Our knowledge regarding agroforestry systems is primarily from tropical forests, whereas multiaged approaches are more common in temperate forests. However, both types of systems are flexible for use in boreal, temperate, and tropical forests. Citing recent reviews about retention forestry (Gustafsson et al. 2012; Lindenmayer et al. 2012) and agroforestry (Bhagwat et al. 2008), Roberge et al. (2013) noted the success of both systems in different environments, supporting the idea that either system probably has wide applicability in tropical, temperate, and boreal forests. Both systems have potential ecosystem benefits because they may more closely emulate natural ecosystem processes. Retention forestry emphasizes retention of structural elements including live and dead trees and patches of uncut forest but does not fully encompass a silvicultural system. Structural elements can also be retained in agroforestry systems. As landscape strategies for enhancing biodiversity, both fully developed multiaged systems and agroforestry systems have great potential. However, they must encompass more than retaining structural elements: a sustainable system needs to account for regeneration and long-term productivity.

Ultimately, either multiaged or agroforestry systems are about managing trade-offs between different strata, species, age classes, or even life forms in the case of agropastoral systems (Figure 6.2). These are complex questions involving densities of overstory trees, understory plants, or livestock. However, like multiaged stocking control, agroforestry systems also are more than managing trade-offs between overstory and understory for timber production. They provide opportunities to enhance diversity and vary the structure of individual land units (stands or agroforestry areas) to meet objectives over larger scales. Roberge et al. (2013) concluded that both multiaged and agroforestry systems have much to gain from the experience of each other, and also the experience and scientific understandings from other biomes.

continued

Box 6.1 *Continued*

Figure 6.2 Two-storied agroforestry system in Costa Rica with eucalyptus spp. in overstory and coffee in understory.

a great many options for management, and may encompass the majority of silvicultural options in contemporary silviculture.

This chapter is organized into systems that result in small openings and systems that result in large openings, rather than the conventional single tree and group selection. The objective is not to develop a new set of terminology but instead to recognize that the present distinction between traditional single tree and group selection systems is fuzzy and largely arbitrary. Although this classification corresponds very closely to traditional single tree and group selection, it is also important to recognize that there are a great many multiaged options for opening sizes, not just two. Also presented are multiaged systems that enhance structural variability, systems that emulate disturbances, and systems that give a great amount of discretion to the manager. It is recognized that the design of stand openings and the stocking control method are related decisions. The stocking control methods are presented separately in Chapter 7. This chapter focuses on ways of organizing openings within stands to achieve larger scale objectives.

6.2 Multiaged systems

6.2.1 Small openings

Systems that use small openings to perpetuate multiaged stands include traditional single tree selection, the European Plenter system, as well as many other variants (Figure 6.3). In these systems, small openings are dispersed throughout a stand. Areas in between openings may be treated to thin, developing cohorts or improve vigor of advance regeneration. In these systems, openings or gaps are small, and edge effects may exist throughout entire stands. Collectively, these systems tend to favor shade-tolerant species, because the light

Figure 6.3 Plenter forest in Switzerland managed using a very small opening sizes.

environments provided by these openings are generally less than maximum. In some forest types, implementation of systems with small openings in mixed forests may exclude the shade-intolerant species. For example, in many forest types in western North America, single tree selection favors tolerant fir species and discourages intolerant pines and larch. However, systems that create small openings in stands of shade-intolerant species can also regenerate these species. Ponderosa pine in the western United States is an example of a shade-intolerant species that is often found in natural and managed, multiaged, single-species stands with small and large openings.

Systems that use small openings are often designed to disperse effects in the stand in many smaller openings. This is often viewed as a milder form of disturbance, as compared to using fewer but larger openings. Small openings are also assumed to be a means to emulate gap phase and shifting mosaic patterns of stand replacement (see Chapter 5). However, natural disturbances typically leave variable-sized openings, total stand area affected is inconsistent over time, and these gap disturbances may not affect much stand area in a given time period (Runkle 1982; Oliver and Larson 1996). In practice, the management systems that leave small openings tend to leave uniform-sized openings and affect a larger area of a stand more consistently than the natural stands they attempt to emulate.

Yet another stand structure that may fit within the small opening category are the simple structures obtained by reducing the density of an overstory and regenerating a new age class. These two-aged structures may resemble the outcome of a shelterwood or seed-tree system with reserves (Figure 6.4). The result is a relatively simple structure with many of the desirable attributes of more complex stands; however, these simpler structures are easier to implement and maintain, and it is easier to project changes in these structure over time.

Single tree selection systems, and all multiaged systems, are usually designed to achieve some consistent level of timber production over time. At the stand level, this is achieved through periodic harvesting of equal-sized areas. Hence, systems that use small openings are not harvesting less area; rather, they are harvesting an equal-sized area as other systems but in a more dispersed pattern. These small openings are less obtrusive in their effects on aesthetics. This advantage may be a justification for using this system. An operational disadvantage is that access is necessary for all areas receiving timber removals. For systems with small openings, this can require extensive networks or roads and skid trails to reach the entire stand without damaging developing age classes of trees.

Regeneration in these systems is usually natural, but planting seedlings is also a viable alternative. In many of the forest types where these systems are successfully used, such as northern hardwoods in North America or in the mixed beech-fir forests of central Europe, natural regeneration is often plentiful in the form of advance regeneration, sprouts, and new seedlings.

Figure 6.4 Two-storied stand formed after a heavy low thin on the Fernow Experimental Forest in West Virginia, US.

6.2.2 Large openings

Silvicultural systems that use large openings may be attempting to accommodate the light requirements of shade-intolerant species, improve operational efficiency over systems that create small openings, or emulate disturbances that form relatively large openings. Group selection systems have been used for centuries (Matthews 1989), and the primary justifications were seemingly based on the light requirements of shade-intolerant species and operational efficiencies. Larger openings decrease edge effects as they increase in size and attain some of the operational efficiencies of clearcut systems when the groups become large.

Group selection systems are generally organized to treat a stand with a series of operations whereby every group is harvested over a period similar to an even-aged rotation with that site and species. This resembles **area control forest regulation,** via which a sustainable harvest is assured by periodic harvests that are equal in size. This is also often the justification for organizing treatments in multiaged stands with smaller openings and the concept of the balanced stand (see Chapter 7). Few places have managed group selection or large opening systems for many cycles, and it is common to see group selection implemented in even-aged stands or old forests, where the objective is conversion or transformation to multiaged structures. Conversion from an even-aged stand, or simply moving an existing multiaged stand into a regulated group structure, involves delineation of units for openings within the stand. The stand is subdivided into small units so that subsequent harvesting will affect a similar area through each cutting cycle. Figure 6.5 shows a group selection layout in a stand where four cutting cycles have been completed, but the stand is delineated for future group harvests through the year 2060.

Regeneration in large openings can be from natural sources or planted seedlings. In some cases, these treatments are part of restoration strategies where previous treatments favored shade-tolerant species and resulted in species conversions. Planting can be an effective means of establishing desired species and obtaining target species mixtures.

6.2.3 Systems to enhance variability

Traditional silvicultural approaches generally homogenized stand structures at the stand level. Stands were traditionally defined as management units with a homogeneous structure (O'Hara and Nagel 2013). Many current management directives attempt to create within-stand heterogeneity. One of these is **variable-density thinning,** which has been used primarily as a means of enhancing stand variability in even-aged stands (Carey 2003;

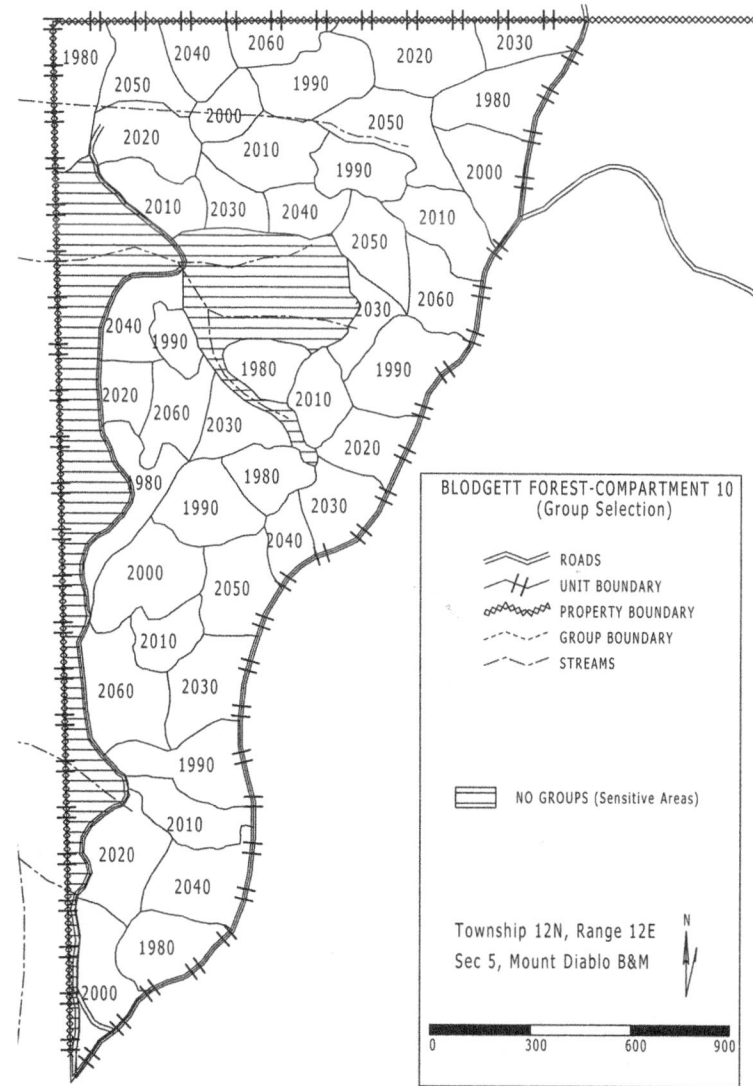

Figure 6.5 Map of previous and planned group selection units in a stand (compartment) in the Blodgett Forest in California, US. Groups are less than 1 ha in size.

O'Hara et al. 2012). However, Pukkala et al. (2011) used variable-density thinning to enhance variation in multiaged Norway spruce stands in Finland. In either even-aged or multiaged systems, the thinning method results in different thinning intensities within the stand. These thinning intensities might range from severe thinning in some areas of the stand to other areas that remain untreated. In multiaged stands, some areas might be harvested and thinned while other areas are left untreated, despite having high stocking levels. The difficulty with treatments, such as variable-density thinning, that attempt to treat a single stand in a variety of different ways is achieving stand-level heterogeneity in a systematic way (O'Hara et al. 2012).

Variable retention is used to retain structure in even-aged harvest units (Mitchell and Beese 2002). By retaining some residual trees, a two-aged or multiaged stand is created, and the posttreatment stand has both a stand structure that provides additional wildlife habitat and is often more aesthetically appealing than a clearcut. These trees are typically left in either aggregated or dispersed patterns (e.g., Aubry et al. 2009), but represent a relatively simple form of multiaged stands. Retention trees (reserves) might be left for a rotation and then harvested or perhaps left uncut to provide future snags or large trees. Variable retention is a key part of emerging strategies of retaining trees for multiple purposes (e.g., Gustafsson et al. 2012). However, at present, there is little guidance, and only limited regulatory requirements, for variable retention strategies. They represent a very flexible means of developing multiaged (two-aged) stands that are both relatively simple in stand structure and operationally simple to implement.

An **irregular shelterwood** or **Femelschlag** is another means to enhance within-stand variability (Figure 6.6). This system involves creating group openings that are expanded with addition entries until the stand is cutover (Figure 6.7). Although disturbance emulation was probably not a goal in the origination of this system in central Europe, it provides a means to form variable-sized gaps and expand these gaps over successive cutting cycles (Seymour 2005). As a silvicultural system, the irregular shelterwood possesses an extremely high level of flexibility. The system varies the opening size, includes some reserve trees, and expands openings over time, thereby providing a large number of stand structure elements for the manager to work with. An irregular shelterwood system could create multiple tree gap disturbances which include some surviving trees and expand over time, emulating many wind events. This might resemble the expanding gaps described by Torimaru et al. (2012) in Japan. Examples of irregular shelterwood or Femelschlag systems are described for the Black Forest

Figure 6.6 Central opening in Femelschlag in mixed forests in Slovenia.

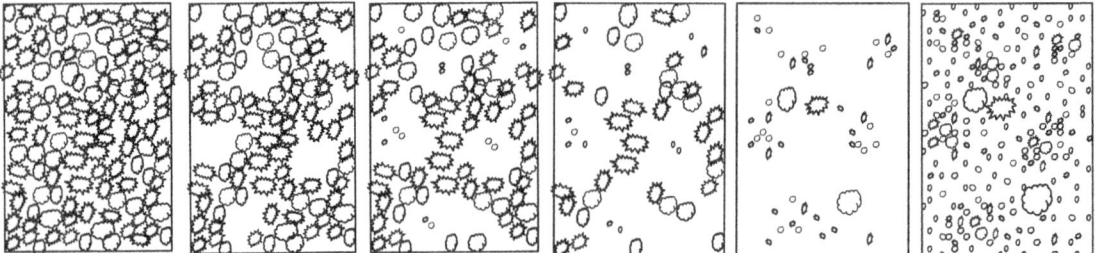

Figure 6.7 Sequence of steps in irregular shelterwood from left to right.

in Germany by Puettmann et al. (2009) and in northeastern North America by Seymour (2005) and Raymond et al. (2009).

Another option is simply creating both large and small openings within a single stand. This might be analogous to a hybrid system of both single tree and group selection in the same stand, or perhaps another way of implementing variable-density thinning. Stands with a range of opening sizes would provide additional structural diversity as well as flexibility to integrate existing structural features into designing the stand. For example, rather than attempting to make all openings the same, the manager could integrate existing openings and other structural features to enhance heterogeneity over the entire stand.

6.2.4 Systems to emulate natural disturbances

Emulation of natural disturbance regimes with silvicultural systems has become an increasingly common objective on many ownerships (Attiwill 1994; Angelstam 1998; Harvey et al. 2002; Franklin et al. 2002; Seymour et al. 2002; Perera et al. 2004; Drever et al. 2006; Geldenhuys 2010; O'Hara and Ramage 2013). Although emulating disturbance is a common theme for contemporary silviculture, it is not without precedent (Box 6.2). Multiaged silviculture should be a primary means to emulate disturbance regimes which result in partial disturbances and residual age classes or cohorts of trees. Disturbances are highly variable events with wide ranges of potential effects on forest stands. Disturbance

---

BOX 6.2 Unit area control

The ponderosa pine–Jeffrey pine forests that are found on the east side of California's Sierra Nevada and Cascade ranges are characterized as low productivity, sparsely stocked stands. The "unit area control" procedure was developed as a means to regulate or "control" spatially heterogeneous stands by dividing them into smaller "unit areas" (Hallin 1951, 1959). A feature of these stands that was noted with the development of the unit area control procedure was the occurrence of variable-sized groups of even-aged trees or two-age classes where senescent older trees exist over younger age classes. This implied a natural small-scale disturbance pattern that resulted in even-aged groups of trees (Figure 6.8 and 6.9). It was also noted that there was a preponderance of overmature groups. The

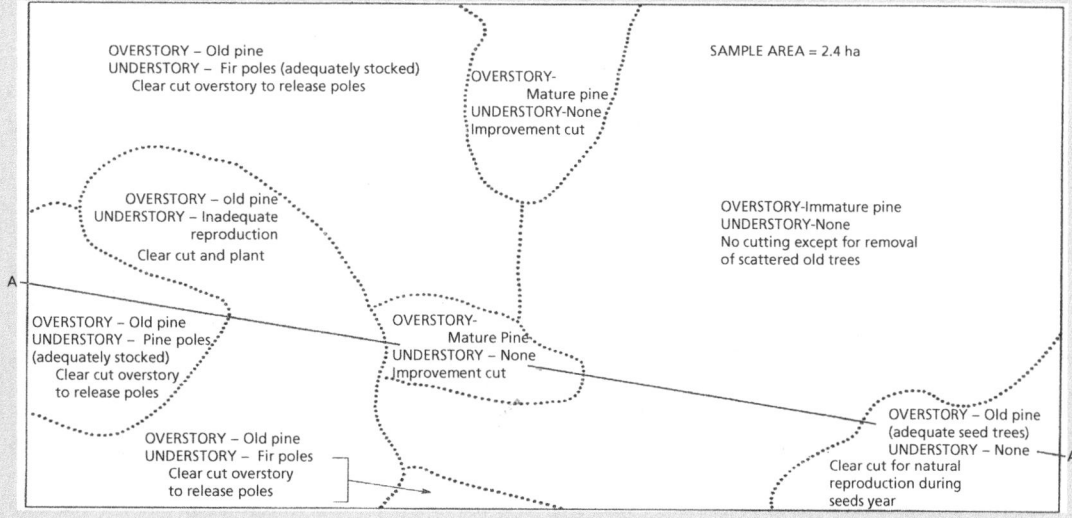

Figure 6.8 Spatial patterns of unit areas under Unit Area Control procedure. Diagonal line shows the location of the transect shown in Figure 6.9 (from Hallin 1959).

continued

Box 6.2 Continued

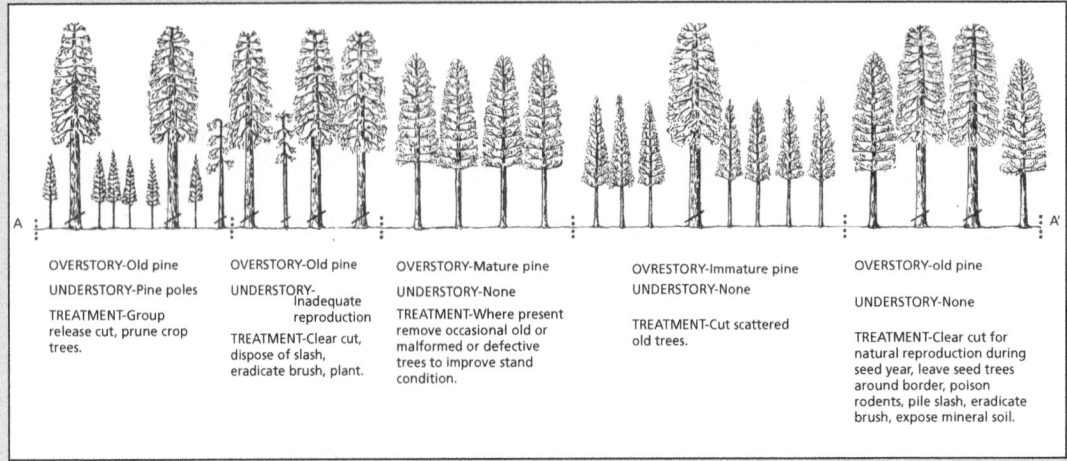

Figure 6.9 Cross-section of forest under Unit Area Control showing variation in stand structure across the transect shown in Figure 6.8 (from Hallin 1959).

primary goal of the unit area control method was to develop regulated stands that were resistant to insects and produced large, high-quality trees. Older trees are particularly susceptible to bark beetle attack, and these trees were removed if they overtopped vigorous regeneration. Since these large trees were also an overrepresented age class in the regulation plan, they were also the first priority for harvest. The regulation issues were largely concerned with reducing the trees in the overmature age classes and developing younger age classes which were relatively scarce.

In application, unit area control worked with the existing patchy pattern of age class/stand structures (Figure 6.8). These unit areas or groups varied in size from well less than a hectar to 2–4 ha. There was some subdivision and consolidation of units to reduce unit size and convert small units into a manageable size. However, unit areas were generally unchanged from the existing spatial patterns.

Designation of unit areas was primarily based on Dunning's (1928) tree classification system. Other criteria for designation of unit areas were age, species, degree of stocking, and presence of seed trees. Overmature trees were more than 300 years old. In this previously unmanaged landscape, old trees were both a risk to bark beetle attack and mortality. Management of unit areas included harvest cuts or overstory removal cuts of three types: 1) removal of overmature trees to release established advance regeneration; 2) removal of overmature trees to favor natural regeneration, a method used when good seed trees were present and when good seed crops could be anticipated; and 3) removal of overstory trees followed by tree planting. The competitiveness of shrub species was a concern on these moisture-limited environments, so quick reestablishment of trees was a priority.

Regulation produced a specified allowable cut. Since the unit areas were managed as even-aged units, even-aged concepts were applied. Rotations were set at 140 years, which was close to the culmination of mean annual increment and would produce dominant trees approximately 50 to 60 cm in diameter. Unit areas were temporally distributed on a 20-year cutting cycle within a stand. The target regulated stand was to have equal areas in each age or condition class. This may not have been achieved in every stand, but over larger scales, the goal was to have an equal distribution of age classes. The allowable cut was then simply the yield of these 140-year-old stands based on even-aged yield tables.

Unit area control was essentially a form of group selection that differed from the typical system because groups or units were highly variable in size and sometimes quite large. There was also a strong focus on even-aged development patterns within units. Unit area control attempted to emulate the natural disturbance regime, but with increased homogeneity of unit sizes to produce consistent volumes. The application of unit area control was an early example of blending the natural spatial and disturbance patterns into a new system of regulated management.

emulation, even within a single stand, has a great range of potential applications and effects on stand structure. It may involve active mechanical treatments or the cautious use of a disturbance agent such as fire.

Disturbance regimes are also not static, and climate change has the potential to modify or completely change disturbance regimes. Disturbance emulation may therefore be a moving target and may require adaptive strategies to adjust to changes in disturbance regimes. Another important facet of disturbance emulation is developing stand structures that are resistant to disturbance effects and resilient enough to recover rapidly from disturbance events. O'Hara and Ramage (2013) argued that multiaged stands provide more resistance and resilience to disturbances and are an important management strategy for adapting to climate change. However, the best strategy will be more than disturbance emulation: managers will also need to anticipate disturbance events, build variable treatment intervals or cutting cycles into silvicultural systems, and integrate salvage operations into management of residual stand structures (Puettmann 2011; O'Hara and Ramage 2013).

Disturbance emulation may take different forms with alternative disturbance regimes. In the fire-prone forests of western North America, fire regimes can be highly variable, leaving a great diversity of residual stand structures depending on stand structure, species, and other environmental variables. Systems that emulate these disturbance regimes should also be highly variable in terms of patterns and shapes of openings, and numbers, sizes, and arrangements of reserve trees (Larson and Churchill 2012; Larson et al. 2012; Churchill et al. 2013).

6.2.5 "Free" systems

There is a need for managers to have increased flexibility or freedom to design stand structures without the constraints of a silvicultural system that may encumber the creativity of the manager. Silviculture has traditionally been defined as both art and science (Helms 1998), but the more systematic it becomes, the less art is involved. One option is to provide managers with greater flexibility to create both heterogeneous stands and heterogeneous landscapes. This might, for example, give the manager greater ability to modify whatever general objectives existed for a particular stand because of unique features in that stand, or even to create unique features. There have been some very successful multiaged management systems that have limited systemization but instead rely on manager discretion. These include what Bončina (2011) described as "freestyle silviculture" in Slovenia, and the "Stoddard–Neel" approach in the southeastern United States (Neel et al. 2010; Chapter 7; Box 6.2).

As more silvicultural systems attempt to create a diversity of stand structures and include a variety of implementation strategies, it may become more difficult to systemize silviculture. This "freedom" may put the concept of "silvicultural systems" in question because it may require a series of treatments or operations that vary with time and cannot be easily planned. This would be a form of adaptive silviculture. However it would not remove the need to look to the future and anticipate the changes in stand structure that result from different treatments. Perhaps all that is needed in these cases are less detailed systems that not only provide general guidance—or what Graham and Jain (2005) described as a vision of desired future conditions—to long-term silvicultural planning but also a great deal of flexibility to the land manager. There will always be a need to systematize silviculture so it is consistent and repeatable from one forester to the next.

Multiaged systems in some forest types may have no previous experience or examples, and few examples in natural stands to demonstrate natural disturbance regimes. These no-analog systems are situations where creativity and adaptive approaches will be necessary to develop and refine new systems. Efforts to apply systems from other forest types should be exercised with great caution.

6.3 Synthesis

Multiaged silviculture offers a great variety of options to manage stands. This variation can come from a number of different gradients. One gradient is the range in opening sizes. Another is the gradient in horizontal heterogeneity, whereby treatments such as variable-density thinning can be used to enhance structural diversity. Yet another is variations in age structure, which can be developed for relatively simple two-aged stands up to complex stands that have many age classes. These are also gradients in silvicultural opportunity that correspond to current management trends on many lands toward more variability in stand structure between stands and across landscapes. No one method is right all the time, and the application of a single multiaged system across landscapes will serve to homogenize that landscape. Hence a diversity of systems—or including methods to diversify systems—will be needed to achieve this large-scale heterogeneity.

The descriptions of these multiaged systems have strayed from the confinements of traditional terminology because the rules that separate these methods into boxes are arbitrary and have constrained our ability to manage forests. As David Smith (1962) articulated: "When the important task of inventing the solution has proceeded far enough the less important step of attaching a name to it can be taken."

The systems presented here represent a fundamental shift in objectives from traditional uneven-aged silvicultural systems that focus on timber volume outputs to systems that focus more on residual structures and meeting multiple objectives, including timber production. This is consistent with the trends in international forestry toward achieving multiple objectives from forests, including not only production of timber but also systems that provide wildlife habitat, maintain natural processes, and produce other ecosystem services. These systems will require greater flexibility to accommodate existing structural features and environmental heterogeneity. These new systems will also present additional challenges. Prediction of stand growth and yield will be difficult. In many forest types, our understanding of stand dynamics is insufficient to guide management of new silvicultural systems to achieve complex stand structures. Nevertheless, it is still important to attempt to manage for complex structures through adaptive strategies that experiment with alternative strategies and provide new insights into stand dynamics.

CHAPTER 7

Multiaged stocking control

7.1 Introduction

Stocking control or **stocking regulation** of multiaged stands is the process of controlling the density, species composition, and sizes of trees through periodic harvest treatments, thinnings, and regeneration treatments. These treatments or operations serve to reallocate growing space within the stand. Stocking control is an integral part of multiaged silviculture because it provides the means for harvest treatments, consistent target stand structures, and long-term sustainability. Alternatively, stocking control is the primary means to transform one structure to another. It also provides the guidance on other structural elements that contribute to stand complexity, such as snags, down woody debris, or species composition. Management objectives are an overriding concern in development of strategies for regulation of multiaged stands. Stand structures fluctuate through a range of conditions between treatments as new trees are recruited and as trees expand into available growing space. There is an implied assumption that stocking control should achieve a relatively constant stand structure over time. Multiaged stocking control therefore attempts to develop a strategy for maintaining a range of stand structural conditions that can be maintained over time and meet management objectives. If stands are overcut, the reduced residual growing stock may be less productive, yielding less volume over time. If stands are undercut, there is a risk of overstocking, inhibition of regeneration, or encouraging features of a more uniform structure. An additional risk with multiaged stand control is that the system may favor less desirable species. For example, a system which maintained a high level of stocking might achieve growth targets but might encourage more shade-tolerant species. This chapter presents basic concepts related to multiaged stocking control and provides examples of different types of stocking control tools.

7.2 Stocking control concepts

A central goal in multiaged stand management or in the development of a stocking control procedure is to achieve sustainability with the management system. Sustainability in multiaged stands is generally a function of three indicators: 1) maintenance of a continuous production of wood volume or biomass from one cutting cycle to the next; 2) maintenance of a consistent stand structure from one cutting cycle to the next; and 3) development of regeneration to serve as replacement trees for harvested trees or trees lost to mortality (Figure 7.1; O'Hara and Gersonde 2004).

Multiaged stocking control involves allocating growing space to different stand components, which can be age classes or cohorts, canopy strata, or species groups (O'Hara and Valappil 1999). It can therefore be viewed as a process of establishing trade-offs between these different components (see Figure 4.3; O'Hara 1998). If an older age class is allocated more growing space, then this would imply a younger cohort has less. Likewise, if a lower canopy stratum is allocated more growing space, then less growing space is available to higher strata. The correct allocation is one that meets objectives and is sustainable. The trade-offs can be viewed in Figure 7.2 as a series of two-dimensional graphs for stands with two-canopy strata (Oliver and O'Hara 2005). Alternatively, Martin et al. (2005) developed a system that assessed understory stocking in relation to overstory density, a procedure that

Multiaged Silviculture. Kevin L. O'Hara.
© Kevin L. O'Hara 2014. Published 2014 by Oxford University Press.

Figure 7.1 Harvest treatment in a Plenter forest in the Black Forest of Germany.

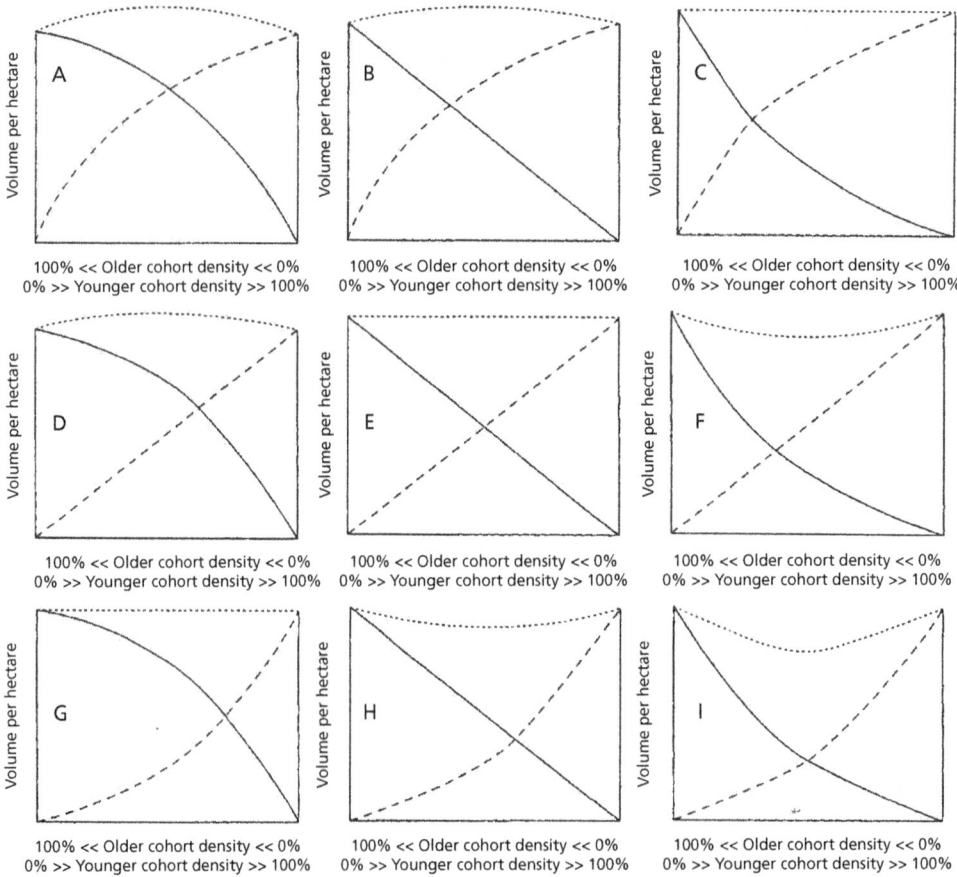

Figure 7.2 Multiple possible trade-offs between simple two-aged or two-strata stands. Where the stand structure produces greater volume, the upper line shows a peak at an intermediate level (A), and where total volume is reduced, the total volume is lowest at an intermediate level (I). All trade-offs are possible, but these only represent those from two-aged stands: with additional age classes, the possible effects on volume production are much greater (from Oliver and O'Hara 2005).

over more shade-intolerant species such as Scots pine and several birch species (Liu and Hytteborn 1991). However, other species also regenerated successfully. Spruce seedlings were found on downed logs, birches on mineral soil, and pine seedlings were more randomly dispersed within gaps. In this particular situation, there may have been sufficient light penetration through the canopy at this high latitude to maintain all species. In tropical forests at 9.25° N, Brokaw (1985a) analyzed responses in gaps up to approximately 700 m^2. In larger gaps, less shade-tolerant species were dominant in size and numbers but were already experiencing self-thinning by year 6. Shade-tolerant species were present across the range in gap size but were most competitive in smaller gaps.

In gaps in New Zealand beech forests (42.22° S), both red and silver beech were present as advance regeneration at the time of gap formation. Silver beech were often larger than red beech at time of release. Larger gaps favored the more shade-intolerant red beech, which also was more competitive in gap centers where the light regime was best and where branch expansion from trees surrounding the gap was less (Stewart et al. 1991; Runkle et al. 1995; Hurst et al. 2012).

In temperate mixed forests in British Columbia (55.37° N) latitude, Coates (2000, 2002) found no spatial patterns or partitioning related to position in artificial openings of various sizes. Tree density increased from north to south, and differences related to opening size were minor. There were few differences based on shade tolerance and regeneration, and growth appeared to be limited by different factors, such as moisture on the north sides of these openings.

Regeneration mechanisms are also important to determine which individuals or species persist. Smaller gaps favored shade-tolerant species in a series of studies in eastern hardwood forests between approximately 35.35° and 42.52° N (Runkle 1982). Although these gaps, which were less than 100 m^2, favored shade-tolerant species, which were often present as advance regeneration, other species were able to persist. In western North Carolina (35.05° N), Phillips and Shure (1990) used artificial openings ranging from 0.016 to 2.0 ha to study upland hardwood responses. Most regeneration in all openings was from stump or root sprouts, or advance regeneration. There were no major changes in species composition before and after treatments or between opening sizes, suggesting that the vegetative regeneration overcame light limitations associated with opening size. In artificial gaps ranging up to 0.04 ha in the Great Lakes region of North America (47.75° N), most trees were present before gap formation and were dominated by shade-tolerant sugar maple (Bolton and D'Amato 2011). The regeneration of yellow birch was related to the presence of large coarse woody debris, which was more prevalent on larger gaps. Coarse woody debris may be important for some species, implying that removing tree stems in harvest treatments may reduce the regeneration potential of some desired species independent of gap characteristics.

Natural gaps were common in several forest types in Japan, comprising as much as 17% of total area in warm temperate evergreen broadleaved forests (Yamamoto 1998). Gaps were small (less than 100 m^2), but multiple tree fall gaps were relatively common and comprised as much as 45% of gaps in the subalpine conifer type. In old-growth evergreen broadleaved forest in Japan (34.15° N), Manabe et al. (2009) found gap expansions from multiple gap-forming events, and large variation in sequences and types of disturbances events. Although differences in species composition were not large across the gaps studied, the variability in gap formation suggests these sequences of gap-forming and gap-expanding events could have a major effect on species composition and gap development.

Trees established in managed groups can be either natural regeneration or planted seedlings. Group openings are often a response to management objectives related to regenerating shade-intolerant species. For example, in northern hardwood forests in eastern North America (44.06° N), group sizes of 2000 m^2 averaged 66%–75% shade-tolerant species, as compared to 92% in smaller single tree selection openings (Leak and Filip 1977). In round group openings ranging from 0.1 to 1.0 ha in California's Sierra Nevada (38.9° N), York et al. (2003, 2004) found all species of planted seedlings grew better in the interior of the group openings and on the north side. Tree growth did not approach the rates of open-grown trees until group sizes approached quantifies these trade-offs for partially cut stands. In designing multiaged stands, three critical points are: 1) stand components must compete with each other, 2) younger components are necessary to replace older ones, and 3) there is an upper limit on how much growing space is available for allocation to the total stand.

7.2.1 Stocking and growth

For even-aged stands, there has been considerable research on the relationship between periodic increment and stocking. Although these studies typically ignore variations in stand structure that can also affect increment for a given level of stocking (e.g., O'Hara 1989), increment generally increases with stocking. For example, Curtis et al. (1997) found increasing periodic increment in coast Douglas-fir through the entire range of stocking (Figure 7.3 left). Likewise, Pretzsch (2005) found a similar relationship in European beech (Figure 7.3, right). The implication from these studies of managed even-aged stands is that greater stocking corresponds to greater leaf area and greater photosynthetic potential. In the studies shown by Curtis et al. (1997), density was held constant to attempt to isolate density effects. The European beech and Norway spruce results shown by Pretzsch (2005) received a variety of thinning treatments. Hence, volume growth tends to increase with stand density, but thinning to control density introduces other variables that confound this relationship. Whether similar patterns occur in multiaged stands is uncertain, but, at present, there is no evidence to indicate otherwise. One exception found similar levels of production from different stocking levels in a 57-year study in northern hardwoods in Michigan, US (Gronewold et al. 2012).

An alternate way to view stand productivity is through the relative **growth efficiency** or **growing space efficiency** of trees, stand components, and whole stands. Efficiency is usually expressed as a ratio of increment to a measure of occupied growing space, such as crown length, crown projection area, or leaf area. If a stand component is efficient, then it is producing volume increment at a higher level per unit of occupied space. For individual trees in even-aged stands, some studies have found that dominant trees were most efficient (e.g., O'Hara 1988), and some have found that suppressed trees were more efficient (e.g., Reid et al. 2003). Some of these differences are simply the result of differences

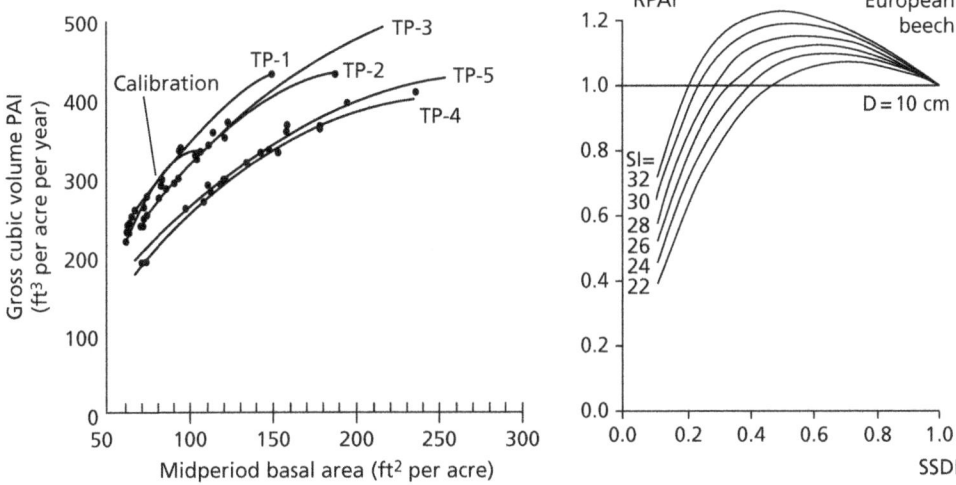

Figure 7.3 Volume increment–density relationships. The Douglas-fir diagram (left) shows the effects of controlling density on volume increment through a series of five treatment periods (TP; from Curtis et al. 1997; with permission from the Society of American Foresters). The European beech figure (right) shows periodic annual increment (RPAI) and stand density (SSDI) for different site indices (SI) when tree diameter (D) is held constant at 10 cm (from Pretzsch 2005; with permission from Springer Science + Business Media).

in species efficiencies (Stancioiu and O'Hara 2006a). In either case, these individual tree relationships suggest that organizing individual trees into stand structures can have a large effect on stand production. However, there is no evidence that these efficiency relationships provide new insights into the basic stand production/stocking relationships shown in Figure 7.3.

Growing space efficiency in mixed species stands is more difficult to interpret because species occupy growing space in different ways. Many of these differences relate to shade tolerance or in the way species allocate carbon to different needs. A shade-tolerant species will have a greater ratio of leaf area to crown projection area than a less tolerant species. However, shade-tolerant species are generally less efficient at converting foliage area to volume or biomass. Efficiencies are typically less for shade-tolerant species, although their actual growth rates may be similar to less tolerant species. Gersonde and O'Hara (2005) compared tree growing space efficiencies of five species in mixed conifer forests of California, US, over a range of projected leaf area (Figure 7.4). Species ranking by efficiency corresponded to shade tolerance ranking, with ponderosa pine being most efficient. There was some change in ranking at large and small leaf areas that may have been due to model fitting. Directly comparing growing space efficiencies for different species is difficult at best, even when they are from the same stand.

For multiaged stands, there are also some confounding results related to efficiency. O'Hara (1996) found trends of increasing growing space efficiency with increasing age of ponderosa pine age classes. This implies increasing the proportion of older trees in a multiaged structure would increase stand productivity. In contrast, in multiaged eastern hemlock–red spruce stands, Seymour and Kenefic (2002) found greater efficiency in upper canopy trees with moderate-sized crowns but that efficiency declined with age when crown size (leaf area) was held constant. They attributed the decline in efficiency, in part, to periods of suppression that in theory were experienced at one time by all trees in multiaged stands. Although this hypothesis suggests that structures with a greater allocation of growing space to upper crown class trees will have greater volume production, it also suggests an age-related decline that O'Hara (1996) did not see in ponderosa pine. Both of these studies indicate that there are ways to affect stand volume production through the design or allocation of trees within stand structures. In the context of the way in which stocking affects volume production, the ponderosa pine study demonstrated how production increases with increasing stocking but that the stand structure—or the relative sizes of trees and the arrangement of leaf area—was also of great importance.

7.2.2 Stocking and regeneration

Regeneration is a major consideration in multiaged systems. Without consistent, timely and well-placed regeneration, the multiaged system cannot achieve the overriding goal of sustainability. Obtaining timely natural regeneration is somewhat due to chance, because natural regeneration may be dependent on inconsistent seed crops, unpredictable weather, or predation by animals and insects. Regeneration will generally occur more commonly with lower stocking because of greater growing space availability (see Chapter 8). Additionally, this regeneration will generally grow more rapidly with greater growing space availability. This represents a trade-off in multiaged stocking control: lower stocking favors regeneration in terms of numbers and growth but reduces stand volume production.

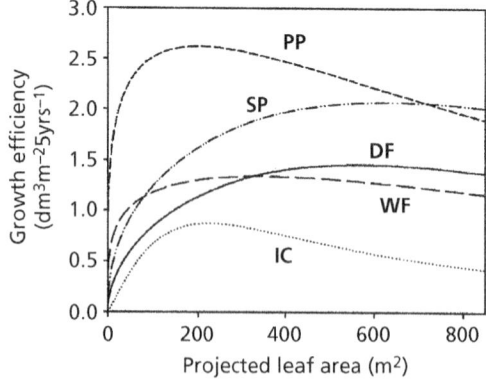

Figure 7.4 Patterns of growth efficiency with increasing projected leaf area in five conifer species in California, US. PP = ponderosa pine, SP = sugar pine, DF = coast Douglas-fir, WF = white fir, IC = incense-cedar (from Gersonde and O'Hara 2005).

Artificial regeneration is an important means of supplementing natural regeneration in multiaged stands. Although multiaged silviculture often carries a label of being natural and relying on natural regeneration only, there is no reason to preclude artificial regeneration. Seedlings can be planted to accelerate the process of developing a new cohort and thereby shortening cutting cycle lengths. Seedlings can also be used to increase certain species, genotypes, or provide trees in specific locations in a stand.

Cumulative stocking affects the light environment of the stand and therefore can have a profound effect on the understory environment. Greater stocking will then result in reduced light availability in the understory. For many forest types, this will determine whether regeneration will occur or not. It may also affect which species regenerates, or which species regenerate most successfully, in the understory. Tolerance to shade may be an important determinant of which species are successful in this environment.

7.2.3 Stocking and stand structure

Stocking control has traditionally focused on representing stand structure through "horizontal" measures such as diameter distributions or stand density measures that are usually based on tree diameters (Curtis 1970; West 1983; Ernst and Knapp 1985). Stand structure is more than the cumulative diameters or basal area on a horizontal plane at breast height. It is a three-dimensional system, with a physical structure and a variety of ecological functions (Spies 1998). These include the distribution of live tree sizes, as well as the vertical distribution of foliage, horizontal patterns of trees and shrubs, and the presence of standing and down wood. Stand structure also affects volume production through location and efficiency of tree crowns in even-aged (O'Hara 1988, 1989; Long and Smith 1990) and multiaged stands (O'Hara 1996; Kollenberg and O'Hara 1999; Seymour and Kenefic 2002).

There have been many approaches used to describe stand structure beyond traditional stand density measures. These approaches attempt to distinguish between stands of different structure or to describe differences in structural attributes (Pommerening 2002). Many have attempted to subjectively describe stand structure based on stand development processes (Oliver 1981; Oliver and Larson 1996; Carey and Curtis 1996), and others describe the stand structure based on the predominant size of the trees (e.g., seedlings, saplings, etc.; Thomas 1979). Diversity indices (e.g., Magurran 1988) are used to describe size class or species distributions (Sterba and Zingg 2006) but without a spatial component. There have also been many efforts to quantify horizontal and vertical spatial patterns and develop indices to represent structure (Pommerening 2002; Sterba and Zingg 2006; Zenner et al. 2011; Gadow et al. 2012). Although there have been major advances in describing stand structure, there is not yet a method that links stand structure to management beyond simple diameter distributions.

7.2.4 Measuring stand density

Stand density is typically measured in an attempt to quantify the amount of competition or level of growing space occupancy in a stand. A long history of work has advanced stand density measurement in even-aged stands. Stand density is usually expressed either in absolute or in relative terms. **Absolute stand density** measures are a simple count or sum per unit area (e.g., trees per hectare, basal area per acre, volume per hectare). **Relative stand density** measures are expressions of the ratio, proportion, or percent of absolute stand density to a reference level defined by a standard level of competition (Helms 1998). Relative density measures take an absolute density measure and make it comparable across different stages of stand development. For example, relative stand density measures provide a common scale to compare the competition level of a young radiata pine stand with an older radiata pine. In addition to providing a measure of growing space occupancy that can be compared for stands in different stages of development, relative density measures also provide an index to the maximum potential growing space occupancy. This provides simple measures of relative competition that are usually assumed to be independent of site quality.

Absolute stand density measures are easy to measure, commonly used, and easy to conceptualize. Relative density measures are more difficult to calculate and conceptualize, and, as a result, are used less commonly. The most common relative stand density measures use average tree diameter or basal area as a reference level (e.g., stand density index (Reineke 1933), tree area ratio (Chisman and Schumacher 1940), crown competition factor (Krajicek et al. 1961), the central hardwood stock guide (Gingrich 1967), relative density index (Curtis 1982), and others). Curtis (1971) and West (1983) found these measures behaved somewhat differently than relative density measures that used volume (e.g., Drew and Flewelling 1977, 1979), or tree height (e.g., Wilson 1979) as a reference level. Overviews of stand density measurement in even-aged stands can be found in Bickford et al. (1957), Ernst and Knapp (1985), Long (1985), and in Chapter 7 of Tappeiner et al. (2007).

For multiaged stands, both absolute and relative stand density are used to describe competition or total growing space occupancy. Unlike even-aged stands, multiaged stands also need a division of growing space among stand components. This is where stocking control becomes more complicated for multiaged than for even-aged stands. Instead of determining an appropriate level of growing space occupancy or a range of competition as in even-aged stands, this growing space must also be divided among multiaged stand components, such as age classes or species.

7.2.5 Density management zones

Silviculturists have long sought an optimal stand density for managed stands. An optimal stand density may maximize wood volume production, or another may achieve the best habitat for a particular species, or yet another may provide the most aesthetic stand. Although there are certainly optimal stocking levels or stand densities for many resource objectives, they will generally not be the same for different objectives. Stand density also represents only one aspect of a prescription: another is the stand structure, which can vary independent of stand density. For example, a level of basal area could describe a stand with a great diversity or little diversity in trees sizes, or a stand with a few large trees or a stand with many. Regardless of the management objective, optimal stand densities

or stand structures change over time as a stand develops. It is impractical, or even unrealistic, to assume a stand could continuously exist at its optimal stocking. Hence, stands are managed to stay within a range of conditions (density or structure) that encompasses the optimal condition. For stand density, this is referred to as the **density management zone** (DMZ). In addition to encompassing the range of desirable density over time, the size of the DMZ also integrates the time period between treatments. For example, a wider DMZ would correspond to greater time between density adjustments.

A DMZ defines the prescribed range in density within a silvicultural prescription (Figure 7.5). For even-aged stands, it represents the density between the minimum density that might occur after a thinning and the maximum that occurs before a thinning or final harvest. This DMZ is prescribed to meet some management objective within operational constraints. For example, a narrow DMZ would require either very frequent, light thinnings or very slow growth rates. Likewise, heavy, infrequent thinnings would require a large DMZ, and a stand with a fast growth rate would avoid frequent thinnings if the DMZ was large.

In multiaged stands, the DMZ is defined by the range of stocking between the beginning and end of the cutting cycle for the entire stand (Figure 7.6). At the beginning of the cycle, or immediately after a harvest treatment, stocking is low. It then increases

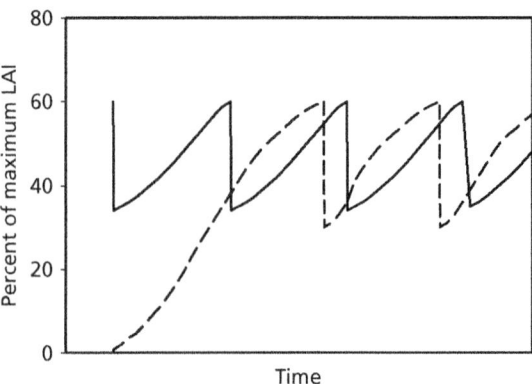

Figure 7.5 Density management zones (DMZ) for even-aged and multiaged stands in terms of leaf area index (LAI). In these hypothetical examples, the even-aged DMZ fluctuates from about 30% to 60% LAI, and the multiaged stand from about 35% to 60%.

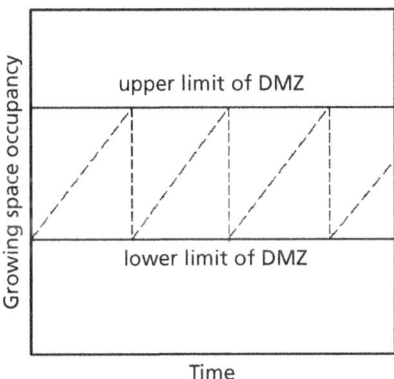

Figure 7.6 Density management zone (DMZ) for a multiaged stand showing multiple cutting cycles fluctuating between the upper and lower limit of the DMZ.

to a maximum immediately before harvest. This pattern would be repeated over each cycle under an idealized system. If competition was causing mortality, stocking might exhibit the opposite pattern if it was measured in trees per unit area rather than using basal area, volume, or a relative stand density measures. Intermediate thinnings between harvest treatments might alter the amount of stocking, as would any episodic mortality events such as windthrow or insect damage.

7.2.6 Cutting cycle length and cutting severity

The **cutting cycle** is the time interval between harvest treatments in multiaged stands. It is an important concept with regard to stocking control because a direct relationship exists between cutting cycle length and stocking removals during harvest treatments (Figure 7.7A and B). If a longer period between harvest entries is desired, then the stocking removals at harvest will have to be severe to maintain the stand in the DMZ. Likewise, a short cutting cycle requires more frequent, light volume harvests (Hansen and Nyland 1987; O'Hara and Valappil 1999).

The DMZ also describes an important aspect of the multiaged stocking control regime. The DMZ should encompass the range of density of the entire cutting cycle. However, this can be higher or lower depending on management objectives (Figure 7.7C and D). For example, a lower DMZ may be more suitable for a dry site prone to bark beetle problems than a site without these potential problems (Figure 7.7D). Or on a site where regeneration is not problematic, stocking could be maintained in a higher DMZ (Figure 7.7C). In both situations, stocking might fluctuate across similar ranges within a DMZ, but the minimum and maximum level of the DMZ may be dramatically different.

7.2.7 Balanced stands and sustainability

A central concept in multiaged silviculture is the balanced stand. The concept of the "balanced stand" appears to have originated with Meyer (1943) and Meyer and Stevenson (1943). They equated a negative exponential diameter distribution to a natural or "virgin" structure capable of maintaining a constant volume over time. Meyer was apparently looking for a standard to insure sustainability and believed this was achieved through maintenance of a constant structure over time and a constant level of cutting that maintained the natural distribution in some sort of equilibrium. Building on de Liocourt's (1898) work in Europe, Meyer (1943) published diameter distributions of forest-level studies in several locations that followed a negative exponential distribution. He then assumed this *forest-level* pattern was a natural uneven-aged structure and an appropriate model to guide a sustainable *stand* structure (O'Hara 1996).

Another view of a balanced uneven-aged stand was provided by Smith (1962), who defined the balance as being achieved with each size or age class occupying equal area (Figure 7.8). This interpretation is analogous to area control forest regulation applied to individual stands (O'Hara 1996). Equal amounts of growing space are allocated to each age class, and each age class is assumed to be harvested sequentially, thereby maintaining a stable stand structure and stable production of wood over time.

These justifications for "balanced" stands were attempts to find a sustainable basis for management. Meyer was trying to justify a sustainable system of regulation for uneven-aged stands, and Smith was trying to justify why a negative exponential diameter distribution might be sustainable (O'Hara 1996). Although the intent was to insure sustainability, there have been few examinations

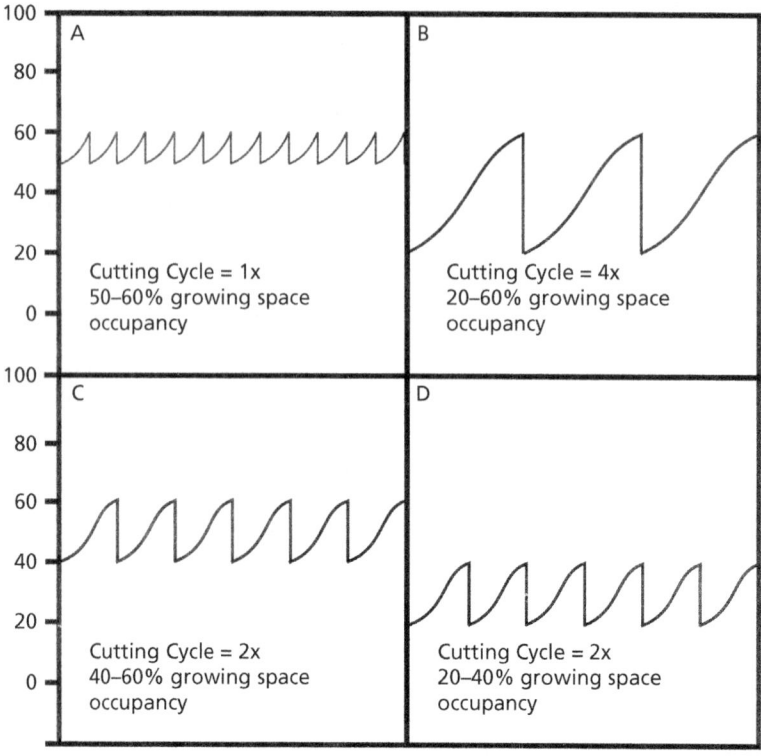

Figure 7.7 Density management zones (DMZs) and cutting cycles are interrelated. When the cutting cycle is long, the DMZ must be large (B), and a short cutting cycle requires a smaller DMZ (A; from O'Hara and Valappil 1999).

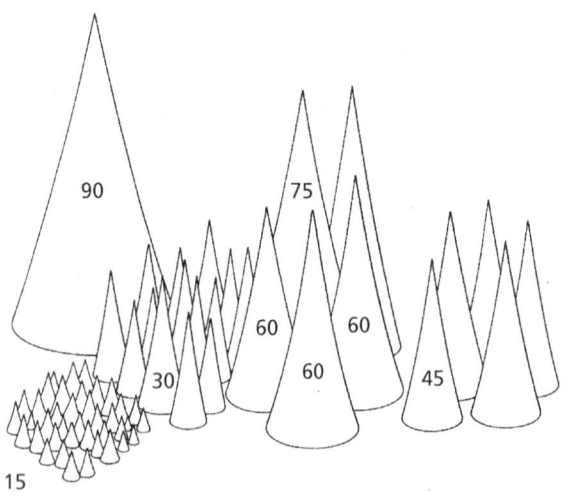

Figure 7.8 Equal growing space for each age class has been one interpretation of a "balanced stand." In this figure, each age or size occupies equal area, which requires that younger/smaller classes have many more trees that older/larger size classes (from Smith et al. 1997; reproduced with permission from John Wiley & Sons).

of these assumptions. By providing for regeneration and maintaining a target structure over time, many methods meet these basic criteria for sustainability. However, merely being sustainable does not assure that a stocking method provides anything approaching potential productivity or meets other objectives. In a simulation analysis of multiaged ponderosa pine, O'Hara (1996) found the negative exponential diameter distributions produced considerably less volume than other structures because so much growing space was allocated to small trees. Stand structures with greater allocations of growing space to larger trees were more productive while also being sustainable. In contrast, Donoso (2005), working with mixed forests in Chile, developed a crown index that suggested balanced structures were productive and provided sufficient regeneration. However, this latter work does not demonstrate that the artificial construct of a balanced structure meets objectives better than other options.

A similar concept is the "equilibrium" stand structure. This equilibrium has been used to describe both a stand where structure and process become relatively constant (e.g., Bormann and Likens 1979), and a target multiaged stand structure that meets some predefined objective (Schütz 2001a). For example, systems in central Europe have used equilibrium diameter distributions, where higher stockings will result in reduced regeneration and lower stockings will result in reduced increment (Schütz 2001a). These multiple states of equilibrium would vary by forest type. The equilibrium curve used in European beech dominated sites would have lower density and standing volume than a stand dominated by central European conifers (Schütz 2006).

7.2.8 Diameter distributions

Frequency distributions of tree diameters have a long history of use for multiaged stocking control. These distributions often serve as target structures for multiaged systems. An important early publication was the work of de Liocourt (1898), who documented a general "reverse-J" distribution for forests in France (Kerr 2014). Examples of stocking control tools based either rigidly or with more flexibility, on diameter distributions include Arbogast (1957), Alexander and Edminster (1977), Guldin (1991), Baker et al. (1996), Schütz (2006), and many others. A reverse-J diameter distribution has declining numbers of trees with increasing diameter and may not necessarily follow the more precise negative exponential form. Meyer (1943, 1952) found the negative exponential form in a variety of forest-level studies and promoted it as a target for multiaged structures. It was believed to be a natural structure and, because Meyer observed it in somewhat natural structures, he believed it demonstrated a sustainable structure. The negative exponential diameter distribution has since become synonymous with uneven-aged structures in many circles, and the more general reverse-J form is often used as diagnostic criteria for determining age structure (Leak 1964). Although even-aged stands often have normal distributions and multiaged stands have distributions that are reverse-J (see Ford 1975; Harcombe and Marks 1978; Kunisaka and Imada 1996; Leak 2002), there are exceptions. For example, even-aged stands can also form reverse-J diameter distributions (Oliver and Larson 1996).

Stand-level analyses of diameter distributions have revealed great diversity in multiaged stand structures. For example, Janowiak et al. (2008) found a variety of diameter frequency distributions in multiaged northern hardwood stands in Michigan (Figure 7.9). In theory, any of these could be used to develop target stocking control regimes for multiaged stands. Goff and West (1975) reported that mixed northern hardwood–conifer stands formed steady-state diameter distributions with a rotated sigmoid form (Figure 7.9). This was assumed to be due to reduced mortality and greater growth in the middle strata. Similar diameter distributions were seen in uncut beech forests in central Europe where a rotated sigmoid form was also common (Westphal et al. 2006). Likewise, others have found the rotated sigmoid form in unmanaged stands (Lorimer and Frelich 1984; Parker et al. 1985; Goodburn and Lorimer 1999). These distributions change over time and are affected by management. For coast Douglas-fir, Zenner (2005) found stands with infrequent disturbances developed rotated sigmoid diameter and then reverse-J distributions as they developed into "old-growth" structures. Stands that experienced a more frequent history of minor disturbance formed reverse-J distributions.

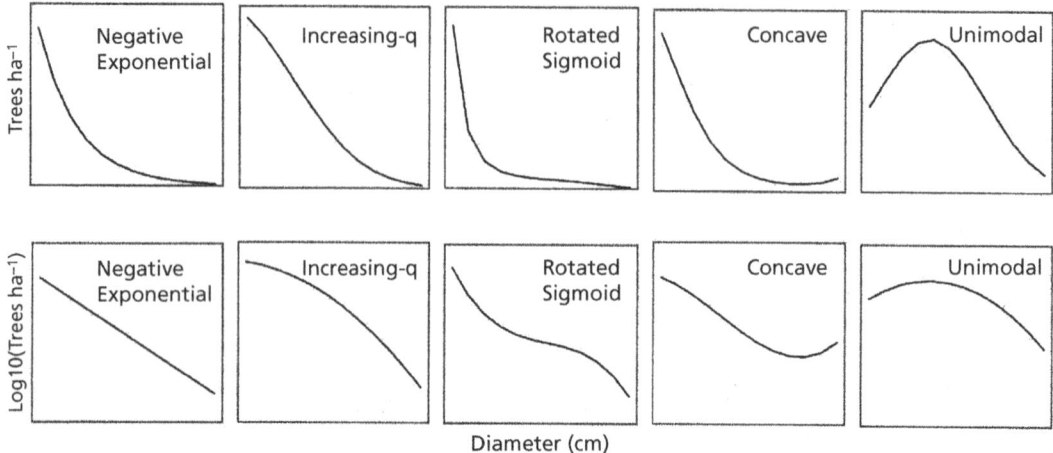

Figure 7.9 Examples of diameter frequency distributions on both linear and semilogarithmic scales (from Janowiak et al. 2008; reproduced with permission from the Society of American Foresters).

Leak (1996) found northern hardwood stands had negative exponential distributions that became rotated sigmoid after partial cutting. In managed northern hardwood stands, Janowiak et al. (2008) found a variety of forms, but they tended toward an "increasing-q" form (Figure 7.9). There are also issues of sampling and scale, which potentially affect the shape of the diameter distribution (Rubin et al. 2006; Janowiak et al. 2008).

Although there are general trends in the form of diameter distributions of unmanaged stands, a valid question is whether these diameter distribution forms should be the target structures for managed multiaged stands. Managed stands are typically maintained at substantially lower densities than those of unmanaged stands. For example, the upper limit of the DMZ for even-aged stands is often only 50% to 75% of the maximum (Drew and Flewelling 1979; Long 1985; Cochran et al. 1994). Multiaged stands would also be managed at similar ranges of density (O'Hara and Valappil 1999). Since natural stands are assumed to be equilibrium structures, it is also assumed that managed stands with similarly shaped diameter distributions are also sustainable, although they may be maintained at a substantially lower range of density. The primary evidence to support this is the presence of abundant numbers of small trees in these various reverse-J diameter distributions and operational experience with multiaged stands maintained with these distributions. However, the sustainability of one structure does not, by itself, indicate another structure is unsustainable. Many structures can meet the criteria for sustainability but may vary greatly in structure, productivity, or appearance (O'Hara 1998).

7.3 Multiaged stocking control

Forest managers attempting to manage a multiaged stand without any previous experience and guidance might struggle with the complexities of this difficult management system. Decisions include determining which trees to remove and which to leave, what are the appropriate numbers of trees of each species, age class, size class, or canopy strata to leave, and the length of cutting cycles. As foresters attempted to manage multiaged stands, they have looked to a variety of examples for inspiration and guidance. Foremost among these has been the expectation that the multiaged stand—or usually what were referred to as uneven-aged or all-aged stands—occurred commonly in nature and were therefore a logical model for emulation. This is seen in both the description and the theory behind the ecological climax model, which, by its very name is assumed to be the desirable endpoint of plant community development. Additionally, self-perpetuating tree replacement was assumed to occur continuously, thereby forming all-aged stands (Oliver and Larson 1996). As a result, the assumption of all-aged forests with individual tree replacement is an underlying concept in the formation of many early stocking control procedures for multiaged stands.

A variety of multiaged stocking control systems have been developed, and these tend to be representative both of the current ecological knowledge at the time of their development and of the local features of the forests. These stocking control systems sometimes may also include an element of economic justification that provides for sufficient volume removals to cover harvesting costs. But many others are simply attempts to meet objectives of providing for regeneration and a sustainable structure. Other systems exist that have not been documented and are based on local knowledge and experience.

7.3.1 Diameter-limit cutting

A common intervention related to multiaged stand culture has been various forms of diameter-limit cutting. In these systems, trees above a given tree diameter at breast height are removed, and the residual trees generally unmanaged (Figure 7.10). These diameter-limit treatments were often assumed to be removing older trees in stands where younger trees would then grow into the canopy, thereby perpetuating an uneven-aged structure. However, stands were often even-aged, and more dominant trees were simply better genotypes or faster-growing species. This removal of bigger trees was also driven by economic justifications rather than purposeful, if misguided, emulation of ecological systems. Hence, the determination of what trees to remove, or the diameter limit, was largely based on merchantability. The best trees—or the high-grade trees—that were cut gave rise to the term **high grading**. As a result, diameter-limit cutting has a poor reputation because it reduces timber quality and is possibly dysgenic. Kenefic and Nyland (2005) provide additional information on diameter-limit cutting in forests in northeastern North America.

In a multiaged stand, diameter-limit cutting that repeatedly removed trees above a given tree diameter or that used a diameter limit that could be adjusted at each entry could be effective. For example, in a single-species stand where distinct age classes are formed and occupy distinct canopy strata, the diameter-limit cutting might closely resemble an individual tree selection cut. Diameter-limit cuttings may also be successful if sufficient quality growing stock was left; however, long-term studies are still needed (Sendak et al. 2003). However, by definition, diameter-limit cutting does not treat the size classes lower than the diameter limit. Alternatively, in a multiaged stand with many species of variable shade tolerance, the removal of larger trees may discourage shade intolerant species and dramatically alter species composition. Or, a diameter-limit cut that was too severe or too frequent might also have undesirable effects on future stand development and structure.

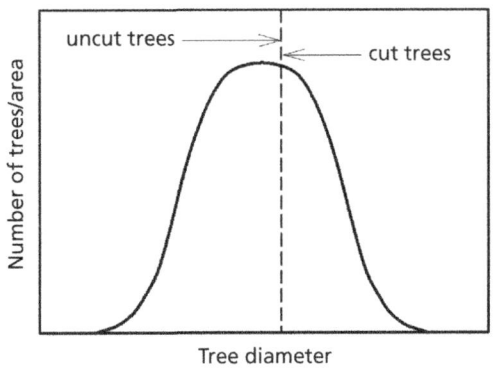 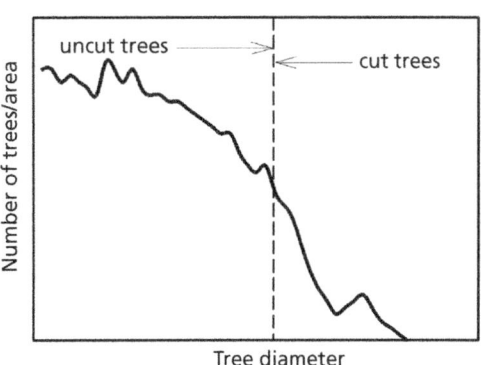

Figure 7.10 Diameter distributions showing possible effects of a strict diameter-limit cutting. Diameter-limit cuts have traditional been applied in a wide variety of stand structures, including those resembling even-aged stands (left) and more complex multiaged stands (right).

In a long-term study on the Penobscot Experimental Forest in Maine, US, diameter-limit cutting was compared to selection cutting (Kenefic et al. 2005). Both treatments involved three harvest entries over several decades. The appeal of the diameter-limit treatments was apparent through greater volume removals and therefore greater initial economic returns to the landowner. By the third harvest treatment, the value of removed trees was several times greater in the selection treatment, and the diameter-limit treatments had favored the less desirable balsam fir over red spruce. These findings demonstrate both the initial attraction of diameter-limit cutting and how the real effects of these treatments become more apparent over time. Differences in treatments were increasingly apparent, resulting in fewer large trees and the greater defect in the diameter-limit treatment.

An alternate form of diameter-limit cutting is to remove trees *below* a diameter limit. These upper diameter-limit treatments have been used on forests in the western United States on public ownerships where restoration is an objective and where maintaining sufficient numbers of large trees are a dominant management issue. The effect, however, is often to make any form of management infeasible, because only small trees can be removed (Stine et al. 2014).

Figure 7.11 Mixed stand of silver fir, Norway spruce, and European beech in Switzerland.

7.3.2 Plenter system

In central Europe, the **Plenter system** has been used since the 1800s to control stocking in multiaged stands. This system attempts to achieve sustainability through the development of an equilibrium diameter frequency distribution. This has also been described as a form of "demographic steady state" (Schütz 2006), where ingrowth equals the volume removed by harvesting and mortality. Sustainability is achieved by maintaining a constant stand structure from one cutting cycle to the next, so that harvest volume equals increment. The Plenter system may be the oldest and most continuously used system for multiaged stocking control. The system originated in the silver fir, Norway spruce, and European beech forests of central Europe (Figure 7.11).

In theory, the Plenter system uses a negative exponential diameter distribution to develop the desired equilibrium structure. An equilibrium is achieved when standing volume remains constant from one cutting cycle to the next and when growth equals harvest, thus providing for long-term sustainability. In reality, however, countless variations on the Plenter structure may exist in which single tree selection is used for stocking control (Burschel and Huss 1987). The stand structure objective of a Plenter system consists of trees of all sizes and ages. However, tree age is not considered an important variable because many central European tree species can remain in a state of understory suppression for many years.

Early Plenter systems used detailed inventory methods—often a 100% sample—to determine cutting levels and specific trees for cutting. These control or check methods were systems which date back to Gurnaud and Biolley in France and Switzerland (Schütz et al. 2012). An inventory of trees by size classes provided data which were compared to the desired structure to determine surplus trees for cutting. Short cutting cycles were used with a repeat full measurement of the stand at each cutting cycle.

Schütz (2001a) described the Plenter structure as being maintained through continuous control of the growing stock (standing volume); a growing stock in excess of the equilibrium would lead to reduced regeneration and recruitment into smaller diameter classes, whereas levels of growing stock below the equilibrium would reduce total increment and reduce the quality of trees. Schütz (1975) developed an equilibrium stocking relationship for Norway spruce–silver fir that varies from the negative exponential diameter distribution. On linear scales, the equilibrium relationship resembles a negative exponential form, but on logarithmic scales, more trees in larger diameter classes are apparent than a negative exponential diameter distribution (Figure 7.12).

Schütz (1997) justified the sigmoid shape (Figure 7.12) as resulting from a nonlinear increase in periodic increment with tree size and the disproportionally greater effect on the remaining stock when large diameter trees are removed. Increment in small diameter classes is relatively low and, therefore, is assumed to require a high number of stems to produce enough diameter class advancement, causing a steep decline in the diameter distribution in these size classes. As trees grow into larger diameter classes, mortality declines, and diameter class advancement increases, causing a slower decline in stem number. Increased tree harvest in the larger diameter classes causes a more rapid decline in the distribution. The equilibrium diameter distribution can be calculated from empirical values of diameter growth and harvesting of trees and tree mortality. The equilibrium stand basal area, or the total growing space occupancy, can be derived from growth and survival of regeneration depending on stand density. A family of diameter distributions can be constructed with estimates of corresponding standing volume. However, only a few curves will produce an equilibrium because of the negative effect of increased growing stock on regeneration and ingrowth. Realistic values for stem numbers in the lowest diameter classes are taken from observations in the field together with standing volume and ingrowth. Finally, the form of the diameter distribution is also a function of the management objective in the Plenter forest. Although different amounts of growing stock can be held at equilibrium through periodic inventory and stocking control, they result in varying yields of small, medium, and large timber, and these outcomes will influence subsequent management decisions by the forest owner (O'Hara and Gersonde 2004).

The development of alternative equilibrium diameter distributions can be applied to other forest types or sites. These alternative diameter distributions may vary in shape due to differences in diameter increment and rates of removal. In silver fir-dominated forests at lower elevations in central Europe, the equilibrium distribution may have a more pronounced sigmoid shape, whereas a more consistently negative slope might be found in high elevation forests because of more uniform diameter increment across all size classes. These differences represent the relative diameter increment across the diameter distribution: equal diameter growth would result in a negative exponential relationship, whereas unequal diameter increment results in the rotated sigmoid form (assuming more rapid diameter increment in larger classes; Schütz, 1997). This resembles the "rotated sigmoid" curve (see Figure 7.9) described by Goff and West (1975), with a greater number of larger trees than indicated for a negative exponential curve.

The advantage of this stocking control approach is that it has some flexibility in developing target

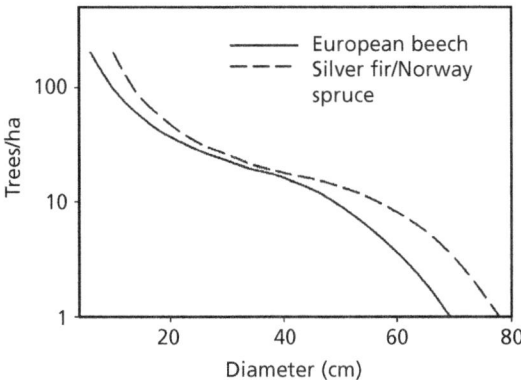

Figure 7.12 Equilibrium stand structures for a Plenter system European beech stand and a Plenter system Norway spruce–silver fir stand (adapted from Schutz 2006).

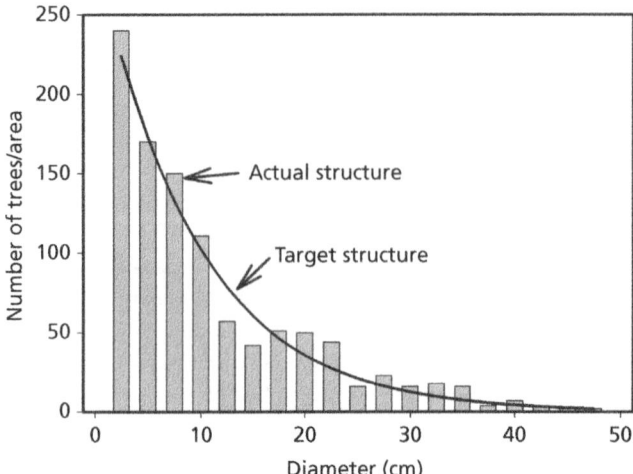

Figure 7.13 Negative exponential curve shown as "target structure," and the "actual structure" showing a stand that might be treated to conform to the target structure.

diameter distributions. Removal rates are dependent on growth rates and competition, thereby providing some assurances of sustainability. There is also a long history of the use of this approach in central Europe and considerable experience with application of the system. However, there are relatively few examples of its application outside central Europe. As with many stocking approaches for multiaged stands, the target structure or, in this case, the equilibrium diameter distribution, is a compromise or trade-off between getting regeneration and maximizing wood production. In practice, small-scale variability creates opportunities to meet multiple objectives. In the central European forests where this method was developed, the three primary species—silver fir, European beech, and Norway spruce—are all relatively shade tolerant and can regenerate in somewhat poor light environments (Stancioiu and O'Hara 2006c). In other forest types where desired species include a greater range of shade tolerance, maintaining this equilibrium or balance is more difficult because tolerant species will be favored by the light, frequent cutting treatments. An analysis of long-term plot data from Switzerland for stands managed with the Plenter system concluded that the system is still evolving, because long-term trends in species composition, stocking, and increment were not constant (O'Hara et al. 2007a). The emphasis on a diameter distribution in any stocking control system also presents some limitations: although the target diameter distributions may show some flexibility, in common usage, they are continuous distributions with a negative slope. They would be more difficult to apply in stand structures with discontinuous diameter distributions or with distributions that have non-negative slopes between some diameter classes, as you might find in a two-aged stand or a stand with many reserve trees (Figure 6.4).

7.3.3 The BDq system

The BDq approach is a diameter distribution-based stocking control tool in common usage in North America. Its development originates primarily with Meyer (1943), who brought concepts from Europe related to using negative exponential or reverse-J diameter distributions as target structures for uneven-aged stands. BDq refers to the three parameters of this stocking tool: B for basal area, D for maximum diameter, and q for the q-factor. The q-factor is the diminution quotient for the negative exponential distribution (Figure 7.13). Because of its importance to the BDq approach to stocking control, the method is also known as the q-factor approach (O'Hara and Gersonde 2004). The q-factor is defined as:

$$q = \frac{n_i}{n_{i+1}}$$

where n_i is the number of trees in the ith diameter class, n_{i+1} is the number of trees in the next larger class, and q is the q-factor. Using 5 cm diameter classes, a q-factor of 1.5 would imply that if one diameter class had 12 trees, the next smaller diameter class would have 18, and the next 27, etc.

The q-factor defines the negative slope of the diameter distribution, but because it is based on diameter classes, the slope is dependent on the size of the diameter classes. If the diameter class size is increased, then the q-factor also must increase to represent an equivalent diameter distribution. This change can be described by a relationship where an increase in the diameter class size by a factor of 2.0 requires the q-factor increase to that power or 2.0. Hence, if the q-factor was 1.3 with 5 cm diameter classes, the equivalent q becomes 1.69 with 10 cm diameter classes. This requires that a silvicultural prescription specifies the q-factor and the size of the diameter classes to provide a complete description of the desired stand structure.

Typically, q-factors vary between 1.0 and 2.0. A q-factor of 1.0 defines a flat diameter distribution with the same number of trees in all diameter classes. A q-factor of 2.0 defines a steep negative exponential distribution with many small trees. The number of trees in a smaller diameter class can be calculated using

$$n_i = n_j \times q^{(j-1)}$$

where n_i is the number of trees in a smaller diameter class, n_j is the number of trees in a larger diameter class, and q is the q-factor. For example, for a diameter distribution with 12 diameter classes, a q-factor of 1.4, and 10 trees in the largest class, the number of trees in the smallest class (n_i) would be:

$$405 = 10 \times 1.4^{(12-1)}$$

For the same situation but a q-factor of 1.2, the number of trees in the smaller class would only be 74; however, with a q-factor of 1.6, it would be 1759 trees. This demonstrates the sensitivity of the diameter distribution—and particularly the number of small trees—to the q-factor.

On semilogarithmic scales, the q-factor defines a straight line. The basal area B defines the area under the curve, and the maximum diameter D defines the intercept on the diameter axis. Mathematically, both of these last two parameters are not necessary, but they make implementation easier. Both the B and D parameters can vary for a given q-factor. For example, a given q-factor can describe stands with a range of basal area levels or a range of different maximum diameters.

For application, the BDq system requires the development of a target diameter distribution and an inventory of current stocking. As with the Plenter system, the target and actual diameter distributions are compared to identify diameter classes with too many or too few trees. This guides the marking of trees for harvest when trees are cut from diameter classes with surplus trees or possibly left to compensate for other diameter classes where tree numbers are deficient (Figure 7.13). Ideally, the diameter distribution will conform to the target negative exponential distribution after cutting. Once a stand is regulated—or conforming to the target negative exponential distribution—the surplus trees in each diameter class would be cut at each cutting cycle. Because they represent the cutting cycle increment, cutting these trees and returning the stand to the target diameter distribution at the end of each cutting cycle assures sustainability.

The foundation of the BDq system is the q-factor, or negative exponential diameter distribution, that defines the stand structure. Unless the q-factor is close to 1.0, it will include many more small than large trees. For this reason, many recent applications of the BDq system have used low q-factors or a segmented diameter distribution, where a two or more q-factors are used to describe the structure. For example, one q-factor might apply to the commercial-sized trees and another to smaller, non-commercial trees.

7.3.4 Stand density index allocation

An alternative approach is to allocate relative density values from an even-aged relative density measure to diameter classes in a multiaged stand. The relative density measure serves as the measure of growing space occupancy with this approach. This represents a more complex approach to stocking control than the Plenter system or the BDq system because the density measure, rather than number of trees, is being allocated to diameter classes (O'Hara and Gersonde 2004).

The central question with this approach is whether the distribution of a relative stand density index by diameter classes provides a better means of allocating growing space than allocating number of trees. This is essentially a comparison of whether a relative density measure is more effective than an absolute density measure for representing competition or growing space occupancy. In principle, the use of a relative density measure would appear to provide the same benefits over an absolute density measure as with even-aged stands. Relative density measures are weighted per unit of tree size and thereby provide an index of density that is more independent of tree size or stage of stand development. For even-aged stands, a relative density measure can be used to compare the competition in two stands that might be at very different stages of development. If stocking allocations are being made for multiaged stands by diameter class, then it probably makes little difference whether a relative or absolute density measure is used. But for larger diameter class groups, or alternate separations of stand components, the relative density measure may have advantages.

How much of the relative density should be allocated to each diameter class? Long and Daniel (1990) allocated stand density index (SDI) to groups of diameter classes for ponderosa pine. Their examples showed how a diameter distribution defined with a q-factor resulted in unequal allocation of SDI and basal area (Figure 7.14). This has implications for how we perceive stands to be balanced (see Section 7.2.7; O'Hara 1996) since, in this example, a stand structure designed with the q-factor and assumed to be balanced produces unequal distributions of growing space occupancy based on basal area and SDI (O'Hara and Gersonde 2004).

In even-aged stands, SDI is calculated as:

$$SDI = N \left(\frac{D_q}{25} \right)^{1.6}$$

where D_q is the quadratic mean diameter of the stand in centimeters, and N is the number of trees per hectare. However, in multiaged stands, or any stand with a nonnormal diameter distribution, an additive equation for individual stand components is recommended. Long and Daniel (1990) proposed

$$SDI_c = \sum N_i \left(\frac{D_i}{25} \right)^{1.6}$$

where SDI_C is the stand density index of a single component, N_i is the number of trees in the component, and D_i is the midpoint of the ith tree in the

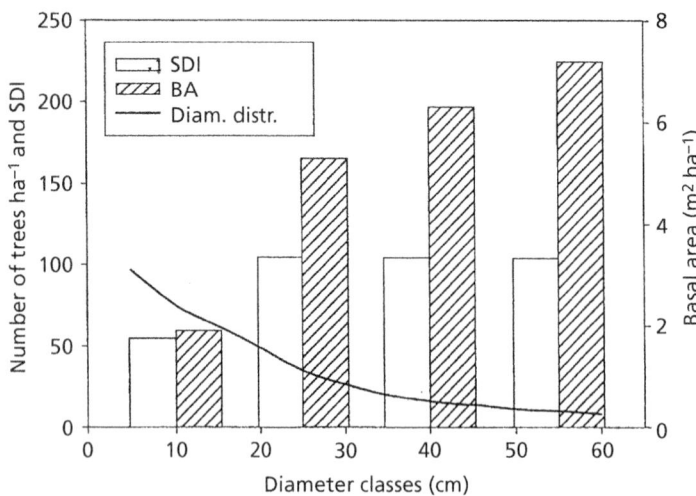

Figure 7.14 Stand density index (SDI), basal area in m² (BA), and trees/ha (Diam. Distr.) for a ponderosa pine stand. This structure shows equal SDI for the three larger diameter class groups but less for the smallest diameter class group. Note that basal area and SDI are not distributed equally. The trees/ha follow a q-factor that ranges from 1.15 to 1.42 (adapted from O'Hara and Gersonde 2004).

component in centimeters. In current usage, diameter classes, or groups of diameter classes, are the most common component. However, similar methods can be applied to other stand components. There is some question about the importance of the additive approach to SDI calculation. Ducey (2009) found that the equations produced divergent values in some structures or when the diameter distribution was truncated to exclude smaller trees. Other references on using SDI in complex stands include Ducey and Larson (2003), Woodall et al. (2003), Zeide (2005), and Ducey and Valentine (2008).

7.3.5 Leaf area allocation

Leaf area is another representation of occupied growing space that can be used to guide stocking in multiaged stands (O'Hara and Gersonde 2004). **Leaf area** is the surface area of a tree or plant's foliage and is expressed either as the amount of one-sided or projected surface area, or as all-sided surface area. **Leaf area index** (LAI) is the sum of all the upper or all-side leaf surface areas per unit of ground area below the canopy. Because it is normally expressed as m^2 leaf area per unit of ground area in m^2, it is unitless.

LAI is a useful measure of growing space occupancy because it generally increases during early stand development in even-aged stands and attains a maximum that is related to site quality (Vose et al. 1994; Margolis et al. 1995). For example, on a high productivity site, a higher LAI would be attained than on a low quality site. Similar limits apply to the maximum attainable LAI in multiaged stands (O'Hara and Gersonde 2004).

The use of LAI has been described as a "first principles" approach to stocking control (O'Hara and Valappil 1999) because LAI involves a more direct representation of growing space occupancy. LAI could represent the opportunity to increase stocking on higher quality sites that can support greater LAI, or the need to adjust opening size due to higher light quality on poorer sites that support less LAI. This approach also allocates growing space to canopy strata, age classes, or species rather than diameter classes. For example, a ponderosa pine guide uses age classes or cohorts (Figure 7.15), and a Norway spruce–Scots pine guide uses canopy strata and species (Figure 7.16). When growing space is allocated to a stand component (cohort, strata, species, etc.), it affects both the amount of growing space left for other components and also their increment (e.g., Figure 7.2; also see Figure 4.3). Leaf area is also a relatively efficient variable for predicting tree growth. The leaf area allocation approach therefore provides some predictions of these trade-offs.

The user designs the desired stand structure by setting a maximum total level of growing space occupancy or LAI, the number of components, and how the LAI is allocated among the components. Maximum LAI can be estimated through plot data, previous research studies, or remote sensing. A spreadsheet model called the "Multiaged Stocking Assessment Model" (MASAM) was used to assess different stocking allocations. In the ponderosa pine example in Figure 7.15, LAI was set at 6.0, and four cohorts were assigned variable amounts of growing space and numbers of trees. The model does not provide the user with a diameter frequency distribution; instead, estimates of a variety of details about the structure being designed are listed under the category "Diagnostic Information." This information is based on age classes or canopy strata, so the number of these components varies with the target stand structure. The diagnostic information includes information for the beginning and end of each cutting cycle for LAI, basal area, and SDI. End-of-cutting cycle values are provided for stand increment and average tree vigor.

The model is constructed so that the oldest age class, or cohort, is removed at the end of each cutting cycle. The second oldest cohort becomes the oldest, and a new cohort is regenerated. If canopy strata are used instead of cohorts, the model assumes the tallest stratum is removed at the end of the cutting cycle, and the second-tallest becomes the tallest stratum during the next cutting cycle. Cutting cycles are assumed to be repeated over time; however, the user can change the stocking parameters to design a series of consecutive structures that might be useful for transforming a stand from even-aged to a stand with two or more age classes.

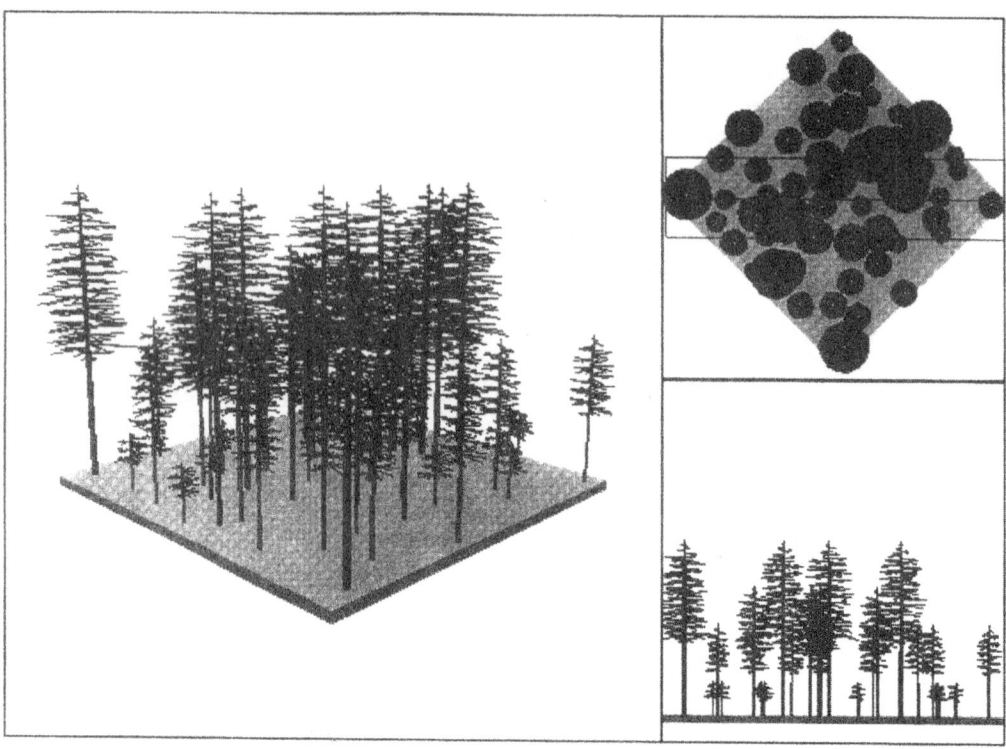

Ponderosa pine MASAM - MONTANA					
	USER-SPECIFIED VARIABLES				
TOTAL Leaf Area Index (LAI)	6				
	Cohort 1	Cohort 2	Cohort 3	Cohort 4	TOTAL
Number of Trees/Cohort/Hectare	46	61	76	91	274
Percent of LAI/Cohort	40	30	20	10	100
	DIAGNOSTIC INFORMATION				
	Cohort 1	Cohort 2	Cohort 3	Cohort 4	TOTAL
Leaf Area Index/Cohort ECC	2.4	1.8	1.2	0.6	6.0
Leaf Area Index/Cohort BCC	1.4	1.0	0.5		2.8
Leaf Area/Tree (m^2) ECC	521.7	295.1	157.9	65.9	
BA/Cohort (m^2/ha) ECC	12.1	8.3	5.1	2.5	28.1
BA/Cohort (m^2/ac) BCC	6.3	4.1	2.1		12.5
Avg. Vol. Increment/Tree (m^3/yr) ECC	0.05	0.02	0.02	0.00	
Avg. Vol. Increment/CC (m^3/ha/yr)	1.5	1.0	0.7	0.1	3.4
Quadratic Mean DBH/Cohort (cm) ECC	51.2	37.0	26.0	16.6	
Tree Vigor (cm^3/m^2/yr)	77.4	80.6	71.2	46.4	
Stand Density Index ECC	145.0	114.2	80.8	47.4	387.5
Stand Density Index BCC	86.1	64.9	39.6		190.6

Figure 7.15 Example of stocking control in a multiaged ponderosa pine stand with four age classes or cohorts using the leaf area approach. A linear increase in trees per acre with younger cohorts is used, but leaf area is allocated with a linear decline with younger cohorts. Estimated production from this structure is 3.4 m³/ha/year (from O'Hara et al. 2003). BA, basal area; BCC, beginning of each cutting cycle; CC, cutting cycle; ECC, end of each cutting cycle; LAI, leaf area index.

Figure 7.16 Depiction of a mixed Norway spruce–Scots pine stand in southern Finland and which had stocking controlled through allocation of leaf area. Norway spruce are shown with longer crowns than Scots pine. LAI totals 1.9, with 1.2 being Scots pine. Basal area totaled 30.5 m^2/ha, and periodic annual increment, 9.5 m^3/ha/year (adapted from O'Hara et al. 2001).

An advantage of this approach is that it can easily be used to design a diversity of stand structures. For example, a user is not restricted to a particular shape of diameter distribution. Figure 7.17 shows a two-cohort ponderosa pine stand that is similar to the example in Figure 7.15 except that fewer cohorts are retained. The model can also be adjusted to retain some trees as reserves or to guide restoration treatments (O'Hara et al. 2003). In this case, the normal assumption that the oldest cohort/largest canopy strata is removed at the end of each cutting cycle is modified to retain these trees.

7.3.6 Stocking control and group selection

Similar stocking control concepts also apply to stands managed with group selection methods (see Chapter 6), but with certain qualifications. Group selection essentially expands the spatial scale of removals so that, rather than leaving individual tree opening, larger groups of trees are removed (Figure 7.18). Single tree and group selection blend together at intermediate opening sizes. Group selection stands are typically controlled using area control forest regulation concepts from even-aged forest management. Within a stand, the area treated is the same for each cutting cycle. Each group is typically managed over a rotation as though it were an even-aged stand. The stand size divided by the number of cutting cycles in a rotation to get the area treated each cutting cycle. Or conversely, the cutting cycle can be determined by dividing the rotation by the number of cutting cycles.

The diameter distribution of a stand managed under the group selection system may follow a reverse-J diameter distribution because of decreasing numbers of trees in older groups. Group selection units may have more constant levels of a relative density index or similar amounts of LAI, regardless of group age. However, the density of trees in groups can be managed to follow any trajectory as they develop. For example, a well-managed stand might have a diameter distribution that approaches a flat, straight line if density were closely controlled and mortality factors were minimal. Likewise, relative density or LAI could be distributed in different patterns depending on the density management limitations and other objectives. Flexibility is important in stocking regimes for group selection systems, as it is for single tree selection systems.

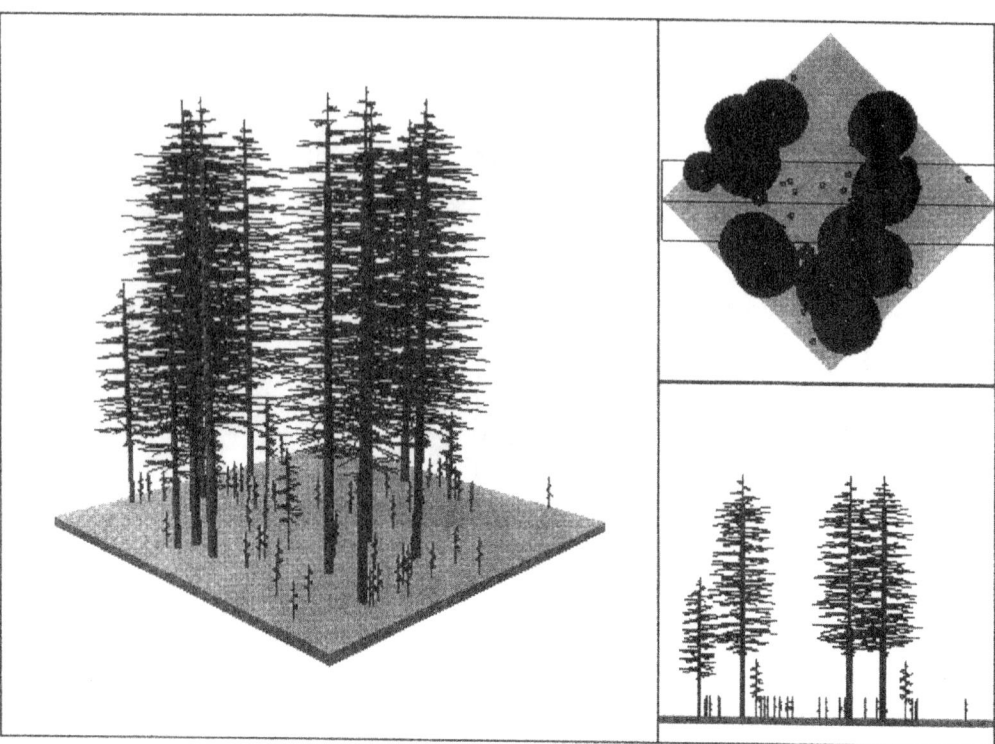

Ponderosa pine MASAM - MONTANA					
	USER-SPECIFIED VARIABLES				
TOTAL Leaf Area Index (LAI)	6				
	Cohort 1	Cohort 2	Cohort 3	Cohort 4	TOTAL
Number of Trees/Cohort/Hectare	50	12	25	40	127
Percent of LAI/Cohort	90	7	2	1	100
	DIAGNOSTIC INFORMATION				
	Cohort 1	Cohort 2	Cohort 3	Cohort 4	TOTAL
Leaf Area Index/Cohort ECC	5.4	0.4	0.1	0.1	6.0
Leaf Area Index/Cohort BCC	1.8	0.1	0.0		1.8
Leaf Area/Tree (m^2) ECC	1000.0	350.0	48.0	15.0	
BA/Cohort (m^2/ha) ECC	26.9	2.1	0.4	0.1	29.5
BA/Cohort (m^2/ac) BCC	8.6	0.2	0.1		8.9
Avg. Vol. Increment/Tree (m^3/yr) ECC	0.13	0.02	0.00	0.00	
Avg. Vol. Increment/CC (m^3/ha/yr)	3.9	0.2	0.1	0.0	4.2
Quadratic Mean DBH/Cohort (cm) ECC	73.4	41.5	12.7	5.7	
Tree Vigor (cm^3/m^2/yr)	99.410	76.925	68.914	48.700	
Stand Density Index ECC	280.0	27.0	8.4	3.8	319.3
Stand Density Index BCC	112.5	4.0	2.4		118.9

Figure 7.17 Example of ponderosa pine restoration using the leaf area approach. A large proportion of growing space allocated to the oldest age class or stratum helps meet the restoration goal of a presettlement type of multiaged stand structure but does not follow the conventional definitions of balanced stands (from O'Hara et al. 2003).

Figure 7.18 Group selection in mixed conifer forests in the Sierra Nevada, California, US. A new group is established in the foreground, and an approximately 15-year-old group is in the middle ground.

7.3.7 "Free" and other alternative approaches to stocking control

Silviculture is evolving to accommodate an increasingly diverse set of objectives, including many that attempt to increase stand and broad-scale diversity in stand structure. Many current stocking tools assume a traditional approach of being focused on stand averages, thereby ignoring variability in stand structure. It is also difficult to integrate the diversity of stand-level elements of structure into a single stocking tool. For example, a multiaged stocking prescription may be concerned with species composition, standing and down wood, and spatial arrangements of elements, as well as the traditional stocking parameters of numbers of trees and size classes. This is a difficult set of parameters to include in a general stocking guide, as well as a specific stand-level prescription. An alternative approach is to give flexibility to the manager to develop a prescription or mark a stand based on the structural features of that stand. A very general stand structure may be recommended that is within broad guidelines or constraints but does not have specific structural requirements. The manager could then integrate existing structural features, such as the presence of large trees, or site variation into design of the stand structure. These "free approaches" are therefore an alternative that do not utilize the specific stocking parameters of many current approaches. Examples include the free selection approaches of Graham and Jain (2005), Mount (2010), and Bončina (2011).

There are many other stocking control approaches for multiaged stands beyond those described here. Some of these have been discussed in the scientific literature; others have not and may only be known regionally. Examples of more formal approaches include the "volume/guiding diameter limit" (VGDL) approach that was used for decades on the Crossett study in southern pines (Reynolds et al. 1984; Baker et al. 1996; Guldin 2011). VGDL sets harvest volume equal to growth over a cutting cycle and uses a flexible upper diameter limit to guide tree selections toward minimum residual stand volume. Hallin's (1959) "unit area control" treated homogeneous areas within a stand as harvest areas, thereby perpetuating an uneven-aged structure through a form of group selection (see Box 6.2). Seydack's (1995, 2012) system for tropical, mixed-species forests, was to use preemptive harvesting of older senescent trees. The "maturity selection" and "improvement selection" systems for ponderosa pine blended financial and biological maturity of individual trees to make tree selections (Munger et al., 1936; Pearson 1942; O'Hara et al. 2010). Two other regional approaches in North America include the long-term selection work in Pioneer Forest (Box 7.1) and the Stoddard–Neel approach (Box 7.2) in the Midwest and southeastern United States, respectively.

Box 7.1 Single tree selection in North American central hardwoods at Pioneer Forest

The central hardwood region of the United States includes large areas of mixed hardwoods growing on moderate-to-low productivity sites. Both tree species and biodiversity can be high in these systems. Various oaks, particularly white oak, are often the most valuable or desired species. The Pioneer Forest in southern Missouri is a 65,000 ha forest managed with a single tree selection approach since the 1950s. The history of Pioneer Forest and its management approach have been described by Iffrig et al. (2004), and Guldin et al. (2008).

The Pioneer Forest was acquired in the 1950s in a somewhat degraded state. One objective was therefore to build volume or growing stock, maintain the more valuable oaks, and produce some income. The single tree selection method that evolved over time on Pioneer Forest removed approximately 40% of the volume on cutting cycles that ranged from 15 to 25 years. Individual tree selections generally removed 30–37 merchantable trees/ha (12–15 trees/ac) based on tree age, species, tree quality, presence of insects or pathogens, and spatial arrangements of trees and tree crowns. These factors may vary in importance depending on local site conditions or management history. Tree markers look at every tree and leave the healthiest trees on the site that have the best potential for future development on that site (Iffrig et al. 2004). Although species vary in value, there are no competitive species that are not valuable.

The stocking control procedure at Pioneer Forest therefore is not well suited to quantification or development of detailed guidelines because it is sensitive to highly variable site and stand conditions. The marking on one site may favor one set of species while the marking on another may favor a different suite because of site variation, existing tree species composition, or possibly for other objectives such as leaving mast-producing trees for wildlife (Iffrig et al. 2004). Resultant structures also do not form consistent diameter distributions at the stand- or section-level (Loewenstein et al. 2000).

After decades of practicing the same general approach to single tree selection, Pioneer Forest has seen a steady increase in standing commercial volume along with periodic volume removals. Figure 7.19 shows an increase of 184% in commercial volume for all species and positive trends for individual species or species groups. For example, the

Figure 7.19 Recently marked mixed pine–oak stand in Pioneer Forest. White arrows show trees marked for removal.

continued

> **Box 7.1** *Continued*
>
>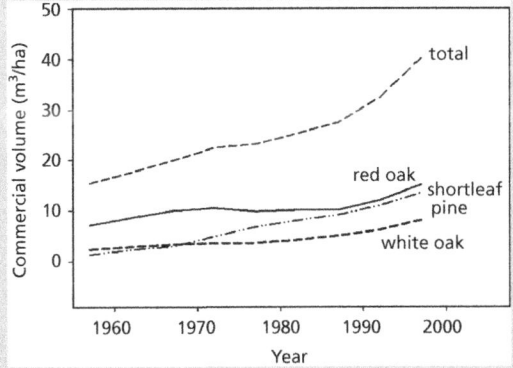
>
> **Figure 7.20** Commercial volume for 1957–1997 from Pioneer Forest inventory plots. Board foot volumes per acre were converted to cubic meters per hectare using a conversion factor of 1 board foot/ac per 0.0125 m³/ha. Original data from Iffrig et al. (2004) included only broadleaved trees greater than 3.9 cm and pines greater than 3.2 cm in diameter at breast height.
>
> approach has resulted in an increase in white oak—one of the more important species—of 340% over the 40-year period represented in Figure 7.20. The Pioneer Forest approach has been in place sufficiently long to demonstrate flexibility to accommodate a variety of site conditions and an increasing inventory over time, a sustainability of species composition, production of wood products, and the capacity to guide both sustainability for longer term and restoration toward a suite of economic, social and economic benefits.
>
> The Pioneer Forest system is successful without having any standard marking protocol or preset stocking standard. Instead, it relies on the ability of managers to consistently mark trees in sufficient numbers to produce viable sales of timber. It demonstrates the importance of manager experience to understand the local forest type and also how providing managers with the flexibility to make decisions with minimal guidance can succeed. However, this works especially well in this forest type where natural regeneration is reliable and all species have some value.

7.4 Synthesis

Stocking control in multiaged stands is an exercise in the allocation of growing space to a finite number of stand components. The basis for dividing growing space among stand components can be to achieve objectives related to sustainability, maximization of timber production, creation of wildlife habitat, or many others. How this growing space is allocated will vary with the objective, the forest type, the site characteristics, and the existing stand conditions. As a result, many different procedures have been developed. Some are commonly known and are described in the scientific literature. Others are local stocking control procedures that are less known. Because stocking control is such an important part of multiaged silviculture, these tools occupy a central place in the management of multiaged stands.

In this current era of managing for multiple resource values over large scales, there are advantages to using stocking control tools that are to a range of objectives and that are meaningful at multiple spatial scales. But there are justifiable reasons to use any method. The Plenter and BDq methods have a long history of use in Europe and North America. Diameter-limit cutting is a means of maximizing short-term benefits but with risky long-term effects. SDI builds on a strong foundation of even-aged density management theory. The leaf area allocation approach provides strong links to other ecosystem values and to a range of target stand structures. Many other systems that are based on local experience are far less well-known but meet objectives and may have been in use long enough to demonstrate sustainability. Approaches termed "free" provide broad guidelines or constraints for managers to design and implement stand structures that are unique to the local conditions and that integrate existing structural feature in the prescription or marking. Ultimately, the question of which stocking control system to use depends on objectives, ease of use, availability of developed systems, and local experience. In an emerging era when structural heterogeneity is sought at multiple scales for a diverse suite of objectives, stocking control in multiaged stands may be evolving to provide a much greater degree of discretion to the manager.

Box 7.2 The Stoddard–Neel approach in longleaf pine

Longleaf pine is a very fire-dependent species that occurs in the southeastern United States. It was once common throughout this region, primarily on the coastal plain, but it is currently found only on a small proportion of its original range. Fire return intervals in longleaf pine stands are extremely short, sometimes as low as a year or two. Longleaf pine is fire resistant and also initiates through a "grass stage" that is also very fire dependent. Both stands with several age classes and open stand structures with great sight distances are common when stands are managed to emulate these natural forests (Figure 7.21).

The Stoddard–Neel approach has been developed to maintain these structures and the extremely high plant and animal diversity found in these ecosystems. The application of this approach in southern Georgia has been described by Mitchell et al. (2006), Jack et al. (2006a), Moser (2006), and Neel et al. (2010). A strong impetus for management in this part of Georgia is bobwhite quail, which are a desired game species and common in frequently burned longleaf pine landscapes. The Stoddard–Neel approach manages for quail habitat, maintenance of biodiversity, and aesthetic qualities in longleaf pine stands in addition to timber values. Prescribed burning is used very frequently because it encourages the understory flora that the quail require while also discouraging the midstory broadleaved vegetation. Trees are often removed in patches that favor regeneration. But the overstory density of longleaf pine interacts with fire by providing the primary fuel—pine needles—necessary to carry a fire. Hence, quail habitat requires the prescribed burning, and the burning requires the longleaf pine in a suitable overstory density. The ecosystem benefits include extremely high native species diversity, regeneration, and control of the broadleaved trees that invade these sites.

Burning schedules are flexible to accommodate the needs of management; these needs can include maintenance of the understory plants required by the quail, avoiding accumulations of fuels, or controlling the development of broadleaved trees. A short fire return interval of several years will result in cooler fires than a longer interval, but each will have different effects that may be desired in some cases and not in others.

The Stoddard–Neel approach emphasizes the residual structure rather than the trees or volume removed. Neel et al. (2010) describe a strong philosophy toward building growing stock and value in the woods: harvesting may remove as little as 50% or as much as 90% of growth in a cutting cycle (Moser 2006; Neel et al. 2010). Increment may also be low: Jack et al. (2006b) presented an example

Figure 7.21 Multiaged longleaf pine stand in Georgia, US, managed with frequent prescribed burning.

Box 7.2 *Continued*

of 50 years of the Stoddard–Neel approach that produced only about 0.5 m^3/ha/year. There are no residual basal area targets or targets for diameter distributions. A tree marker has to be capable of evaluating a tree's current value, potential value, and its contribution to the diversity of the system. Hence some good trees will be left for future value, and some poor trees will be left for their aesthetic or diversity value. Regeneration is encouraged by creating gap openings which may be expanded in future markings, and typically these canopy openings are located to release advance regeneration rather than to establish new age classes. This approach demonstrates the importance of frequent burning to provide control of competing broad leaved oaks and other trees. With frequent burning and no measured residual stocking targets, the Stoddard–Neel approach is flexible to provide regeneration and sustainability through the continued development of new age classes. However, with little quantification or "formula" for this method, it is difficult to understand for persons unfamiliar with longleaf pine forests, and very difficult to train new tree markers.

CHAPTER 8
Regenerating multiaged stands

8.1 Introduction

Regeneration is an integral process in multiaged silviculture. Without regeneration at the right time and place, the multiaged system is not sustained. As a result, the success and sustainability of multiaged systems is often a function of whether timely and sufficient regeneration occurs, rather than whether the stands meet some prescribed targets for overstory stand structure. Much of our knowledge of forest regeneration relates to even-aged stands. Several books have been written or compiled on this subject (e.g., Lavender et al. 1990; Duryea and Dougherty 1991; Hobbs et al. 1992), but coverage specific to multiaged stands has been limited to brief presentations on selection silvicultural systems. Much of what we know about even-aged stand regeneration is applicable in some way to multiaged stands. This even-aged stand information also provides a basis for understanding important differences between stand structures and extensions to regeneration in environments with existing overstory competition for light or moisture.

Regenerating managed multiaged stands is often assumed to be exclusive to natural forms of regeneration. This apparently comes from long-standing perceptions that multiaged systems are more natural than even-aged systems (Lamprecht 1989; Matthews 1989; Larsen 1995) and therefore well suited for natural regeneration. Much of our early knowledge and experience in multiaged silviculture comes from central Europe, where successful natural regeneration is relatively common. Similarly, regeneration in many broadleaved stands in eastern North America has also been generally reliable in multiaged stands. The success of natural regeneration in these two regions provides support for the perceived naturalness of these systems and the use of natural regeneration. However, nothing precludes artificial forms of regeneration such as tree planting or direct seeding in multiaged stands. Perhaps the term "artificial" has negative connotations that discourage alternative means of regenerating multiaged forests. However, artificial regeneration is an important silvicultural option for gaining adequate restocking, insuring forest practice regulations are met, securing desired species composition, shortening cutting cycles, adjusting spatial patterns of regeneration, and anticipating the effects of climate change. Regeneration options for multiaged stands are presented in this chapter, along with operations necessary to achieve regeneration success.

8.2 Natural regeneration

In multiaged forests, important types of natural regeneration may include seedlings, advance seedling and sprout regeneration, stump sprouts, suckers, and many others (see Oliver and Larson 1996; Greene et al. 1999). The limiting factors for each of these may vary at different times during the regeneration process. For artificial regeneration, there are similar limiting factors that vary with space and time. The overriding factor in multiaged systems that affects regeneration is the presence of existing overstory competition that affects the light regime, moisture regime, or both. The overstory in the multiaged stand is generally much heavier than what might be found in even-aged stands with reserve trees. This effect of the overstory is to reduce light to the understory. However, it also provides seed sources, protects again heat and cold, and affects moisture and nutrient availability.

The three primary needs for natural regeneration have been expressed as the environment, the seedbed, and the seed source (Roe et al. 1970). This traditional regeneration triangle is focused on natural seedling regeneration because of the inclusion of the seedbed. For vegetative regeneration of quaking aspen, Shepperd (2001) described a different regeneration triangle that included the environment, protection, and hormonal stimulation. Other guidelines also exist and indicate that many factors may be essential for successful natural regeneration. Although we may be able to identify three factors that are most important, other factors may also be more important at different times, locations, or situations.

8.2.1 Seed dispersal and availability

Seed dispersal is generally not a concern in multi-aged stands. At one extreme, individual trees are removed, which results in very small gaps; at the other, group selection and other systems, such as variable retention or two-aged systems, may form relatively large gaps. It is only in the latter case that seed dispersal may present problems. Generally, if parent trees are sexually mature and vigorous, seeds will be produced. Exceptions occur, of course, as many species experience very inconsistent annual seed production, with intervals between good seed years that might be as much as 10 years long. For example, in southern Finland, Saksa (2004) reported similar seed production in even-aged and uneven-aged stands dominated by Norway spruce. Since most wind-dispersed seed falls near the parent tree, and fewer seeds fall with increasing distance from the parent tree or stand edge (Figure 8.1), the distances are generally not problematic in multiaged stands. A two-hectare circular opening would have a radius of approximately 80 m; a typical wind-dispersed seed such as that from a ponderosa pine or Norway spruce would easily cover this distance, although seed would be more plentiful near edges (Figure 8.2). For seeds dispersed by animals and birds, it is generally assumed that these vectors will spread seed through large gaps. However, little information exists to document this. Seed may also be present in the "seed bank," either as buried seed in the forest floor or aerially in the canopy. This seed bank may include tree species, such as cherry or

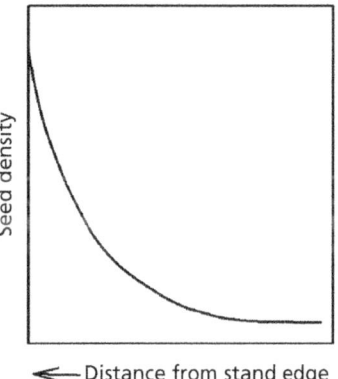

Figure 8.1 Seed dispersal patterns for most species follow a exponentially declining relationship from the stand edge or seed wall. The shape varies with seed weight, wind resistance, height of the seed source, and dispersal mechanism.

basswood, and shrub species, such as gooseberry or currant. Since these dormant seeds may require heat or light to stimulate germination, multiaged systems may not favor germination of these species.

Although seed dispersal may be sufficient, seedlings still require light and other resources for survival. Page et al. (2001) found that the density of regeneration increased but growth decreased with increasing overstory basal area in Sitka spruce in Scotland. Seed dispersal is therefore enhanced by the very same conditions that impede growth of seedlings. This adds another dimension to the trade-off between managing resources for different age classes or canopy strata in multiaged stands.

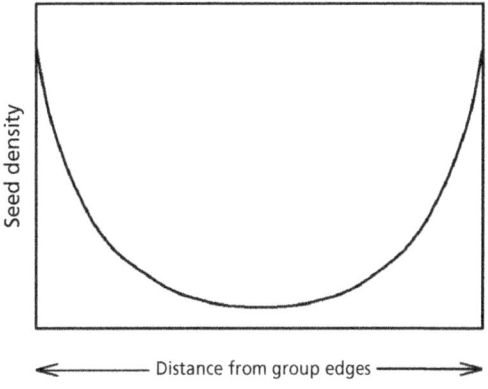

Figure 8.2 Seed dispersal in group opening may come from all edges.

In complex stands where sexually mature individuals may occur at low frequencies or where certain species may be scarce, pollination may be a greater concern than seed dispersal. For example, large distances between trees of one species, or between male and female members of dioecious species, may reduce pollination, thereby reducing numbers of viable seeds of desirable species or contributing to inbreeding. Likewise, seed production problems may occur in temperate forests where some species produce abundant seed crops only infrequently or when production of seed does not coincide with cutting cycles. For example, coast Douglas-fir typically produces heavy or medium seed crops only every seven years on average (Isaac 1943). A monoecious species, such as a pine, in a multiaged stand sparsely stocked with sexually mature trees might also be prone to inbreeding because of distances between pollinating and seed-bearing trees.

8.2.2 Seedling regeneration

Seedling regeneration includes natural seedlings that become established after a disturbance. Typically, the disturbance—whether a natural disturbance or a harvest treatment—provides conditions suitable for establishment by reducing competition, increasing light availability, and providing a more suitable substrate for seed germination. However, seedlings can be established at other times, just less commonly. The age class of seedlings that develop following a disturbance are competing with each other as well as with other understory vegetation and overstory trees. As seedlings become established and grow, the overstory and other understory vegetation are also recovering from the disturbance. Competition therefore may increase dramatically during the cutting cycle or in the period after a disturbance.

Numerous studies have looked at the growth of understory seedlings in relation to percent above canopy light or a similar measure of light intensity or canopy competition. Seedling height and diameter growth increase with increasing light intensity, although the shape of these relationships varies (see results for coast Douglas-fir (Mailly and Kimmins 1997; Drever and Lertzman 2001), western redcedar (Drever and Lertzman 2001), Norway spruce (Chrimes and Nilson 2005; Stancioiu and O'Hara 2006c), European beech (Collet and Chenost 2006; Stancioiu and O'Hara 2006c), and other conifer and broadleaved species (Oliver and Dolph 1992; Chen et al. 1996; Wright et al. 2000)). A primary variable affecting the shape of these relationships is shade tolerance. For example, Gratzer et al. (2004) looked at a variety of species in the Himalayas and found radial growth in the shade-tolerant Himalayan hemlock increased rapidly with increasing light as compared to intolerant species such as Sikkim larch (Figure 8.3). A shade-tolerant tree will typically have a steeper

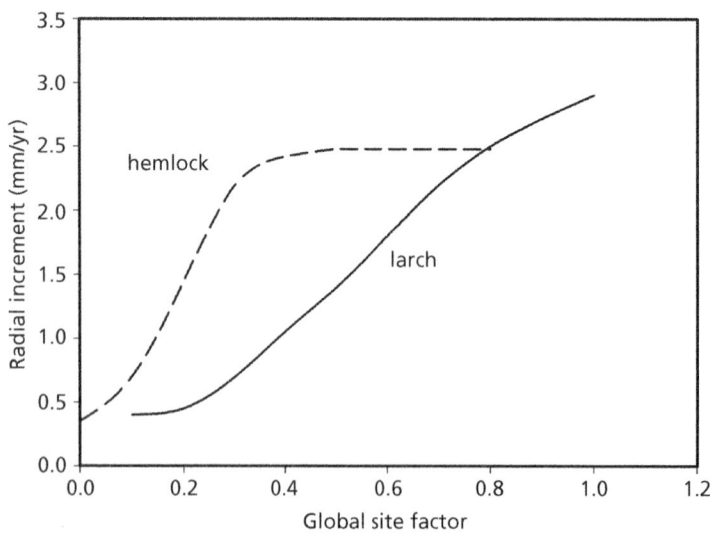

Figure 8.3 Radial growth patterns of two contrasting species in Bhutan: Himalayan hemlock and Sikkim larch. The "global site factor" is an index of diffuse and direct light received (adapted from Gratzer et al. 2004).

response to increasing light intensity as compared to a shade-intolerant tree (Figure 8.3). In the Romanian Carpathians, Norway spruce appeared to have the greatest range in relative height and volume increment as compared to European beech and silver fir (Stancioiu and O'Hara 2006c). However, this was a much narrower range in species' shade tolerance than studied by Gratzer et al. (2004). On harsh sites, seedling performance in relation to the light regime may be confounded by facilitative effects of shrubs and other plants. For example, Keyes et al. (2009) found ponderosa pine in central Oregon, US, had greater survival under shrubs.

A simpler and possibly more relevant expression of the same trends is through the relative performance of seedlings under different regeneration methods or overstory treatments. These studies have compared a variety of overstory treatments that are less easily quantified than actual light measurements. In the Sierra Nevada of California, US, McDonald (1976) compared the growth of five conifer species, three broadleaved trees species, and two shrub species with five different overstory treatments. All species were tallest in the clearcut and decreased in height with reduced cutting intensity, with several species failing to reach 25 cm in height after nine years in the single tree selection treatment (Figure 8.4). However, seedling density was greater in lower quality light regimes, with several species having the highest density in the selection system. Although light studies have indicated which light regimes achieve the greatest growth, there has been less study of the effects on seedling establishment and survival. In other forest types, there may be little effect on species composition from overstory treatments. For example, in southeastern Alaska, US, a range of overstory treatments had little effect on the regeneration of Sitka spruce and western hemlock, the two primary species (Deal and Tappeiner 2002).

8.2.3 Advance regeneration from sprouts

Vegetative reproduction or coppice reproduction—from advance regeneration sprouts, stump sprouts, seedling sprouts, or root suckers—is an important source of regeneration in some multiaged systems. With the exception of root suckers, dormant buds and adventitious buds are the source of these vegetative sprouts. These forms of regeneration often have an advantage over seedlings because they arise from root systems that may be well developed. For example, an advance regeneration oak may have a root system that is considerably older than the above-ground tree (Heggenstaller et al. 2012). A stump sprout has the resources of the parent stump system that include food reserves as well as the resource acquisition ability of the root system. They therefore have the potential to grow

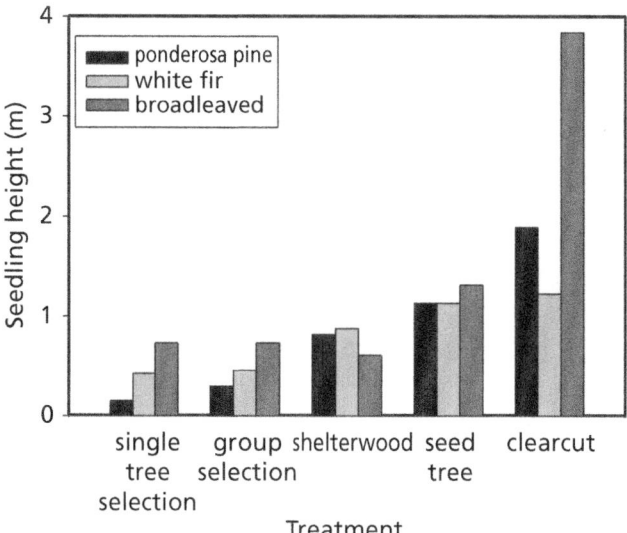

Figure 8.4 Nine-year height growth of seedlings in various silvicultural treatments in California, US. Conifer seedlings were from direct seeding; broadleaves were primarily of sprout origin and included California black oak, tanoak, and Pacific madrone. All species showed sensitivity to light as represented by treatment. Ponderosa pine was more sensitive than the more shade-tolerant white fir (adapted from McDonald 1976).

very quickly and typically outcompete seedlings of the same species (Bond and Midgley 2001; Johnson et al. 2009). Although vegetative sprouts inherit an intact root system, there may be delays before they can support this system. As a result, the root system may initially deteriorate after harvesting before sprout leaf area is sufficient to support it. In full sunlight, this "functional equilibrium" (e.g., Brouwer 1983) would occur sooner that for a sprout in partial shade (Figure 8.5). It is also possible that the root system might die before the sprouts can develop sufficient photosynthetic capacity to support it. The light regime for sprouts is therefore an important consideration in coppice multiaged systems that affects both sprout growth and viability of the parent root system.

Several processes occurr in a stump sprout system that also apply in various degrees to other forms of vegetative reproduction. A more vigorous root system generally produces a more vigorous sprouting response. Additionally, vigorous trees are more prolific sprout producers than less vigorous trees. However, increasing age and tree size negatively affect sprouting success (Johnson 1977; Dey et al. 1996; Weigel and Peng 2002), suggesting that maximum sprouting may occur at an intermediate tree size. For example, Solomon and Blum (1967) found red maple sprouting was most prolific from tree diameters of approximately 25 cm and declined for smaller and larger trees. Dey et al. (1996) found similar relationships for height growth from five oak species but that the peak varied somewhat with different species (Figure 8.6). This trend was not observed in other non-oak species in the same study. Similarly, larger advance regeneration trees also sprout more vigorously than smaller advance regeneration (Sander et al. 1976). Older trees often have a reduced ability to produce sprouts (Johnson 1977). This may be due to reduced viability of dormant buds or possibly reduced vigor of the parent tree. Although these confounding relationships tend to complicate preharvest assessments of sprouting potential, some general guidelines can be followed:

1. Sprouting varies by species regardless of size and vigor.
2. Sprouting increases with parent tree vigor. This applies to various forms of vegetative reproduction, including root suckers.
3. Older trees experience reduced sprouting potential.
4. Larger trees experience reduced sprouting, but this variable may be confounded with age.
5. There are genetic differences between individuals that affect sprouting ability.

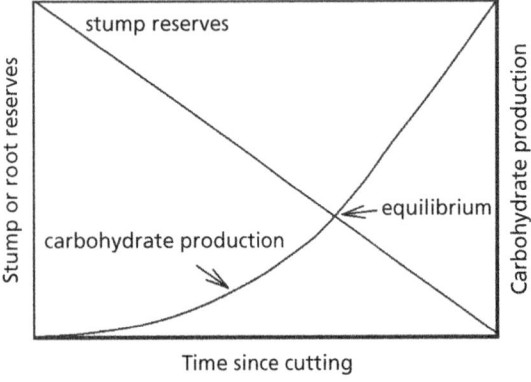

Figure 8.5 Hypothetical production relationships for a single stump after cutting. The stump contains carbohydrate reserves at the time of cutting that support the sprouting and development of a photosynthetic capacity. However, the reserves are also needed to support the stump root system, so consequently, they become depleted. The equilibrium is the point where production of carbohydrates by the new sprouts exceeds the losses in maintaining the system. A stump system with low reserves may not reach the equilibrium and would die.

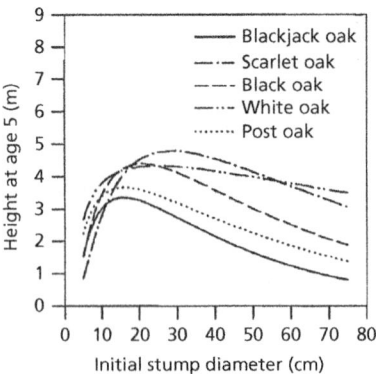

Figure 8.6 Estimated sprout height in relation to stump diameter for selection of broadleaved species from eastern North America (from Dey et al. 1996, Modeling the regeneration of oak stands in the Missouri Ozark Highlands, Can. J. For. Res. 26, 4, 573–583, © Canadian Science Publishing or its licensors).

For vegetative reproduction in partial shade environments, similar processes are observed for seedlings. In coast redwood, O'Hara et al. (2007b) found a similar light response relationship for stump sprouts as others have found with seedling regeneration. Hence, a saturation function whereby sprouts perform better with increasing light but at a decreasing rate will apply to most species (see Figure 4.2). Likewise, in water oak (Gardiner and Helmig 1997) and several other oak species (Dey and Jensen 2002), sprout response also increased with increasing light. A variety of oaks in eastern North America experienced a 2% decline in sprouting for every meter increase in overstory basal area (Atwood et al. 2009). For oaks in southern Missouri, Larsen et al. (1997) developed a probabilistic model for regeneration (primarily sprout origin) that indicated dramatic reductions in the likelihood of obtaining regeneration at higher overstory stocking levels (Figure 8.7). They recommended maintaining overstory basal area at less than 14 m²/ha. The most competitive sprout regeneration will be in better light regimes. There may also be an interaction between site quality and the light regime: oak regeneration may occur on dryer sites, where competitive species are less successful, rather than on the wettest sites (Hodges and Gardiner 1993; Johnson et al. 2009).

The process of developing a self-sufficient root system where the sprouting tree supports the preexisting root system may take longer in multiaged systems where growth of sprouts is reduced in partial shade (Figure 8.5). This relationship would be complicated by clonal patterns or networks of sprouting trees that exist in species such as quaking aspen that sprout from extensive root systems that may cover many hectares. Likewise the sprouting response of smaller groups of clonal trees or nonclonal trees linked through root grafting would also be more complex. In these cases, the entire system would be affected by the presence of shade and would therefore take more time to reach a level of self-sufficiency or functional equilibrium (e.g., Brouwer 1983; Powers and Wiant 1970) than a system in full sunlight. However, the networks formed by systems connected through root grafting may have weaker internal sharing than purely clonal systems.

8.2.4 Advance regeneration from seedlings

Advance regeneration also includes existing seedlings and saplings that live in the understory and do not resprout after being cut, burned, or damaged. These are generally shade-tolerant conifers and broadleaved trees not damaged by overstory removal operations or other disturbances. In managed multiaged stands, advance regeneration may form at any time, but regeneration events and growth surges are likely to coincide with cutting cycle entries, or disturbance events in unmanaged

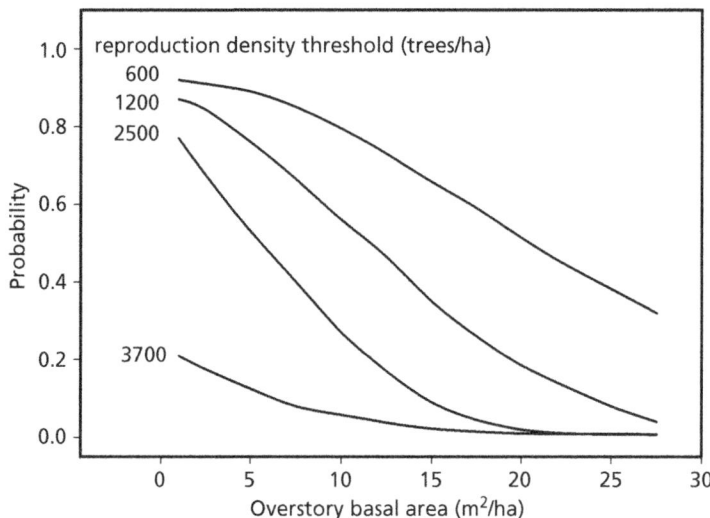

Figure 8.7 Estimated probability of occurrence of all species of regeneration greater than 2 m in height for the Missouri Ozark Mountains, US, in relation to overstory basal area. "Reproduction density thresholds" are target densities: hence the probability of having 600 trees/ha is very large when the basal area is less than 10 m²/ha but very low when there are 3700 trees/ha at the same overstory density (adapted from Larsen et al. 1997).

stands. Advance regeneration may be quite old (see Figure 4.10): for example, Parish et al. (1999) found many subalpine fir trees over 200 years old but less than 10 cm in diameter in old forests in interior British Columbia. In coastal British Columbia, Parish and Antos (2006) reported periods of suppression and release dating back centuries in a slow gradual process of height development.

Silviculturists regenerating stands with advance regeneration have the option of using these trees or replacing them with planted or natural seedlings. Advance regeneration trees have a height advantage if left standing and a root system if they are cut or damaged, and they may also have sprouting potential. For even-aged silvicultural systems, advance regeneration may shorten the rotation length and reduce site preparation and planting costs. For multiaged stands, advance regeneration has similar advantages and may be well adapted to the partial shade environment following release. In either case, there are good arguments for either retention of advance regeneration or not; ultimately, the decision requires a good understanding of postrelease growth of advance regeneration as compared to seedling growth.

The regeneration in multiaged stands usually exists in some state of suppression, and its release—or ability to respond to increased resource availability—is a critical aspect of regeneration success in multiaged silviculture. There have been many studies of advance regeneration release beneath even-aged overstories for both conifers and broadleaved species, and results have been highly variable. This variation may be due to genetic differences in individuals (St. Clair and Sniezko 1999), or difficulties in quantifying either the light regime or the vigor of the tree. Most of these studies have examined trees released into full sunlight. However, the same general principles apply to multiaged stands where released trees have an improved light environment but possibly with only partial sunlight. Most species will experience growth rates following release that correspond to their light regime but possibly with a delayed response (Figure 8.8). As a result, tree growth following release of advance regeneration will be less in poorer light regimes. This applies to both conifer and broadleaved advance regeneration and both sprouting and nonsprouting regeneration.

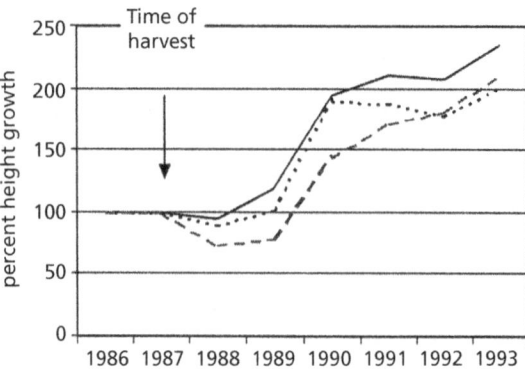

Figure 8.8 Delayed release in interior Douglas-fir in British Columbia, Canada (from Kneeshaw et al. 2002, Patterns of above- and below-ground response of understory conifer release 6 years after partial cutting, Can. J. For. Res., 32, 2, 255–265, © *Canadian Science Publishing or its licensors*).

Advance regeneration trees can be assumed to move toward a functional equilibrium where biomass is allocated to needs such that growth is maximized (Brouwer 1983; Poorter and Nagel 2000). This model implies that if, for example, a tree required more moisture, it would allocate biomass to root development. If tree stability were limiting, the tree might allocate carbon to stem girth. The transition from a shaded light environment to an improved light environment would theoretically result in trees experiencing increased light intensity, transpiration losses, temperature fluctuations, and physical stress on the tree stem through wind. Tree reactions would be highly variable depending on the species, pre- and postrelease environments, and the characteristics of individual trees. The height growth response would therefore be delayed in a tree that required additional root allocation or in a tree that had to expand its foliage area.

A multitude of variables affect the amount of growth response and how quickly it occurs after increased light exposure. These include the level of improvement in the light regime, the species involved, the shade tolerance of the species, the moisture regime, the vigor of the trees prior to release, the history of release and suppression in the subject trees, and the type of release measured. In many cases, these variables produce confounding effects, or the effects may not be consistent along a resource gradient. For example, taller advance regeneration trees may appear to be desirable because of their size. However, questions exist about whether taller trees respond more quickly following release or if smaller trees will respond more quickly and gain a size advantage. Studies have documented reductions in growth responses with increasing advance regeneration height in Scots pine (Vaartaja 1952), western hemlock (Oliver 1976), grand fir (Ferguson and Adams 1980), lodgepole pine (Murphy et al. 1999; Kneeshaw et al. 2002), interior Douglas-fir (Kneeshaw et al. 2002). Others have found taller advance regeneration, including that with Norway spruce (Örlander and Karlsson 2000; Glöde 2002), interior Douglas-fir (Claveau et al. 2006), balsam fir (Stadt et al. 2005), subalpine fir (Claveau et al. 2006), and white spruce (Comeau et al. 2003; Stadt et al. 2005), experienced more rapid growth following release. Differences in these results may be attributable to simple differences in species or perhaps the prerelease light regime. Oliver (1976) attributed poorer response in larger western hemlock advance regeneration to increased respiration requirements but also noted larger trees in a better light regime responded immediately. Part of the response may be related to leaf morphology in sun and shade foliage and the tree's prerelease light regime. If a tree resides in a better light regime before release, it may require no foliage expansion or conversion, and its response may be immediate. Tucker and Emmingham (1977) reported that western hemlock trees lost half their foliage after release, thereby requiring a major replacement of foliage to adjust to the new light environment. Trees that are released from a poorer light regime will therefore have a greater adjustment than trees in better prerelease light regimes.

Shade tolerance is another factor that affects the response of advance regeneration to release. Some studies have noted a more rapid response in shade-tolerant species than in shade-intolerant one, and, among conifers, the rapid response of firs (Helms and Standiford 1985; McCaughey and Ferguson 1988; Oliver and Dolph 1992; Wright et al. 2000). Mechanisms that explain this pattern may be related to greater adaptation to shade, greater allocation to foliage, or simply a greater ability to convert from a shade-endurance mode to a full- or partial-light mode. Results are not consistent: Claveau et al. (2006) reported only weak response relationships with shade tolerance in interior spruce, interior Douglas-fir, and subalpine fir. The fact that these trees were released from a deciduous overstory suggests that prerelease light levels may have been greater than levels that would have favored only the subalpine fir. A study that examined trees that had experienced cyclic release and suppression found that release in shade-tolerant species was less sensitive to prerelease history than that in shade-intolerant species (Wright et al. 2000). Only the most shade-tolerant trees can be found in the poorest light regimes, which explains some of the range of results obtained in studies of advance regeneration release.

Below-ground resources are an often-overlooked factor affecting advance regeneration response to release. If a tree's root biomass was in functional equilibrium with the above-ground moisture and nutrient requirements before release, then above-ground onset of rapid height growth may indicate a below-ground limitation. For example, Claveau et al. (2006) observed an increase in root allocation following release, but the response alone could not be connected to a specific growth limitation. Mitchell (2001) found in four overstory treatments that neither moisture nor nutrients were limiting growth but that the limitation was light. Kneeshaw et al. (2002) found moisture-related stress reduced initial height growth following release.

Although the research on advance regeneration response to release is contradictory in many ways, there are some generalities that emerge. One is that tree release is positively correlated with prerelease growth rates: vigorous trees respond more rapidly and with greater growth than less vigorous trees. The degree of prerelease suppression is therefore a key variable. Shade-tolerant species have a greater capacity to endure a poor light regime, and they also have the greatest ability to respond to release. The sequence of release is often seen first in a root growth response and only later followed by a diameter and then height growth response. The most apparent form of response—height growth—is therefore the one that is only seen after other needs are met. Although moisture and nutrients are critical to growth and vigor of advance regeneration trees both before and after release, light is the key variable to manage and monitor for advance regeneration success.

8.2.5 Physiological and morphological adjustments of advance regeneration

Advance regeneration trees often develop morphological characteristics that distinguish them from similar trees grown in full sunlight. These characteristics include both above-ground and below-ground changes. For conifers, less root biomass for trees in poor light regimes than trees in better light regimes has been reported (Magnussen and Peschl 1981; Chen 1997; Kneeshaw et al. 2002; Claveau et al. 2006). This is a form of adjustment that probably aids in light reception when water is in less demand. The result is lower root-to-shoot ratios for advance regeneration trees as compared to similar trees at better light regimes. There is apparently a shade tolerance interaction with root-to-shoot shoot plasticity. Chen (1997) found the more shade-tolerant Englemann spruce and interior Douglas-fir had much greater plasticity than ponderosa pine over a light gradient.

Many broadleaved species that are capable of sprouting will develop large root systems that are capable of supporting new sprout growth after a release that damages the above-ground part of the tree. Thus the root-to-shoot ratio would be much larger for a resprouting broadleaved tree when compared to either a conifer or a broadleaved tree that was either younger or that one that did not have sprouting capability. Recognition of trends in allocation to roots in broadleaved advance regeneration will be more difficult than for conifers.

An alternative expression of crown morphology changes with light regime in conifers is the silhouette area ratio (Leverenz 1981) that compares shoot silhouette area to the surface area of the detached foliage. For advance regeneration, a greater silhouette area ratio in poorer light environments is typical, and following release, a lower ratio would occur (Tucker et al. 1987).

In addition to variations in crown morphology with light regimes, foliage morphology will also vary with light regime for many species. Much of the plasticity in crown morphology has been attributed to improved reception of light while minimizing respiration. Advance regeneration trees in shade typically have greater **specific leaf area** (SLA)—greater leaf surface area per unit of leaf biomass—than comparable trees in better light regimes. Both conifers and broadleaved tree species have greater SLA when grown in poorer light regimes (Mailly and Kimmins 1997; Chen 1997; Beaudet and Messier 1998); however, for the extremely shade-intolerant species, there may be no range in SLA over a light gradient. Chen and Klinka (1998) reported no plasticity in SLA for the deciduous western larch over a light gradient. For Norway spruce, Metslaid et al. (2007) described increases in needle mass, width, and thickness with increasing light.

There are a variety of morphological changes that can be seen in advance regeneration across a light gradient. Both broadleaved and conifer trees exhibit these characteristics, but shade tolerance plays a role in how these characteristics are expressed. Some characteristics are more apparent in shade-tolerant species and others in shade-intolerant species. Many of the studies of advance regeneration over light gradients have included the open stand as one extreme. For multiaged silviculture, the range in light conditions will often be less, and often considerably less, than in many even-aged stands.

8.2.6 Guidelines for advance regeneration

Although there is no universal trend for whether larger advance regeneration responds better than smaller advance regeneration, there are other indicators of response that are useful for making evaluations. The potential growth of advance regeneration conifers is related to prerelease vigor. Because of the importance of predicting postrelease growth when making decisions over whether to retain advance regeneration, there have been a variety of efforts to develop indices to assess prerelease vigor and release potential. Many of these are based on the characteristic of more shade-tolerant conifers to form a flatter crown in shade as compared to a more conical crown in full sunlight (see Figures 4.10; Parent and Messier 1995; Grassi and Giannini 2005). The length of the live crown of the advance regeneration tree has therefore been a critical variable for predicting postrelease growth in some species, including several true fir species, coast Douglas-fir, and mountain hemlock in western North America (Seidel 1983, 1985; Helms and Standiford 1985;

Oliver 1986). Prerelease height growth is a good indicator (Helms and Standiford 1985; Tesch and Korpela 1993) for these same species. More complex ratios of terminal to lateral growth may also be good indicators of release potential (Page et al. 2001). Whereas these indices vary somewhat by region and species, all are oriented toward shade-tolerant species that exhibit pronounced differences in crown morphology in different light regimes. For boreal species in Canada, Messier et al. (1999) proposed a "maximum sustainable height" that would enable understory species to persist and still respond to increased growing space. It assumed that height negatively affected carbon allocation, with taller trees having greater respiration costs. In an understory environment, height above some threshold would correspond to inadequate reserves for growth. These thresholds were greater with increasing shade tolerance.

A simple equation for comparing the potential height growth of advance regeneration and planted seedlings was developed by Seidel (1980) for conifer species in the interior of the Pacific Northwest. This equation was

$$A = \frac{B-E}{D} + C$$

where:

A = annual height growth of planted seedlings needed to equal height of advance regeneration during a specified time period,

B = average height of advance regeneration at time of release,

C = average annual height growth of advance regeneration after release,

D = length of growth period in years,

E = average height of seedlings at time of planting, and

all height measurements were in equal units.

By this equation, a planted seedling would have to grow more than 0.36 m/year to be larger than a 2 m advance regeneration tree growing 0.2 m/year over 10 years when the seedling was 0.4 m tall at planting. This method makes no assumptions about relative growth rates after the comparison period.

For some eastern North America oak species, Sander et al. (1976) found a strong relationship between prerelease tree size and growth following release, and developed guidelines for advance regeneration requirements for even-aged systems. Release of these oaks was best achieved with cutting or removal of the above-ground portion of the tree so the tree would resprout. Basal diameter was a reliable indicator of vigor, and having a sufficient number of oak advance regeneration stems between 1.25 and 2.5 cm provided a reasonable compromise between sprout growth and not having too many sprouts (Sander 1971; Johnson et al. 2009). The growth patterns of these sprouts will often surpass the growth of undamaged regeneration but the postregeneration growth in complex stands is primarily the result of the light regime.

Ultimately, the decision on whether to retain advance regeneration or plant seedlings should consider more than growth rates or predicted growth rates. In some cases, the advance regeneration may be a different species than the predominant overstory species, in which case a decision to shift species composition will be required if the advance regeneration is relied upon for the next stand. Cost of planting and spatial patterns of advance regeneration trees are other considerations. Harvesting overstory trees may create excessive damage to advance regeneration. Some species of advance regeneration may be prone to infection by stem decay fungi. Others may be too weak to withstand snow and wind stress or have stem-form defects (Jaeck et al. 1984). Conifers with terminal damage will survive but may have a lateral branch or branches turn upward and assume dominance (Tesch et al. 1993), possibly forming forked stems. Alternatively, in places where ungulate damage is a problem, they may be less vulnerable to browsing than planted trees because of their size.

8.3 Artificial regeneration

The planting of seedlings in multiaged stands is an important means of assuring regeneration of desirable species in the desired locations is present. This is analogous to the process of **enrichment planting** in tropical forests (Lamprecht 1989; Helms 1998), where seedlings may be planted to augment species composition or density. Similar terms are **reinforcement planting** or **improvement planting**. (Helms 1998). The numbers of seedlings will typically be much lower than for a comparable even-aged reforestation project. Seedlings can be targeted for areas where they will receive sufficient light or where they will not be damaged by future felling of older trees. For example, larger gaps, such as group selection openings, are a logical place to plant because sufficient light is usually available for tree growth, including the growth of species intolerant to shade. In smaller gaps, intolerant species may need to be planted in areas with more light. For example, the north side of gaps in the Northern Hemisphere would receive a better light regime than on the south side. Likewise, better microsites due to factors other than better light regimes are also good places to plant seedlings. These might include places where shade is provided by a stump or log, depressions where soil moisture is more available, or places where mineral soil and organic matter are mixed. These microsites follow the same needs as in even-aged systems except for the obvious differences in light regimes. As a result, desirable planting sites in multiaged stands will generally be those with superior light regimes, whereas in even-aged stands, shaded microsites will be most desirable on some sites.

For situations where successful regeneration is typically established as advance regeneration, planting prior to an overstory removal or thinning may be necessary. Many North American oak species and other deciduous broadleaved species are more competitive as stump sprouts or advance regeneration. In cases where these trees are not present, planting is a viable option. For example, Paquette et al. (2006) reported that northern red oak and black cherry in eastern Canada could be planted prior to release and respond after only three years. In central Europe, Ammer and Mosandl (2007) found similar tree sizes and growth rates between planted and natural European beech seedlings growing beneath a Norway spruce canopy. For conifers, and probably many broadleaved species, planting can occur at the time of canopy removal.

Studies indicate planted seedlings follow similar growth patterns with light intensity as natural seedlings. As in these studies with planted seedlings, it was possible to control seedling age and position in the understory, they have an experimental design advantage over studies using natural regeneration. Mason et al. (2004) found strong relationships between seedling growth and light intensity for several temperate conifers in Scotland. Their relationships resembled those for natural seedlings with asymptotic growth at higher light intensity (see, e.g., Figure 4.2). In British Columbia, several species were planted under variable light levels and showed increasing growth with increasing light (Vyse et al. 2006). The authors recommended reducing the basal area to less than 15 m^2/ha to provide sufficient light for those sites. Similar patterns of increased growth of other species planted under partial overstories have been reported with similar recommendations.

Seedling requirements for planting in multiaged stands will probably be similar to those for even-aged stands. Larger seedlings will probably be better competitors but will also cost more to produce and to plant. Additionally, the competition from many grass and shrub species may be less in multiaged stands when these competing species have less relative shade tolerance, so that more soil moisture will be available. Alternatively, the competition from overstory trees may make moisture less available in partial shade (Vyse et al. 2006). The shade from the overstory may therefore provide positive or negative effects, depending on the site. Another possible advantage for multiaged stands is the reduction in evaporation and transpiration stresses on seedlings; that may compensate for increased competition for soil moisture from established trees. The result may be a similar type of seedling is appropriate for either even-aged or multiaged stands based on shoot-to-root ratios.

Decisions over seedling stock types will involve variables such as relative costs, ease of planting, and relative performance. As with even-aged stands, container seedlings will be easier to plant, but bare-root seedlings may have greater survival. One of the objectives of site preparation is often to increase the ease of planting. This objective will be more difficult to meet in multiaged systems with small openings, and thus the ability to plant trees with large root systems will be limited. The presence of established root systems in multiaged stands will also provide an obstacle to planting trees and may favor planting of container seedlings. Generally, the decision on seedling stock type will involve many variables, as in even-aged stands. However, key differences in multiaged stands include the need for seedlings that can compete in more shaded environments, more difficult planting through existing tree roots, and more competition for soil moisture.

Direct seeding is also an option in multiaged stands. Aerial applications will typically be infeasible because of costs of seed and difficulty getting seed through an existing canopy. However, application of seed directly to desirable microsites may be an important option in multiaged stands where modification of species composition is an objective. These microsites can be the result of either natural factors or through human activities such as site preparation or logging effects. The same drawbacks to direct seeding that apply to even-aged stands apply in multiaged stands. For example, seed predation by rodents and birds may occur more or less in multiaged stands but will probably vary by region and forest type.

8.4 Cultural treatments

8.4.1 Competition control

Although overstory competition is a major constraint on understory growth in multiaged stands, competition from other trees and other vegetation may reduce understory and overstory tree growth. Balandier et al. (2006) reviewed the forest vegetation management options focusing primarily on even-aged stands. The concepts and tools are similar regardless of stand structure. Several studies have compared understory tree growth with and without vegetation control. Mitchell et al. (2007) planted Pacific silver fir and western hemlock in British Columbia under a variety of overstory canopy densities. Regeneration treated with a manual removal and subsequent herbicide treatments produced significantly larger seedlings of both species in all levels of overstory competition. Likewise, Harrington (2006) found increased volume response to vegetation control in planted coast Douglas-fir, western hemlock, and western redcedar in western Washington State, US. These increases in growth with vegetation control were apparently additive because of reduced overstory competition. In both studies, the authors concluded that in addition to the light regime, competition for below-ground resources was also limiting growth of understory trees. However, on the British Columbia site, nitrogen limitations were a factor, while on the Washington site, the limiting factors included both nitrogen and soil moisture. In a similar study with planted conifers in the Pacific Northwest, Brandeis et al. (2001) found vegetation control was a significant factor affecting the growth of grand fir and western hemlock but not of coast Douglas-fir or western redcedar. The growth of all four species was positively affected by decreasing overstory density. This site was dryer than either the Washington or British Columbia sites, and this fact indicates that site interactions may affect seedling response to competition.

Competition for light and below-ground resources such as water and nutrients can reduce both understory tree survival and growth. Actual effects of competition will vary with site characteristics as well as the types of competing vegetation. For example, shrub species that can grow taller than planted seedlings will compete for both light and below-ground resources. Understory vegetation control will therefore be an important treatment to assure success of regeneration on some sites.

8.4.2 Site preparation

Modification of the microclimate, logging slash, soil, or controlling competing vegetation are common needs when reforesting multiaged stands, whether using natural seedling, planted seedling, or vegetative regeneration. However, the need for site preparation is generally less in multiaged stands than in even-aged stands, due to several factors. One is that with a partial overstory or a stand that is only disturbed in certain patches, the need for regeneration is on only a fraction of the stand area. Hence many forms of site preparation may not be feasible. Harvesting activities are concentrated in areas that will require regeneration: hence the site disturbance in multiaged stands is in the areas needing regeneration. Although this is also true in even-aged stands, it is more commonly an acceptable substitute for site preparation in multiaged stands. Finally, since multiaged stands are often associated with a less intensive form of management, various forms of natural regeneration are often used, and there is greater acceptance of the species mixtures or delays in immediate reforestation that result from natural regeneration.

All three categories of site preparation—chemical, mechanical, and fire (Helgerson et al. 1992)—have potential for use in multiaged stands. Chemical site preparation is used to control competing vegetation, as in even-aged stands. However, many of the shrubs and grasses that are problems in even-aged stands are less shade tolerant than the tree species they compete with, and thus the need for chemical control is reduced. For example, in western North America, shrubs such as deerbrush and manzanita are shade intolerant and are weak competitors in multiaged stands (McDonald 1976). However, in eastern North America, the most challenging competition may be from shade-tolerant tree species such as American beech and red maple. The need for chemical control of competing vegetation during regeneration will therefore vary from what is necessary in even-aged stands and will also vary from one ecosystem to another. Systems that rely on larger openings may have a greater need for chemical treatments than single tree systems.

Whether the chemical treatment is for site preparation or postregeneration species control, the target species will generally be in the understory. Herbicide application methods will vary from targeted hand applications to treat individual species, to broad applications using mist blowers. Aerial applications will generally be precluded because of the presence of an overstory. Hand applications may include backpack sprayers, granular applications, or hack and squirt (frill) treatments.

Mechanical site preparation involves modifying slash or, more commonly, the soil. Objectives related to soil are often related to exposing mineral soil, altering drainage, or soil temperatures. Mechanical site preparation that involves heavy machinery such as bulldozers pulling choppers or disks, or excavators will be relatively difficult in many multiaged stands. In larger openings, such as in group selection or low-density two-aged systems, large equipment will be more effective. In systems with smaller openings, these larger machines will be difficult to maneuver. Less intensive mechanical site preparation is accomplished with hand tools such as planting hoes and shovels. Here, the same types of soil modification may be achieved as with larger equipment, but on a smaller scale. For example, small microsites might be prepared to achieve a depressed soil level, or the soil surface may be scalped to remove competing vegetation. These small microsites may then be used as sites for planting seedlings, direct seeding, or simply left to improve the chances of natural seedfall and regeneration.

Prescribed burning can also be used as a site preparation tool in multiaged stands (Figure 8.9). Fire

Figure 8.9 Mixed conifer stand where prescribed burning was used to create seedbed for new age class of trees; in the Sierra Nevada of California, US.

has the potential to reduce slash levels, expose mineral soil, and control competing vegetation thereby meeting multiple site preparation objectives. However, its use will be limited because of the variability in fire resistance of different tree species. In some systems, fire may be used as a broadcast tool without harming desired species. In others, fire will only be possible as a tool to reduce slash concentrations in piles away from desired trees. In southeastern North America, prescribed burning timed to coincide with seed crops of loblolly pine and shortleaf pine and soil disturbance from harvest treatments represented a potential regime for uneven-aged stands (Cain and Shelton 2002). Although this treatment was not viewed as effective as periodic herbicide treatments to control competing vegetation and move desired pine trees into larger size classes, it does demonstrate a case of reintroducing fire as a natural process into a multiaged stand management regime. Similar regimes may be effective in other multiaged systems that focus on fire-tolerant species.

8.5 Synthesis

The dominant factor affecting regeneration in multiaged stands is the presence of an overstory canopy stratum. The overstory dominates the light regime, creates a unique understory microclimate, affects moisture availability, and affects other below-ground resources. It also affects seed production and dispersal. Natural forms of regeneration are traditionally the most common in multiaged stands. However, artificial regeneration is a viable option to modify species composition or accelerate the regeneration process. Managing the various forms of natural regeneration can be difficult, particularly for advance regeneration. Advance regeneration provides the advantage of being established and adapted to an understory environment. It also may be poorly formed or incapable of release. Hence, the decision to retain or use advance regeneration can be very difficult. Likewise, decisions regarding the viability of regeneration from sprouts or stumps can be confounded by the vigor of the root system and sprout responses in a partial-light environment.

Multiaged silviculture is often associated with more natural forms of management. However, artificial regeneration and cultural treatments, such as site preparation and vegetation control, are appropriate for multiaged stands. The common interpretation that multiaged silviculture is a more natural approach to forest management may be valid, but this does not preclude more intensive management treatments to help meet objectives.

CHAPTER 9
Tending multiaged stands

9.1 Introduction

Many multiaged stand objectives may be best achieved with what are called "intermediate operations" or treatments that are traditionally associated with even-aged systems. These treatments, such as thinning, pruning, or fertilization, are opportunities to alter the stand structure, growth rates, forest health, or the quality of individual trees in multiaged stands. Intensive forest management, which often includes multiple intermediate treatments, is usually associated with even-aged systems, particularly plantations. Multiaged stands are often viewed as a more natural alternative to even-aged stands, to the extent that multiaged stands may be assumed capable of taking care of themselves (Gibbs 1978). However, the use of intermediate operations is an important option for meeting objectives in multiaged stands and can be very intensive (Maguire 2005). This chapter discusses how treatments such as thinning, pruning, and fertilization can be used to improve multiaged stands.

9.2 Thinning treatments

Thinning treatments may be used to achieve objectives related to overall stocking control, or to achieve localized objectives independent of overall stocking. For example, thinning is used to reduce the numbers of trees in certain age classes or strata to make the stand conform to stand-level stocking goals. Or, localized spacing can be adjusted to favor more promising individual trees or to favor trees on superior microsites (Figure 9.1). Thinning can also be used to favor certain species or adjust spacing in group openings. A manager may even anticipate future harvest activities and thin to favor trees that can be avoided in future tree-felling or skidding operations. In multiaged stands, thinning treatments will most likely be part of periodic harvest treatments that occur at the end of each cutting cycle. This is particularly true for merchantable trees, from which stems will be removed; but these entries represent an opportunity to thin non-commercial trees as well (Figure 9.2).

Although even-aged and multiaged stands differ greatly in stand structure, the reasons to thin either type of stand are about the same. The experience with thinning even-aged stands therefore has much to share with multiaged thinning. Thinning treatments are presented relative to two primary objectives: adjusting density or adjusting species composition.

9.2.1 Adjustments in density

The basic thinning methods for even-aged stands (Smith et al. 1997; Nyland 2002) can also be applied to understory cohorts or strata of multiaged stands. Each of these methods—commonly classified as low, crown, selection, and geometric—could achieve some purpose in a multiaged stand. However, these traditional thinning methods are typically intended for even-aged stands that are relatively homogeneous in horizontal space with normal diameter distributions. The structures of multiaged stands are more complex, as are the methods for thinning. In the most simple multiaged stands, those with two age classes, there is an older age class or stratum that interacts with a younger and lower stratum. Either of these strata can be thinned, or, alternatively, the older stratum will be removed at the end of a cutting cycle while the younger stratum will be thinned. In stands with more than two

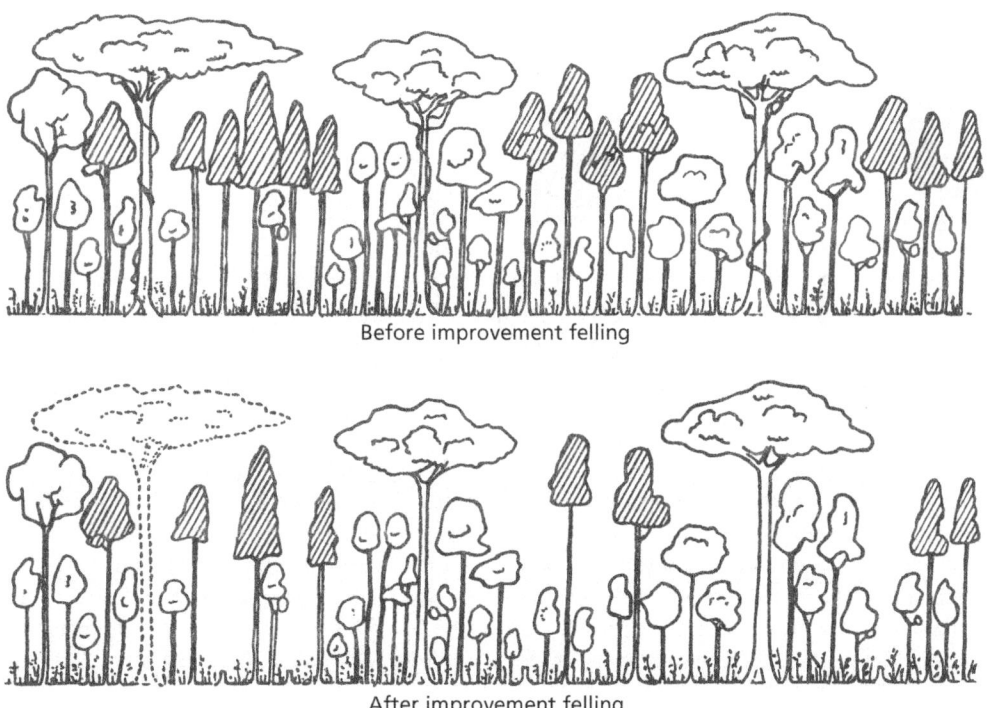

Figure 9.1 Before and after diagrams of improvement cutting in a mixed tropical forest in Equatorial West Africa. The shaded trees are gaboon, which are thinned to give residual trees more space. The emergent tree defined with a broken line on far left is assumed to be poisoned to release trees beneath it (from Lamprecht 1989; with permission from Deutsche Gesellschaft für Technische Zussammenarbeit).

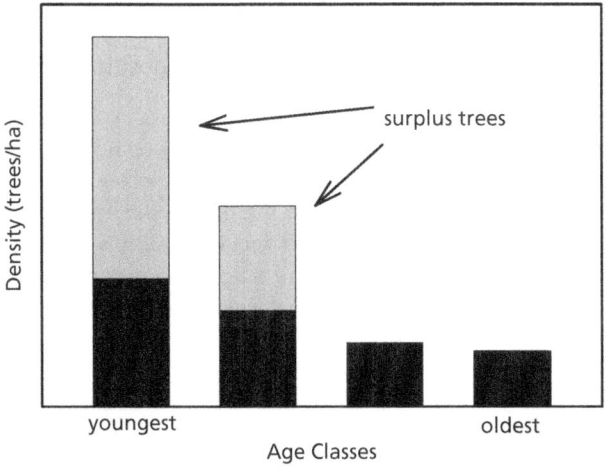

Figure 9.2 Hypothetical four-aged stand with surplus trees that might be removed in a thinning.

age classes, there may be reasons to thin all of the age classes. Perhaps the most applicable of the traditional even-aged thinning methods are forms of "free" thinning, which can combine other thinning methods and focus on providing selected trees, or crop trees, with additional space. For example, the younger cohort in a two-aged stand may benefit from the removal of smaller trees or from thinning from below, as well as from thinning trees around crop trees or a crown thinning (Figure 9.3).

Figure 9.3 Diagram of an understory thinning in a two-aged stand. Trees with hatch marks would be removed to favor larger trees, space trees and remove trees directly below large trees from the older age class.

There may also be spacing objectives that override objectives related to vertical structure. For example, it may be useful to remove trees around larger overstory trees, under which nothing will grow well, or perhaps to remove trees in the felling area of larger trees, as understory damage is unavoidable there. Variable-density thinning (Carey 2003; O'Hara et al. 2012) can also be used in multiaged stands where objectives are to enhance structural variability. Understory trees can be thinned with different objectives or methods in different parts of the same stand. This would enhance both vertical and horizontal stand heterogeneity.

Thinning treatments can be part of a conversion or transformation process to treat stands in ways that will hasten the development of multiaged characteristics (see Chapter 13). Even-aged stands can be thinned to encourage a new age class in some areas, in a manner that is much like a variable-density thinning but with a regeneration objective. Thinning in understory age classes of multiaged stands can encourage additional regeneration and add structural complexity.

Metrics to guide thinning treatments in multiaged stands can include trees or basal area per hectare for different components (e.g., age classes, species, canopy strata, etc.). Relative stand density measures can also be determined for individual stand components and used to guide overall stand stocking levels (Long and Daniel 1990). Tree stability may also be an issue with understory trees in multiaged stands due to wind, snow, or other weather events (Box 9.1).

Height-to-diameter (H:D) ratios are used to assess individual tree stability as well as vigor. Individual tree photosynthate allocation patterns allocate resources first to height, then to diameter (Oliver and Larson 1996; Waring and Running 1998). A tree with less resources will allocate less to diameter and thus the tree will have a high H:D ratio. Although species vary in threshold H:D ratios that indicate stability, ratios from 80–100 are commonly used when both height and diameter are measured in equal units (Cremer et al. 1982; Lohmander and Helles 1987; Wonn and O'Hara 2001; Vospernik et al. 2010; Chapter 15).

9.2.2 Adjustments in species composition

Instead of a density objective, thinning in multiaged stands can focus on encouraging certain species. Species may be encouraged for the economic value of the timber, more rapid growth rates, stand health, greater species diversity, or simply to select for species that are more adapted to certain microsites. For example, one species may be favored in a drier part of a stand, whereas another might be favored in wetter areas.

Thinning can be used to modify species composition to improve stand resistance to disturbance events such as storm damage, insects, or pathogens. Many insects and pathogens, such as many defoliators, bark beetles, and dwarf mistletoes, have strong host species preferences. Favoring certain nonsusceptible species, or encouraging a diverse mixture of species through thinning treatments, is a means to reduce undesirable effects of these agents.

Box 9.1 Intensive stand tending in central Japan

In central Honshu, Japan, the Imazu forests in Sekigahara are intensively managed in mixed, multiaged stands of sugi and hinoki. These forests have been managed using similar methods since about AD 1600. Although there is a long tradition of using multiaged silviculture and refinement of these methods over time, these traditional methods are threatened by changing markets that prefer larger harvests. In the present form of management, many age classes are maintained at any given time in single tree selection systems. Stands are maintained at high densities with total stocking levels that may exceed 500 m³/ha. Individual trees may be over 300 years old, and the primary goal is timber of very high quality. Periodic harvests remove the largest trees, which are designated for high value uses such as in temples and shrines. Replacement seedlings are replanted following the removal of large trees.

Tending of these stands includes intensive pruning treatments in multiple lifts up to as high as 6 m. These pruning treatments are done by laborers who begin when the trees are small. When the trees are large, they are climbed to remove branches. Pruning is relatively frequent, resulting in lifts of only a meter or two. The result is clear tree boles up to 6 m, very high quality wood production (Figure 9.4), and an improved light environment for developing trees in the understory.

Another form of tending that demonstrates the intensive nature of this management system is focused on the stability of young trees. The high stocking levels used in these management systems leave younger age classes vulnerable

Figure 9.4 Large hinoki tree pruned with many lifts and to a total height of approximately 12 m. The diameter of this tree approaches 60 cm.

to storm damage, particularly from heavy snows. To assist young trees in densely spaced groups after snow damage, they are tied to larger trees, leaving a spider web of lines (Figure 9.5). As trees age, they enter better canopy positions where they attain more stable height-to-diameter ratios and no longer need any support.

Figure 9.5 Dense group of multiaged hinoki and other tolerant understory conifers. Lines stabilize trees damaged by snow.

Figure 9.6 Understory silver fir, Norway spruce, and European beech in Austria. These trees already show evidence of a vigorous response to a release that took place several years before this picture was taken (note the expanded internodes on the silver fir on the far right) but may need further release via thinning to favor more desirable species or simply to provide more space.

9.2.3 Free thinning approaches

There are many variations in methods to adjust the density or species composition of a forest stand. With increasing structural complexity, the potential options may increase exponentially. With each additional age class, canopy strata, or species, there is yet another stand component to assess and possibly treat. Each of these components may be thinned with a different objective. There is much potential for "free thinning" approaches that are flexible to the unique conditions of each stand or part of a stand. This may require an experienced forester or operator to make decisions on removals based on broad guidelines that only loosely constrain practices.

9.3 Release treatments and vegetation control

There are situations where desirable trees or species in multiaged stands need to be released from less desirable trees. **Release treatments** free young trees from undesirable, usually overtopping competing vegetation (Helms 1998). Vegetation control may involve mechanical or chemical treatments to favor certain species. These treatments may be more common in simpler multiaged stands, such as two-aged stands, than in more complex stands. However, complex stands, with many age classes or species, can also require interventions to select species (Figure 9.6). Undesired trees can be cut to remove them. Or, if they are species that are prolific sprouters, they can be treated with chemicals to kill the root systems.

In many ways, release and vegetation control are extensions of thinning treatments, and all of these objectives could be met with a single treatment. This provides some efficiencies that are common in even-aged precommercial thinning treatments, where stand density is reduced and undesirable trees may be removed. For example, a multiaged stand can receive a treatment to reduce understory density, favor certain species, and remove select trees in older age classes to provide a better light regime.

9.4 Pruning treatments and wood quality

High quality timber is often an objective of multiaged systems, and greater wood quality is sometimes cited as a specific benefit of multiaged stands (Matthews 1989; Guldin and Fitzpatrick 1991; Leak and Sendak 2002). Trees in multiaged stands typically grow from successively lower states of suppression until they reach the upper canopy (see Figure 4.14). These stages of suppression may keep branches small and less persistent. For species with decurrent growth forms, this competition during early stages also helps maintain single-stemmed trees. Branch size and retention is primarily a function of available growing space and its effect on crown size (Larson 1969). If stands are maintained at lower density levels in either even-aged or multiaged structures, individual tree growth rates will be faster, and there will be less inhibition of branch growth. At either high or low densities, including multiaged stands, **pruning** lower branches can be useful to increase clearwood production.

There are many similarities between pruning regimes in even-aged and multiaged stands. In even-aged stands, pruning strategies generally prune only select trees, and they prune when trees are small to maximize clearwood production (O'Hara 1991). These same strategies are important in multiaged stands. In multiaged stands, only a small number of trees from any age class may be selected for pruning. However, the total number of pruned trees in a multiaged stand may be similar to the number in an even-aged stand. In the multiaged stand, these trees will be of different ages. The selection criteria for selecting trees for pruning are similar in even-aged and multiaged stands: desirable species, individuals of good form, and trees that are expected to be retained through consecutive cutting cycles should be selected for pruning (Figure 9.7).

Pruning methodologies are also similar regardless of the age structure of the stand. Pruning should occur when trees are small so that clearwood

Figure 9.7 Pruned and unpruned silver fir trees following Plenter selection treatment in Germany. The tree on the left has a straighter bole than the tree on the right and is a better candidate for pruning.

production is greater. This means identifying trees to be retained through all subsequent cutting cycles when they are in the early stages of their development. Pruning should retain approximately 40% to 50% of the stem in live crown if possible. There is no advantage to retaining dead branches, as they result in loose knots, and dead branch wounds occlude more slowly than live branch wounds. If crowns are receding quickly, then pruning severity can be increased without adversely affecting tree vigor. To maximize clearwood production, branches should be severed close to the stem (O'Hara 2007).

9.5 Fertilization and forest nutrition

Multiaged stands probably have similar nutrition requirements as even-aged stands. Unfortunately, forest nutrition research on multiaged stands is almost nonexistent. On the same site, the multiaged stand would have a similar potential leaf area index (LAI) and utilize similar amounts of water as the even-aged stand (O'Hara and Nagel 2006). Nutrient losses from leaching, volatilization, and tree removals would probably be offset by inputs such as deposition, weathering, and nitrogen fixation in similar ways in both types of stand structures. Nutrient deficiencies that are present in even-aged stands will also likely be present for multiaged stands on the same site. Differences between these structural extremes that do exist would most likely be related to variations in total growing space occupancy, or LAI, as a result of differences in management regimes. These can affect processes such as nutrient uptake, decomposition, and stand growth.

During even-aged stand development, nutrient supply varies with stand age (Madgwick et al 1977; Binkley 1986). When stands are young, nutrients are largely available, and young tree nutrient requirements are less. After full crown closure, nutrient demands are greater and supplies limited. Later, as stands age, nutrient pools become more fully stored in stand biomass and may limit growth. Multiaged stands may exist in this latter stage, where nutrients are largely held in living biomass, but periodically experience the release of nutrients associated with a cutting cycle harvest.

The thinning and final harvest treatments in an even-aged system may be assumed to be about equal in number to the cutting cycles in a multiaged system (see Figure 7.5). The primary differences are the complete removal of vegetation during the final harvest of the even-aged system as compared to the more constant level of forest cover or LAI in the multiaged stand.

This complete harvest in the even-aged stand leaves the site exposed, thereby raising soil temperatures. This may result in more rapid decomposition and greater nutrient availability if adequate moisture is present. However, on drier sites, inadequate moisture may result in decreased nutrient availability. The increase in forest floor temperatures can therefore have different results depending on the moisture environments of the site. The presence of a continuous canopy may slow or accelerate forest floor decomposition depending on the moisture environment. In colder and wetter environments, such as many boreal or wetter temperature forests, decomposition may be slowed in multiaged stands, and overall nutrient availability may be less than comparable even-aged stands. However the opposite effect can occur in dryer forests. On some sites, such as wet tropical forests, more rapid decomposition can lead to nutrient losses through leaching. There may be differences in decomposition rates and nutrient cycling with even-aged and multiaged systems, but these differences are complex due to the interactions of other variables, most notably forest floor moisture.

Total foliage or LAI levels are an indicator of site occupancy and the use of soil nutrients (Binkley 1986). Multiaged stands may maintain more LAI over time because removals are generally lighter in total than the removals in even-aged stands (see Figure 7.5). This may affect total nutrient requirements. However, in most situations, stands will be maintained at less than maximum potential LAI to prevent overcrowding and maintain high tree vigor (see Chapter 7). Additionally, maximum LAI in the density management regime will be about the same for both even-aged and multiaged systems. Nutrient demands to maintain foliage through a rotation or through multiple cutting cycles may be greater in multiaged stands because they maintain higher overall LAI.

Fertilization attempts to replace nutrients limiting growth during stand development. Usually fertilization consists of nitrogen with phosphorus, potassium, or some other nutrient applied less often. Fertilization is usually intended to address a nutrient deficiency and increase growth, but it may also be intended to improve stand vigor or health, thereby having a nontimber objective. Fertilization may affect stand growth in a number of ways. A common response in even-aged stands is an increase in growth that is limited in duration. For example, a 10-year acceleration in growth might occur, followed by growth rates dropping to the pre-fertilizer levels. This has the effect of shortening the even-aged rotation because it takes less time to achieve a certain tree size or volume per unit area with fertilization than without. Alternatively, fertilization can result in a prolonged growth trajectory that increases the total biomass potential of a site (Binkley 1986).

In combination with thinning in even-aged stands, fertilization often results in an additive growth response where the combination of thinning and fertilization produces larger trees and more stand increment than either treatment individually (Omule et al. 2011). The mechanisms of these responses are not fully understood. Perhaps fertilization accelerates canopy regrowth following thinning, or perhaps fertilization allows the attainment of a higher level of LAI and a higher level of sustained overall productivity.

It is the interactions between fertilizer and thinning of even-aged stands that may provide the best insights into the growth responses of multiaged stands following fertilization. Multiaged stands are most similar to older even-aged stands, where a high proportion of the nutrient capital is stored in the above- and below-ground biomass of living trees and LAI is high. The end-of-cutting cycle harvest in the multiaged stand may resemble the thinning in an even-aged stand. A fertilization treatment therefore can increase nutrient availability and increase stand growth rates. This effect may be temporary but can provide a means to increase sustained rates of stand productivity or as a means to enhance forest health.

Fertilization in even-aged stands can guide similar treatments in multiaged stands where increases in increment or stand health are objectives. These treatments will likely result in increases in LAI and increases in tree growth. Fertilization approximately 10 years before expected harvests may be most effective at providing sufficient response time before the next harvest and for providing trees with sufficient light and moisture to benefit from the increase in nutrient availability. For short cutting cycles, fertilization may be effective in conjunction with cutting cycle harvest treatments.

9.6 Prescribed burning

Prescribed burning is usually not considered as an intermediate operation in multiaged stands because of the presence of regeneration. However, in some ecosystems with very high fire frequencies or fire return intervals, prescribed burning can be a viable option in multiaged stands. One of the best examples of this is the management of longleaf pine stands in southeastern North America (Mitchell et al. 2006). In these forests, natural burning may have occurred every year or two, providing a highly diverse and fire-resistant ecosystem. The dominant longleaf pine are extremely resistant to fire and exist in a "grass stage" as seedlings where the terminal bud is protected from fire as they develop a root system (Figure 9.8). Prescribed burning in these stands reduces fuels and the chances of unplanned fires. The burning also controls oaks and favors the diverse flora this ecosystem is known for (Mitchell et al. 2006) and thus is an integral part of controlling stocking and stand management (see Box 7.2).

This example of prescribed burning in longleaf pine is probably not typical of multiaged stands. Cain et al. (1998) found prescribed burning in loblolly–shortleaf pine stands in the southeastern United States had few effects on the trees but resulted in changes in understory flora. Prescribed burning may be a useful treatment in other multiaged systems, such as those in which ponderosa pine or other fire-adapted species are grown. In other cases, it may represent too great a threat to the regeneration or to other trees with low fire resistance. It is a useful treatment to aid regeneration in group openings, where it can be used as a broadcast treatment or to burn piles (see Chapter 8).

Figure 9.8 Longleaf pine seedling (left) two weeks after prescribed burning in southeastern Georgia, US. The longleaf pine seedling is alive because the inner bud is insulated from the fire. The oak on the right (arrow) was killed by the fire.

9.7 Synthesis

There is limited justification for multiaged stands to receive the same intensity of management as some even-aged stands, particularly plantations. Tending operations in even-aged stands gain considerable efficiency by treating stands on cleared sites, or trees of only one age class. These same types of intermediate treatments can be used in multiaged stands to enhance stand productivity, wood quality, or stand health. The point is not that multiaged stands should receive these treatments, but that they can. Another significant point is that our understanding of these "intermediate operations" in even-aged stands is much more advanced than for multiaged stands. However, these treatments have much in common, and multiaged silviculture can benefit greatly from the use of this knowledge.

Thinning is a particularly important tool for reducing density of younger age classes in situations where natural regeneration is too plentiful in some age or size classes. Thinning treatments can be integrated into stocking control plans that require fewer young trees. Thinning can also be useful to adjust species composition or encourage trees in some locations over others.

Multiaged systems may be viewed as a more natural stand management alternative to even-aged systems. However, this should not preclude the use of intermediate operations. There is a misconception that since multiaged stands may be more natural, they should be managed with a "hands-off" approach. A similar notion is that chemicals to control unwanted tree species or pests should be excluded. These should be separate decisions: one decision relates to the management system or the desired stand structure, and other decision is selecting the tools necessary to produce it. An assortment of intermediate treatments or tools can provide a means of more fully meeting management objectives in multiaged stands.

CHAPTER 10

Transformations to multiaged stand structures

10.1 Introduction

Transformations, or conversions, from even-aged to multiaged stand structures have become a common objective for many forests. These transformations are driven by a variety of objectives. International accords are encouraging more diverse forests. For example, a greater emphasis on multipurpose forests was a directive of the Helsinki Process. Multiaged forests are also encouraged in some forest certification protocols (McMahon 1999; Hartsfield and Ostermeier 2003) or by forest laws (McGinley et al. 2012). Many plantations have little diversity in tree size, species, or stand structure, so they may be targets for transformation. Other planted forests may be considered for transformation because they lack sufficient diversity or are desired as multipurpose forests. The explicit objective therefore may not be to develop multiaged or uneven-aged stands, but instead to develop stands with greater amounts of structural heterogeneity or complexity which, in turn, may provide other benefits for wildlife, aesthetic values, resistance to abiotic disturbances, or resistance to insects and pathogens. Another justification may be to simply avoid the unpopularity of clearcutting treatments (Kimmins 1993). Transformation is not new as might be implied by the events and trends noted above: Schütz (2001b) noted the long history of transformation in the forests of central Europe. As forestry has shifted from emphasis on even-aged or multiaged systems, there have been efforts to transform one system to the other for many decades.

Both "**transformation**" and "**conversion**" are commonly used to describe the process of directing even-aged stands toward multiaged structures. The term transformation will be used in this book because it implies a less complete total change in stand structure. This is consistent with the premise that a great many different silvicultural systems exist on a continuum rather than as a limited number of options or discrete categories. Additionally, the more subtle term "transformation" fits the concept that sometimes minor changes in management regimes can result in large changes in stand structure. The term transformation was also used in the title of the 1999 and 2004 conferences on this topic (Cameron et al. 2001; Pommerening 2006). Conversion is often used to describe management that directs changes in species composition, such as the conversion of Norway spruce to mixed-species compositions in central Europe (e.g., Speicker et al. 2004).

The objectives for both transformation to more complex age structures, and conversions to more complex species compositions are driven by similar demands for greater structural heterogeneity or "naturalness" in stand structure. Hence both objectives are very compatible and may be pursued at the same time. Transformation treatments of even-aged to multiaged stands may not be greatly different than restoration treatments described in Chapter 11. Multiaged stands may be a "natural" structure in many places that have been traditionally managed as even-aged. Hence, the transformation regime may be a form of restoration. Here, the discussion is focused more narrowly on a specific type of restoration, where even-aged stand structures are directed toward multiaged structures.

Multiaged Silviculture. Kevin L. O'Hara.
© Kevin L. O'Hara 2014. Published 2014 by Oxford University Press.

10.2 The silviculture of transformation

Forestry has traditionally held a narrow of view of the possible range of multiaged stand structures. Transformation can therefore be perceived as moving stands from a great many possible even-aged stands or stages of even-aged stand development to a relatively few multiaged options (Figure 10.1). Instead, there is a broad assortment of possible multiaged stand structures and many potential pathways for transformations. This broad assortment of target multiaged stand structures is compounded by a great diversity of initial structures or possible even-aged structures from which the transformation process may be initiated. Additionally, there may also be multiple options or pathways for transformation from one initial structure to the same target structure (O'Hara 2001). As a result, there are a great many variations on what a transformation might entail, and no simple procedure that is widely applicable to all of these situations.

The **target stand structure** is the structure that best meets management objectives. In the case of transformation, the target structure is a moving target because the multiaged structure changes over time as it moves through cutting cycles. The target structure is therefore a range of conditions that exist within the multiaged regime. A typical objective of any silvicultural regime is a sustainable stand structure. In the case of transformation, the objective of the target structure is a sustainable management regime over the long-term.

Management decisions should always be sensitive to the initial state of the forest. This is especially true for transformations. Kenk and Guehne (2001) described several initial conditions that might affect transformation in the Black Forest in Germany, such as site conditions, susceptibility to storm damage, and root disease. Each of these is suggestive of a different transformation strategy. The constraints of existing stand structures or financial resources are likely to limit both the range of target stand structures and the speed of achieving these transformations. Land managers may therefore pursue simple or complex target stand structures for transformation. Simple multiaged structures might be two-aged stands or two-strata stands, whereas a complex structure may have many age classes, canopy strata, or species. Managers will require the freedom to pursue a variety of transformation strategies based on the initial stand structure, the target structure, and the available resources for the transformation. The availability of resources to support transformations will also determine outcomes: in situations with plentiful resources, intensive transformations may result; but if resources are limited, the transformation might be more extensive.

The many different possible initial states and target stand structures result in a multitude of potential alternative treatment regimes leading to great uncertainty, particularly at later stages of the

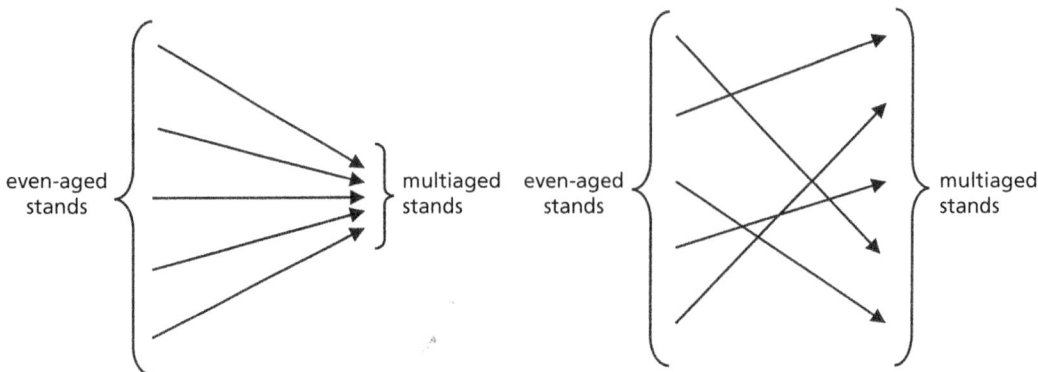

Figure 10.1 Conceptual diagrams displaying alternate ways of viewing options for transformation from even-aged to multiaged stands. The left is a conventional view, where there are many options for even-aged stands but not for multiaged stands. The right displays the idea that there are very many possible multiaged stand structures.

transformation process. This is exacerbated by the general shortage of experience with transformation to multiaged stand structures. Adaptive management (e.g., Yousefpour et al. 2012) is therefore highly appropriate for transformation regimes. There needs to be a general acceptance that these transformations, particularly the more complex regimes, will have to be modified with experience and new information. Transformation regimes and adaptive silviculture should be compatible partners in the transformation process.

General multiaged management regimes consist of descriptions of the number and sizes of trees to cut, and the spatial patterns of trees and openings. Transformations are concerned with the same variables. However, transformations are typically presented as a sequence of treatments that direct the stand toward the desired stand structure. This desired structure may be simple or complex, and the sequence of treatments will usually correspond to this continuum of complexity by being short or long in duration. Transformation treatments are described here as a dichotomy between directing stands toward simple or complex stand structures.

10.2.1 Transformations to simple stand structures

The obvious first step in any transformation from an even-aged stand is to regenerate a new age class or cohort of trees. However, there may be reasons to proceed cautiously, as is sometimes the case with shelterwood systems. With these systems, the shelterwood harvest may be preceded by a **preparatory treatment** to increase windfirmness of residual trees, develop advance regeneration, or enhance vigor and seed production potential of trees (Smith et al. 1997; Nyland 2002). A similar treatment may be useful as part of a transformation regime, whether the intention is to create a relatively simple or a complex stand structure.

The shelterwood harvest treatment is often an excellent way to regenerate a second cohort, and if the overstory or shelterwood trees are retained, a two-aged stand is the result. Similarly, a seed-tree cut with residual or retention trees results in a two-aged stand. In eastern white pine, Wetzel and Burgess (2001) showed that a shelterwood can successfully regenerate white pine, provided seedlings received sufficient light; a minimum of 50% of above canopy light (PAR) was required to maintain sufficient seedling growth. Site preparation and weed control further enhanced growth of eastern white pine. Broadleaved species, such as European beech or lime, can be planted beneath Scots pine stands in Germany to transform stands with minimal overstory treatment (Kenk and Guehne 2001). Retaining some Scots pine results in a two-aged stand; and if they are removed, the stand is converted from pure Scots pine. Malcolm et al. (2001) recommended irregular shelterwood and group selection systems for transformation of conifer stands in the United Kingdom. A variable retention harvest would also result in a two-aged stand. Regardless of what this initial treatment is called, the treatment needs to regenerate a new age class while retaining some of the previous stand.

The creation of canopy gaps or openings is another simple approach to transform an even-aged stand to a more complex structure. These gaps can instantly transform the structure and with regeneration, will result in a two-aged stand. This approach may be particularly useful with shade-intolerant species. If the objective is only to achieve structural complexity, then gaps are a quick means to achieve this. However, if the objective is to also produce a consistent level of timber production while maintaining a two-aged structure, then a regulated approach to the entire stand would require half the stand be harvested in group cuts. This is probably excessive for most objectives. Gap creation may be best in a more complex stand management strategy where gaps can be spread out spatially and temporally in the stand. If a regulated level of timber production is not the objective, then creation of small numbers of gaps is a potential method of adding complexity in a relatively simple way.

An alternative to simple multiaged stands are stratified, mixed-species stands or simple modifications to even-aged regimes. Stratified mixtures are even-aged stands but with stratified canopies because of the variable growth rates of different species. Oliver and Larson (1996) discuss these types of stands in detail (also see Section 4.5). These even-aged stands may meet many of the objectives that are driving transformations. They provide stratified canopies, have mixed-species compositions, and may be easy to manage. Hence they are a simple alternative means of developing a complex stand structure. Extending even-aged rotations can also enhance stand structural diversity and, at least if using coast Douglas-fir, will result in only minor reductions in production (Curtis 1995). However, these even-aged stands typically begin and end with a clearcut, which may be an overriding constraint.

10.2.2 Transformations to complex stand structures

Transforming even-aged stands to complex structures requires a more detailed sequence or system of treatments. These complex structures may include several age classes, multiple canopy strata, many species, or a full range of diameter classes. Achieving this complexity generally involves a sequence of treatments rather the single treatment that may be used to achieve a two-aged stand.

The first treatment is usually to regenerate a new age class. Nyland (2003) stated this point as "silviculturists should have little trouble envisioning how to start a transition, or even plan a second cut and what it should accomplish." Nevertheless, there may be good justification to begin the process with a treatment that enhances the stability or regeneration potential of the stand. Older even-aged stands may be at high stocking levels and vulnerable to wind and snow damage. Advance regeneration that is necessary for forming the new cohort may not be present or of sufficient vigor.

Treatment sequences attempt to gradually direct the stand toward the target stand structure. There may be multiple options or pathways for achieving the target structure. For example, Nyland (2003) diagrammed two alternative management pathways for transforming a hypothetical even-aged stand to the same multiaged stand (Figure 10.2). These pathways may include treatments such as commercial harvests, non-commercial thinnings or improvement cuts, regeneration operations, or others. Kenk and Guehne (2001) described transformation regimes in Germany that included treatments such as precommercial thinning, underplanting, direct or artificial seeding, wedge cuttings to reduce wind damage, and Femelschlag treatments to release understory trees. The silvicultural "toolbox" for transforming even-aged to multiaged stands is quite large and may even include small clearcuts when wind damage is an issue (Heinrichs and Schmidt 2009).

Stocking control is an important part of transformations. In a mixed oak—shortleaf pine forest in Missouri, US, Loewenstein (2005) concluded that transformation from degraded, even-aged stands could be accomplished in three steps. The first entry might regenerate a second age class at age 30 by leaving an overstory density of 40%. At age 60, a third cohort could be regenerated by reducing overstory density to 30%, and the second age class density to 20%. Both of these age classes would be treated. The third age class would eventually be allocated 10% of the growing space. In this example, the target distribution of growing space or density in the transformed stand would be a 3:2:1 ratio among the three cohorts and would take approximately 80 years.

In mixed loblolly and shortleaf pines in the southeastern United States, Baker et al. (1996) described a 30-year, four-step transformation process. They used stocking parameters from the BDq method (see Section 7.3.3) to guide the treatments (Figure 10.3). Whereas Loewenstein (2005) concluded the stocking guide for the fully transformed stand was inappropriate to guide the transformation process, Baker et al. (1996) used it in their hypothetical example. Both cases use a BDq method, but the differences were primarily related to the rigidity with which the guide was followed. The basal area or "B" parameter guides total residual stocking at each entry, but the maximum diameter and q factor are only used later in the process.

Rehabilitation is similar to transformation in that it tries to move a stand from one structure to another. However, rehabilitation is used to move a stand from a degraded condition to an improved condition. Baker and Shelton (1998) used the term to describe a series of experiments in mixed shortleaf and loblolly pines in the southeastern United States where degraded stands were transformed into multiaged stands. On better sites, they showed that stands could be transformed back to productive, well-stocked multiaged stands in only 15 years. Their rehabilitation was evaluated based

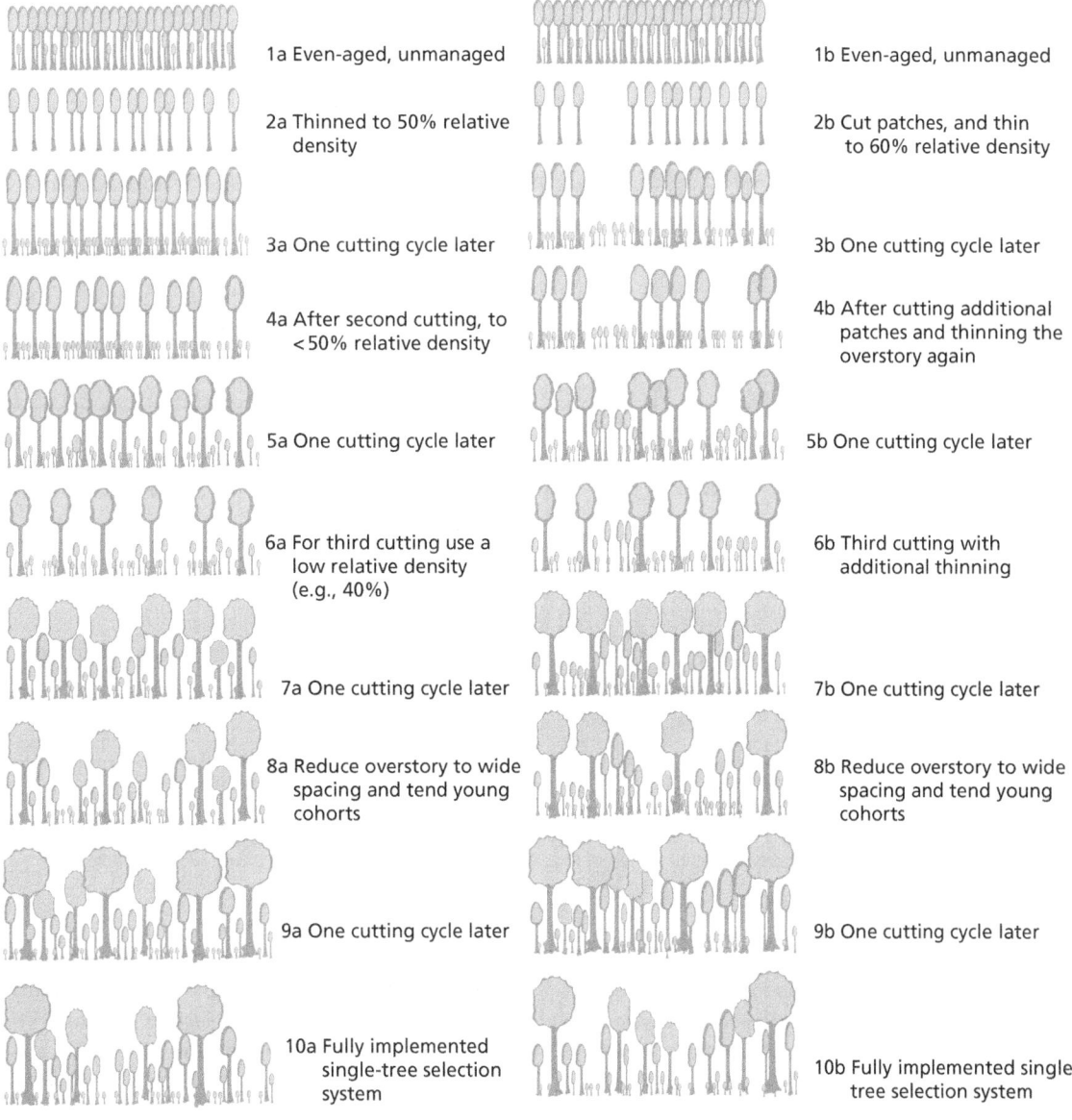

Figure 10.2 Diagram of alternate pathways for transformation from an even-aged to a multiaged stand. The pathway on the left uses a uniform cutting, and the one on the right a patch cutting system. Other alternative pathways are often possible as well (from Nyland 2003; reproduced with permission from Elsevier).

on an overall stocking target rather than a structural target. Nevertheless, these examples demonstrate that transformation may assume pathways from a variety of initial conditions to a wide variety of final stand structures. Treatment options are highly varied, and so are the tools used to guide transformations.

10.3 Sustainability and transformation

When a stand is transformed, there is the general expectation that the new multiaged stand regime will be sustainable over the long-term. This means the new regime will successfully maintain the stand structure, and the treatments necessary to maintain

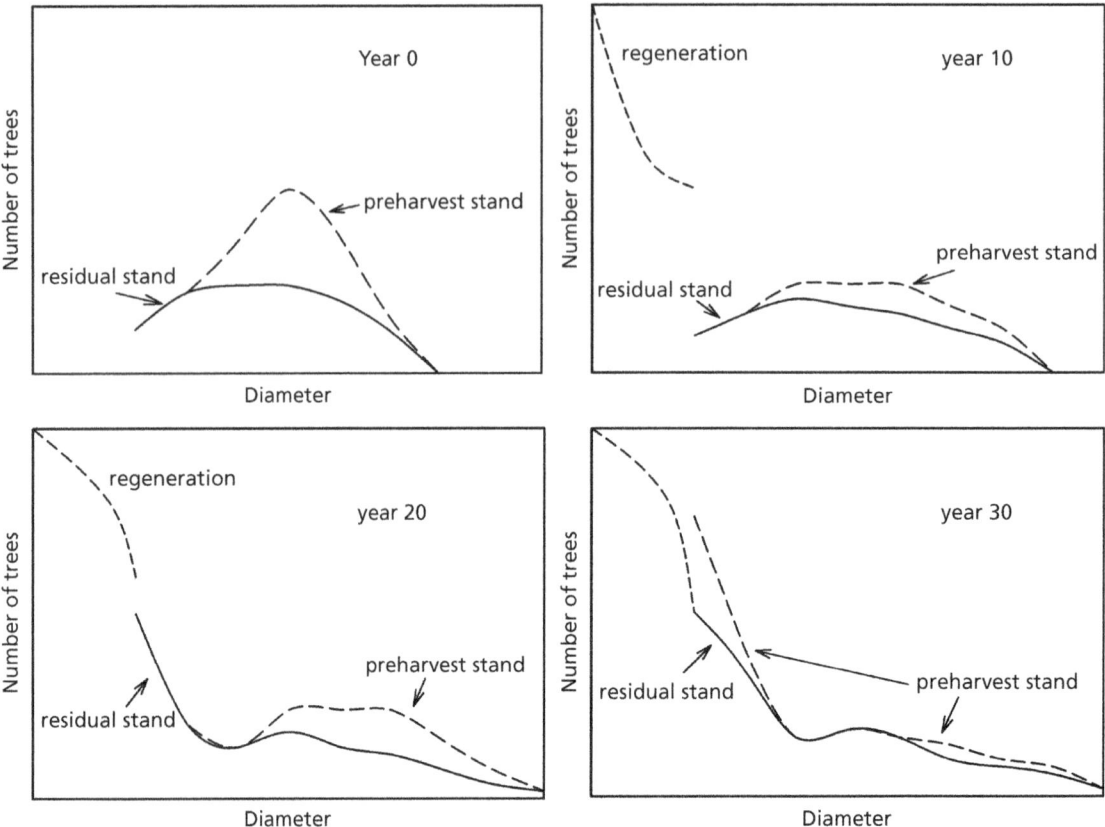

Figure 10.3 Steps in transformation of a hypothetical loblolly–shortleaf pine stand from even aged to multiaged. This transformation takes a normal diameter distribution and transforms it a distribution resembling a reverse-J, with three cutting treatments over 30 years (adapted from Baker et al. 1996).

the structure will be feasible and environmentally sound. However, the transformation treatment itself is simply a transition exercise that need not be a sustainable regime in isolation from the intended multiaged regime that will eventually result. For example, some intermediate steps in the transformation process may not be feasible, or the intermediate structures many not be sustainable.

Among the potential pitfalls of transforming even-aged stands to multiaged stands is the potential for using this process as a justification for removing only the large trees or using a treatment analogous to traditional diameter-limit cutting (see Section 7.3.1). In addition to the negative effects on value and tree size diversity in the new stand, when the best parent trees are removed, these treatments may also have dysgenic effects (see Chapter 14).

If natural seedling regeneration is expected, then there is a potential for loss in genetic quality in the stand. The risk of dysgenic selection is reduced if vegetative regeneration is used or if a seed bank exists. However, if natural seed fall is needed, then the treatments may remove the most valuable trees, reduce the genetic quality in the stand, and also remove the best seed producers. An exception to this, where removal of the largest trees was expected to reduce production, but did not, is described in Box 10.1. In addition to the potential genetic effects and removal of more valuable trees, there is also the potential to adversely affect species composition by not only removing more desirable species but also creating environmental conditions that favor the regeneration and persistence of these undesirable species (Kelty et al. 2003).

> **Box 10.1 Transformation of even-aged Norway spruce in the Schlägl Forest**

The Sonnenwald Forest Management District in Austria provides a unique case of transformation to multiaged. The Schlägl Monastery managed the Sonnenwald Forest using clearcutting and even-aged management until the late 1970s, when a transformation to a multiaged forest began. The forest is nearly pure Norway spruce with some European beech (Figure 10.4). Stand treatments have been described as "target diameter harvests" that removed larger trees from the even-aged stands as an ongoing means of transformation. The objective was a diameter distribution resembling an equilibrium curve (Schütz 1975; Sterba 2004) in the form of a rotated-sigmoid curve. A 3.5 ha plot was established (the Hirschlacke plot) and mapped in an even-aged, 120-year-old stand undergoing the target diameter harvest regime.

The initial reaction to these treatments was a concern they would reduce stand growth by removing the larger, faster growing trees (Sterba and Zingg 2001). Also, as a diameter-limit harvest, the treatments may have been dysgenic and would certainly remove the more valuable trees in the stand, thereby leaving a high graded stand.

After two harvest treatments and two thinnings over less than 25 years, stands were approaching the target equilibrium diameter distribution (Figure 10.5). The change in the di-

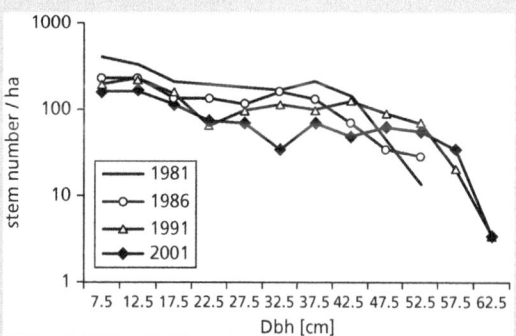

Figure 10.5 Changes in diameter structure from 1977 to 1997 at the Hirschlacke plot (from Sterba 2004; with permission from METLA). DBH, diameter at breast height.

ameter distribution is primarily in the increase in numbers of small trees, but the removal of larger trees has not affected the stocking in the larger diameter classes. The estimated stocking for the equilibrium is 475 m²/ha. The tree data from the Hirschlacke plot were used to examine individual tree growing space efficiency. Growing space efficiency provides the ability to assess different components in a stand such as age classes, canopy strata, or species (see Section 7.2.1);

Figure 10.4 An even-aged stand at Sonnenwald Forest prior to transformation.

continued

Box 10.2 *Continued*

when Sterba and Zingg (2001) used a diameter-based measure of available growing space, they found that small trees with long crowns were the most efficient.

The Schlägl example demonstrates the flexibility and resilience in central European Norway spruce forests, where reliable regeneration provides several options for managers. Despite the potential dual genetic and economic risks, a transformation that removed the larger, more valuable trees was successful at moving stands toward the target multiaged structure. This example also demonstrates the utility of using individual tree efficiency to assess management activities. Analyses of growing space efficiency revealed the target diameter harvests did not have an adverse effect on volume production because the small trees were unexpectedly efficient. Sterba's (2004) simulations indicated a number of alternative equilibrium curves could be met in 100 years. The Sonnenwald Forest Management District had been previously managed as a regulated even-aged forest; at the time of initial transformation, it consisted of even-aged stands of a variety of ages. The complete transformation process for the entire forest will therefore take many decades.

10.4 Constraints on transformations

Tree stability is a major constraint affecting transformations from even-aged to multiaged stands. Tree stability is often represented with ratios of height to diameter (H:D in equal units), which is also a simple measure of stem taper. Trees will approach critical stability thresholds in either even-aged or multiaged structures if growing space is limiting (Faber and Sissingh 1975; Cremer et al. 1982; Wonn and O'Hara 2001). Schütz (2001b) developed a flowchart to aid decision making for stands with potential stability issues (Figure 10.6). Even-aged

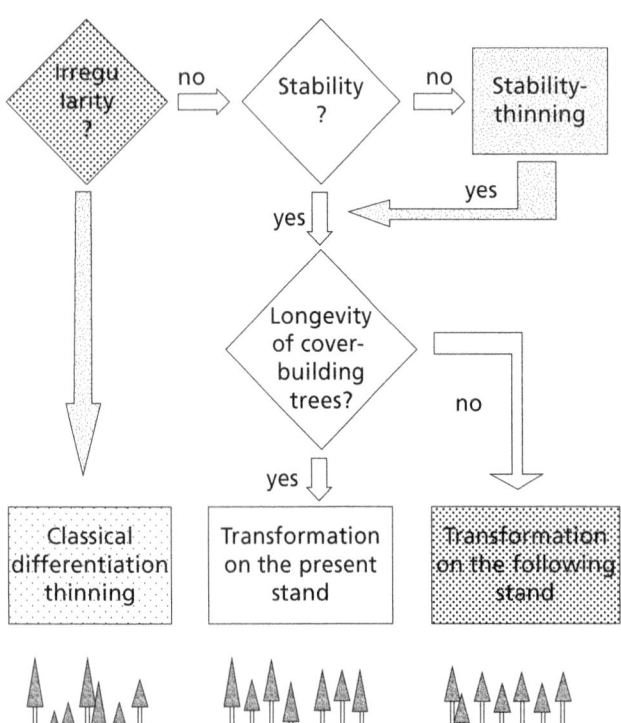

Figure 10.6 Flow chart displaying a decision process for transformation when windthrow is a concern (from Schütz 2001b; reproduced with permission from Elsevier).

increase in H:D ratios can be slowed with early and heavy thinning. In Norway spruce in the Czech Republic, heavy thinning was able to keep H:D ratios below critical thresholds, thereby creating opportunities for transformation to multiaged structures or conversion to mixed-species structures (Slodicak and Novak 2006).

A related concern is with inadvertent species conversions during the transformation process. In some ecosystems, the partial harvests that are used to regenerate new age classes of trees will create environments more suited to shade-tolerant species that may be less desirable. In these situations, more severe treatments or group openings may be necessary. Planting to encourage desired species may also be a useful strategy.

Operational constraints may make it impossible or infeasible to achieve certain structural qualities. Tree spatial patterns of transformed stands may not fully match those of the natural stands they attempt to target. For example, a restoration objective may have a target stand structure that includes a particular spatial pattern of trees in addition to a target vertical stand structure. The difficulty of harvest operations on steep slopes may limit harvests to infrequent but relatively severe treatments. Specific spatial patterns may also be difficult to achieve. In the Black Forest of Germany, distinct differences in spatial patterns were evident between even-aged and uneven-aged stands, and stands undergoing transformation (Hanewinkel 2004).

There may also be limitations on transformation because of tree age. Once a transformation begins, there is generally an expectation that some residual trees from the even-aged stand will be retained for several cycles of the transformation process. Depending on the regime, this may require trees that can be retained for 100 years or more. Schütz (2001b) included tree longevity in his flow chart outlining the decision process of whether to begin the transformation process with the current stand or wait until the next (Figure 10.6). In mixed-species forests where species have variable life expectancies, such as the central hardwoods in North America, having trees of greater longevity provides more flexibility in designing the transformation regime (Loewenstein 2005). Tree longevity is therefore an important factor in making tree selections, or possibly a factor for selecting tree species during the transition process.

As with much of silviculture, economic constraints may also limit the options in transformation regimes. Some transformation regimes may occur at considerable financial costs to the landowner. Economic analyses of transformation regimes are complicated by the many assumptions that must be made. A transformation is assumed to move an even-aged stand to a multiaged stand. The age of the even-aged stand can be quite variable, and this affects the financial returns from the transformation. For example, Knoke and Pluscyzk's (2001) comparison assumed a Norway spruce stand at age 47 would be transformed. This provided some immediate returns for the transformation regime because of the harvest of merchantable timber that would be delayed in the even-aged regime. A similar analysis of a fully mature even-aged stand at the time of harvest would result in a different conclusion because of the immediate returns of the clearcut harvest in the even-aged regime. Economic analyses of transformation regimes are therefore very difficult, and the implications of their results are limited. Other discussions of economics of transformation regimes have been completed by Buongiorno (2001), Knoke et al. (2001), Tarp et al. (2005), and Knoke (2012).

A significant economic concern may be the relative costs associated with the new and old regimes when considered in their final form. For example, if the multiaged management system results in a less expensive natural regeneration strategy as compared to tree planting in the even-aged system, there may be substantial economic advantages for the multiaged system. Or, increased harvesting costs with operating among many standing trees in the multiaged stands may create advantages for the even-aged stand. A more conventional economic comparison between the two alternative systems (e.g., Tahvonen 2009) may therefore be more useful than the analysis of the transformation itself. The expenses of a transformation are important, but should probably be looked at as a simple cost (if any) of transformation, not the sole justification for selecting a new regime (Knoke et al. 2001).

10.5 Synthesis

There is strong international interest in greater use of multiaged forest management (O'Hara 2001). It is part of the surge in interest in more complex stand management strategies. This surge is part of a cyclic swing in forest management (see Chapter 2). There may be a temptation to let the term "transformation," or even "conversion," define the intentions of land managers attempting to meet objectives for complex stand structures. A single transformation regime that was widely applied would probably achieve stand-level complexity but may not be well suited over many stands because it might homogenize landscapes. Instead, it should be recognized that there are many forms of complex stand structures that may vary greatly within a forest type and will certainly vary between forest types. The complex stand structure of a Norway spruce forest may have little in common with the complex structure of ponderosa pine. Likewise, the tools used for transformation should be highly variable and encompass a full range of potential treatments. The analogy of the "silvicultural toolbox" is appropriate here: there are many tools for transformation, from patch cuts to thinning to shelterwood treatments, and many more. There are also many target stand structures that can be the objective of transformation. Some alternative structures may be easier to attain and more economical than others. Transformation silviculture should also be highly adaptive rather than assuming preestablished plans will suffice without adjustments later in the transformation process.

A reoccurring theme in this book is the need for flexibility and that there are many options for multiaged stand structures. However, most transformations have tended to set diameter distributions as their targets. But any stand structure is possible, and some will be more likely to meet management objectives than others. Transformation targets should be flexible, as should the treatment regimes that are used to achieve these targets.

CHAPTER 11

Managing multiaged stands for diverse objectives

11.1 Introduction

Complex stand structures provide a different mix of benefits and ecosystem services than simpler stand structures. The production of these benefits may increase or decrease as stand complexity increases. However, no stand structure can maximize production of all benefits. As a result, optimal combinations of benefits may be best achieved with an assortment of different stand structures. For example, the stand that provides one benefit, such as wildlife habitat, may be compensated by a different combination of benefits from another stand with a different structure. For some benefits, there is a need to examine trade-offs at scales greater than the stand. This chapter describes how some ecosystem benefits or services are provided in relation to stand complexity. Multiaged silviculture, as a means to enhance stand complexity, can develop stands that provide a range of benefits and services. Although many benefits are best assessed at scales larger than the stand, the focus here is primarily the stand, with some references to larger scales. Because there are so many potential benefits and services, only a limited number are presented. The format will be to discuss how different benefits vary with stand complexity.

11.2 Timber—quantity and quality

The production of timber is a common objective of many multiaged systems. Relative production of even-aged and multiaged stands is discussed in Chapter 12. However, timber production involves both the quantity of production and the quality of timber produced. Wood volume production is maximized when growing stock or stand density is maintained at high levels. Maximum production at high stand densities is a well-known feature of even-aged stands (e.g., Long 1985; Curtis et al. 1997; Pretzsch 2005), where maintaining high levels of density and leaf area index (LAI) results in high levels of production. LAI is strongly related to gross production, and net production is the production remaining when respiration is subtracted from gross production. The same processes are at work in multiaged stands: high levels of production will occur in multiaged stands with high levels of LAI.

In multiaged stands, maximizing stand-level wood volume production at high densities results in smaller individual tree sizes. This trade-off was described by Long (1985) for even-aged stands. Similar processes work in more complex stand structures, where both total volume production and average tree size cannot be maximized in the same stand. Stocking control must therefore include an upfront decision on which objective to pursue.

High quality timber is usually based on a combination of tree size and the size and number of branches on the merchantable bole. Other factors such as stem taper and straightness are also important. Tree size is a function of individual tree growth rates, which are maximized when density is maintained at low levels. Branch size and branch retention on the bole are increased, and stem taper is decreased, with lower density. Hence, there is a trade-off between maximizing tree size with low stand densities and maximizing wood quality characteristics, which are enhanced at higher densities.

There are several ways to mitigate this trade-off. Trees in multiaged stands can be pruned to remove

Multiaged Silviculture. Kevin L. O'Hara.
© Kevin L. O'Hara 2014. Published 2014 by Oxford University Press.

Figure 11.1 Butt end of Norway spruce log from a multiaged stand in Black Forest, Germany. Pruned branch stubs are visible developing from the pith, with large amounts of clearwood outside the defect core.

branches on the lower bole and thus improve quality (Figure 11.1; see Chapter 9). Maintaining stand density at higher levels around the most desirable trees can also be effective at reducing the number and size of branches. These trees can later be released to increase wood formation on the desirable bole section. This is a modification of the even-aged stand treatment "crop-tree release" (Heitzman and Nyland 1991), where high-value species are selected and density is managed to improve wood quality. The presence of multiple age classes or canopy strata in multiaged stands makes this more complicated. However, this same canopy competition generally serves to reduce branch sizes and numbers on understory trees without further manipulation. In stands managed to form large openings, these even-aged methods should provide a means to enhance wood quality.

Timber production objectives are therefore more complex than simply maximizing wood volume. The trade-off between wood quantity and quality results in the inability to maximize both at the same time. Additionally, stand density levels affect other stand processes such as regeneration and stand health. Multiaged stands are managed through a range of densities between the beginning and end of a cutting cycle. However, there are also some benefits to managing multiaged stands through different ranges of densities to manage trade-offs between different resource objectives.

11.3. Biodiversity and wildlife habitat

Providing for diversity on managed forests is often an objective that is secondary to other objectives, particularly at the stand level. Maximizing biodiversity becomes more important at large scales because they include provide a greater range of species and habitat. Hence, measuring and providing for biodiversity is very scale dependent (Turner et al. 1989; Magurran 1988, Whittaker et al. 2001). Scale and structural diversity interact to affect biodiversity: greater structural diversity is often correlated with higher biodiversity. Biodiversity is often measured as a function of species richness or the number of species. Species often have different requirements or niches. Hence, different structures have different levels of diversity, and the species that are found in one structure are often quite different from those in another structure (e.g., Neumann and Starlinger 2001; Denslow and Guzman 2000; Franklin et al. 2002; Mani and Parthasarathy 2006). The greatest variety of species is therefore achieved with a variety of stand structures over broad scales.

Wildlife habitat refers to those places where an animal finds cover, food, or water. Managing for wildlife habitat is a common and important objective in many forests. Habitat features can include down and standing dead wood, nesting cavities, mast production, presence of multiple canopy vegetation layers (including shrub and forest floor layers), hiding cover, and many more. Because habitat is highly varied between wildlife species, there is no single stand structure that meets these needs. Instead, landscapes with a varied assortment of structures provide the most habitat. Using historical conditions as models for large-scale arrangements of structures may provide this habitat in proportions that maintain native species. This is a **coarse filter approach** to management that attempts to provide a broad range of habitats that are necessary to maintain the natural diversity of species, ecosystems, and processes (Hunter 1990). A **fine filter approach** focuses on individual species by providing their habitat needs when they are not being provided by broader-scale management or when the species is in a critical population status.

Complex structures are distinguished by their diversity and they provide important forms of wildlife

Figure 11.2 Thinned even-aged jack pine stand in eastern Canada. Young even-aged jack pine stands are essential habitat for the Kirtland's warbler.

habitat. However, some species are dependent on simple stand structures, and particularly those associated with young even-aged stand development may have few surrogates in contemporary unmanaged forests (Figure 11.2). Providing an assortment of habitat is therefore best accomplished by providing an assortment of stand structures ranging from complex to simple. Large landscapes with only multiaged stand structures would have relatively low levels of diversity as compared to landscapes with a large assortment of different structures or silvicultural systems to produce them.

Multiaged stands typically have a large level of structural diversity (O'Hara 1998). Through having a great range of tree sizes, multiple canopy strata, and often with mixed-species compositions, multiaged stands also typically have a large level of habitat diversity. Whereas the greater structural diversity of multiaged stands is well documented, there is less known about overall biodiversity in different structures. In old boreal forests in Finland, Lähde et al. (1999) found high levels of structural diversity in single tree and group selection treatments, and these levels were intermediate between those of the low thinning and the uncut old forest. In French Guyana, Favrichon (1998) modeled diversity in tropical forests and concluded that although more severe cuttings altered species composition, maintaining native assemblages, including high-value broadleaved trees, would best be achieved with light selection treatments. Kerr's (1999) review of silvicultural alternatives in Britain concluded that alternative systems to even-aged plantations would enhance biodiversity. There are many other studies that have documented the high diversity in complex stand structures (e.g., Fries et al. 1997; Bergeron et al. 1999; Bagnaresi et al. 2002). Indeed, the correlations between structural diversity and biodiversity are a strong impetus for expanded use of silvicultural systems that result in complex structures.

It is logical that high structural diversity corresponds to high species diversity. It is also logical that multiaged stands provide a different suite of habitats than other structures. When scales are expanded to include many stands, greater diversity is achieved when a diversity of stand structures are included. Hence, the stand structure that maximizes diversity at the stand level may only be a piece in the puzzle of maximizing diversity at large scales. A mix of multiaged stands with other stand structures will likely achieve the greatest diversity. The structural variation provided in multiaged stands can be enhanced by developing variations in stocking control through procedures such as variable-density thinning. Longer cutting cycles result in greater variation in structure because longer cutting cycles correspond to greater range in stocking. Structural features, such as snags, coarse woody debris, or unique trees or species that provide wildlife habitat, can also be integrated into multiaged systems to enhance habitat diversity.

11.4 Climate change

Climate change threatens the health and sustainability of forests all over the globe. Potential effects include greater disturbance frequency and severity, species extirpations, increased stand health problems, and more (Vose et al. 2012). The role of forestry will be to mitigate undesirable effects and develop adaptation strategies to reduce the likelihood of others (Seidl et al. 2007; Bonan 2008; Bolte et al. 2009; Bolte and Degen 2010). This section focuses on these efforts in the context of complex stand management, and the next section discusses carbon storage.

Proposed adaptation strategies for forestry involve assisted migration of species through relocation or assisted colonization, or modifying seed transfer guidelines (Ste-Marie et al. 2011), restoring structures and processes in highly altered ecosystems (Millar et al. 2007), and renewed efforts to prevent stand replacement disturbances, which increase risk and reduce investment options for adaptation work (Butler et al. 2012). Ecosystems may be more resistant to change if species that are most adapted to expected future conditions are favored (Janowiak et al. 2011). Silvicultural activities that, for example, favor drought resistant species would lead to stands with greater resistance. Nonnative species, such as coast Douglas-fir in central Europe, may be important alternatives when native species are poorly adapted to anticipated future conditions.

Assisted migration strategies are not without their limitations. Aubin et al. (2011) highlighted the ethical issues associated with focusing on some species that might be commercially valuable but not on others. Park and Talbot (2012) argued that climate change will require an intensification of silvicultural practices and may be well beyond our logistical or financial resources. Indeed the intensive nature of large-scale establishment and maintenance of new populations of trees is formidable in any forest setting. In locales where forest management is traditionally more extensive, other strategies that require fewer resources will be necessary.

Because of the long time horizons of forest management, climate change presents a moving target for establishing populations of trees that are adapted to a site. Trees adapted at one time may not be adapted at another. A species that may be adapted to exist in a climate may not be adapted to regenerate in that climate, or vice versa. Multiaged stands offer the ability to overlap generations of trees with different levels of adaptations. If one generation fails because of a difficult regeneration environment or a climatic fluctuation, multiaged stands have additional age classes to occupy the site. Likewise, if an older age class cannot persist because of environmental change, a younger age class may be more resistant to these changes. Multiple age classes provide different levels of adaptations that provide resiliency in multiaged stands. O'Hara and Ramage (2013) described this as a form of resilience to climate change that gives multiaged stands an advantage over other stand structures. Implementation of multiaged silvicultural strategies to develop complex stands will therefore be part of broader strategies to develop forests that are resistant and resilient to climate change.

11.5 Carbon

Carbon storage is an increasingly common forest management objective supported by international agreements and developing carbon markets (Box 11.1). Forests have great potential for storing carbon. However, as with many forest management decisions, there are trade-offs associated with increasing stand-level carbon stores. For example, greater carbon storage in fire-prone forests often results in greater fire hazards (Stephens et al. 2009), lower tree vigor, and susceptibility to insects and pathogens because of higher stocking levels (Fettig et al. 2007) or reduced structural complexity (D'Amato et al. 2011). The incentives to increase carbon storage are a singular objective that should be considered in a broad management context. Current incentives to increase stand-level carbon storage require demonstration of **additionality**, or that an activity or project provides

Box 11.1 Carbon storage in tropical forests

Carbon sequestration in managed forests is an important part of global efforts to reduce the effects of climate change. Forests can store large amounts of carbon, but, within a given ecosystem and forest type, the effects of forest management on stand structure can have a major effect on stored carbon. If multiaged stands can store greater amounts of carbon (Figure 11.3), then they may represent part of global strategies to mitigate climate change effects.

Tropical forests have been managed with "selective logging" systems that either attempt to emulate natural disturbance dynamics or use the selective cutting rubric as a justification for removing the most valuable timber over some lower diameter limit (Sist et al. 2003). In either case, the result is often a mismanagement that results in considerable damage of residual trees and negative effects on carbon storage (Pinard and Cropper 2000). However, although terminology for multiaged systems is highly variable in tropical silviculture, multiaged systems are often a part of sustainable forest management efforts (see Section 11.7) in tropical forests. The intent in these systems is often to use harvest treatments to emulate single tree openings. These treatments favor the establishment of valuable species, reduce fire hazards, and reduce the negative effects on wildlife habitat. Reduced impact logging is a key component of these efforts to minimize damage to residual trees by pre-planning landings, marking valuable retention trees, planning skid trails, cutting lianas that bind trees together before tree felling, and directional felling of trees (Putz et al. 2012).

A recent additional justification for improved management of tropical forests is the potential to greatly increase carbon storage. Putz et al. (2008) estimated large increases in stored carbon from multiaged silvicultural approaches that minimized logging damage to residual trees using the elements of reduced impact logging. Examples reduced carbon emissions by approximately 30% over conventional harvest systems in Malaysia and Brazil. Putz et al. (2008) estimated a global carbon retention of 0.16 gigatons per year from managed forests. Well-designed and thoughtful multiaged treatments in tropical forests therefore have the potential to increases carbon storage and improve management practices on a large-scale basis.

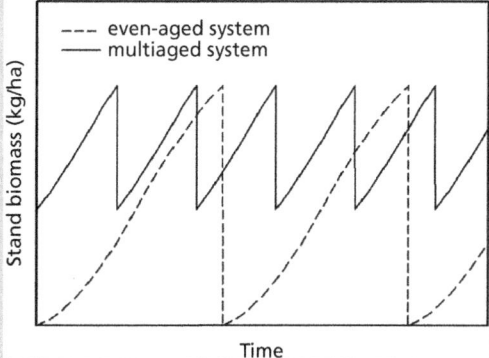

Figure 11.3 Above-ground biomass over time in idealized even-aged and multiaged systems. Biomass storage, and therefore carbon storage, maybe greater in the multiaged system and is approximated by the area under the biomass curve.

additional carbon storage beyond the storage that would have happened without the activity (Kollmuss et al. 2008).

Multiaged silvicultural systems may have potential for additionality over even-aged systems because of greater average stand-level storage when comparing over multiple cutting cycles and rotations. Whereas the amount of carbon storage in a classical even-aged system using the clearcut method will range from being very low after harvest to being high at the end of the rotation, carbon storage in the multiaged system varies in a smaller range at the high end (Figure 11.3). Variations in the management regime—such as with thinning regimes, cutting cycle lengths, etc.—will affect these comparisons. There is also potential for management regimes to influence carbon storage by affecting tree growth rates and wood properties that affect carbon density (Jones and O'Hara 2012). A modeling comparison of silvicultural regimes in Austria (Seidl et al. 2007) found that a multiaged strategy had greater carbon storage than an even-aged system but less than the unmanaged control.

11.6 Restoration

Forest restoration is a broad area within forestry encompassing many objectives related to redirecting stand development or counteracting the effects of converting forests to other uses (Stanturf and Madsen 2002). Restoration objectives can involve many different resources or be designed to enhance many different ecosystem services. These objectives may include timber production, wildlife habitat, carbon sequestration, water quality, or restoration of basic ecosystem processes. Restoration can involve intensive operations, such as mechanical site preparation, or rely on natural processes as a more subtle means to re-establish a suitable forest that meets certain objectives.

Restoration can direct degraded stands on trajectories toward multiaged structures. This is similar to the transformations from even-aged to multiaged presented in Chapter 10. However in this case, the stands may be of any age structure damaged by overcutting or restored from a non-forested state. Another form of restoration is associated with managing stands with invasive species and re-establishing native species. Restoration can also direct stands toward old structures (e.g., old-growth; Bauhus et al. 2009) that usually have complex age structures.

Stands with low stocking due to overcutting can either be restored with treatments to develop a more healthy and productive structure or cleared to initiate a new stand. In either case, these stands can be directed toward multiaged stand structures. In the southeastern United States, Baker and Shelton (1998) developed a replicated study to examine the merits of restoring multiaged loblolly pine–shortleaf pine stands with growing stock reduced from 10% to 50% of full stocking (0.9 to 31.7 m^2/ha of basal area). Rehabilitation treatments included herbicide treatments of undesirable broadleaved species. The option to rehabilitation was to replace the existing stand using site preparation and tree planting. They found stands down to about 20% stocking could be successfully rehabilitated based on growth responses to treatment. This work demonstrates that stands with low stocking have the resiliency to recover and retain a multiaged structure with minimum intervention and investment.

Restoration of multiaged stands should do more than simply develop multiple age classes of trees. Stand structures consist of complex vertical and horizontal spatial arrangements of trees that comprise important structural features. Although vertical structural patterns are the most obvious feature of multiaged stands, the horizontal spatial patterns of trees are also an important consideration when emulating natural patterns. In the Black Forest in Germany, Hanewinkel (2004) found variable spatial patterns in Norway spruce–silver fir–European beech stands undergoing transformation. Large trees exhibited regular spatial patterns in multiaged stands, and smaller trees more random patterns. Hanewinkel concluded that treatments that created gaps in the transformation process might result in more clumpy distributions than would be desirable. However, working in a variety of ecosystems in western North America, Larson and Churchill (2012) concluded that studies of global patterns of spatial heterogeneity obscured patterns of local variation. They recommended that managers needed to integrate local patterns of spatial heterogeneity into tree-marking protocols as a way of integrating spatial patterns into management regimes. Spatial pattern analysis reveals that these patterns are important structural features because they provide resilience and emulate certain aspects of disturbance behavior and regeneration. However, emulation requires some knowledge about the disturbance processes and their effects on stand structure. It is also important to use caution when generalizing about disturbance patterns or residual stand structures from one ecosystem to another.

For deforested sites, restoration begins with establishment of a new cohort. Depending on the condition of the site, this may involve intensive efforts common to plantation establishment. Herbicides or intensive site preparation may be necessary to establish certain species. "Nurse crops" that modify the local environment, or species that fix nitrogen may be helpful in establishing tree species that are sensitive to full sunlight exposure or where nitrogen availability is low, respectively. Dey et al. (2010) found that nitrogen-fixing species, and nurse crops in the form of other tree species could aid the establishment of oaks on former agricultural sites in the southeastern United States. Additional age classes at later stages of the restoration process may also require planting before the first age class has reached sexual maturity. At this point, the restoration process begins to resemble the transformations presented in Chapter 10.

11.7 Sustainable forest management

Sustainable forest management (also known as SFM) is the management of forests to maintain their biodiversity, productivity, regeneration capacity, vitality, and potential to meet the future relevant ecological, economic, and social functions, without causing damage to other ecosystems (Helsinki Declaration 1993). Various forms of multiaged silviculture are often part of sustainable forest management programs. Achieving sustainability with any management system is an important objective and benefit of management. Criteria and indicators are tools used to evaluate and guide sustainable forest management systems. They guide management treatments and evaluate processes. Criteria and indicators are the subject of considerable ongoing research and several books (Prahbu et al. 1999; Raison et al. 2001; Sato 2009). They exist at local and national levels and guide forest certification systems and initiatives that cover an assortment of boreal, temperate, and tropical countries (International Model Forest Network 2012).

For multiaged stands, sustainability is often assumed to be achieved with maintenance of a consistent stocking control regime over time. Schütz (2001b) proposed a concept of demographic sustainability that referred to stands maintaining an equilibrium stocking regime over time. A consistent stocking control regime provides a confined range of variation in stand structure from beginning to end of the cutting cycle. However, sustainability refers to far more than maintaining a consistent stocking regime. Sustainability refers to the capacity of forests to maintain their health, productivity, diversity, and overall integrity over time (Helms 1998). Many of these factors transcend the stand to encompass many stands and broader areas. Stand-level criteria and indicators include changes in species composition, growth rates, and the presence of structural features such as large trees, snags, etc.

O'Hara et al. (2007a) examined several sustainability criteria for multiaged stands managed for nearly 100 years with the Plenter system in Switzerland. Although these stands had been managed with the same basic methodology for many decades, there were significant changes in tree species diversity, size class diversity, stocking, and increment over the study period (Figure 11.4). The changes were subtle, but surprising given the consistency in management over time. Similar small changes in stands managed with shelterwood and selection systems were observed in Spain (Bravo et al. 2010; Tiscar Oliver et al. 2011). This work suggests careful monitoring may be necessary with even the most consistent multiaged systems, and similar criteria

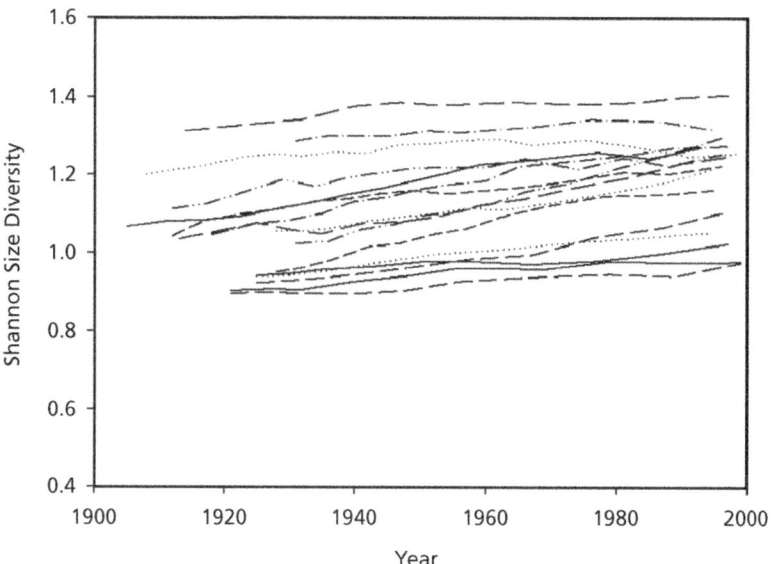

Figure 11.4 Changes in tree-size diversity of multiaged stands managed with the Plenter system in Switzerland. The Shannon diversity index was calculated on diversity in tree sizes. Each line represents a separate plot measured for many decades. Plots trended upward, although management was relatively constant over time (from O'Hara et al. 2007a).

may be helpful in assessing the sustainability of multiaged systems.

The deviation of a stand from an expected pattern may indicate the stand is not following a sustainable path, but more than likely is an indication of the normal perturbations caused by climate, disturbance, or management activities. Developing expected patterns is an integral part of adaptive management approaches to sustainability; however, individual stands need not explicitly follow these patterns. Instead, groups of stands should be evaluated to monitor their collective development. Broad-scale or landscape level collections of stands are a logical unit to use for assessing sustainability.

11.8 Other ecosystem services

There are a great many ecosystem services or benefits derived from multiaged stands. In addition to those discussed in this chapter, multiaged stands provide water, aesthetics, non-forest products, protection functions, and other ecosystem values. Many of these benefits are also provided in simpler stand structures and in varying amounts that may be greater or less than in more complex structures. The strong associations between different types of wildlife habitat and stand structure demonstrate how an assortment of structures are required to achieve the greatest diversity in habitat. This is also true of other ecosystem services, which are derived from a range of stand structures.

Whereas the benefits of timber production are easily quantifiable and assessed on an economic basis, the others are not. As a result, comparisons of the relative values of benefits from different stand structures are difficult. Zhou et al. (2012) compared discounted and undiscounted valuations of several different nontimber benefits from multiaged coast Douglas-fir–western hemlock dominated stands in the Pacific Northwest of North America. Although there is some debate about discounting nonmarket values (e.g., Howarth 2009), Zhou et al. (2012) found that many decisions would be unaffected by discounting. However, the greater issue is simply providing meaningful valuations to nonmarket values.

11.9 Synthesis

Multiaged and other complex stand structures provide many ecosystem services and other benefits. The capability of these structures—and the multiaged systems that implement them—to provide many nontimber benefits is a common justification for their expanded use. The relative production of any benefit varies with the stand structure and the management regime. Any single structure does not have the capability to produce all potential benefits. Management strategies to enhance different ecosystem benefits or services will rarely be the same. Some benefits may be enhanced by treatments that reduce other benefits. However, there is also the potential for joint production of some resource values that allows for multiple benefits, and possibly some synergy, from the same management treatments.

In a forest managed with multiaged systems, there is potential for all stands to provide a similar suite of benefits. Where a diverse set of benefits are desired, a diversity of stand structures will be necessary. Variation in multiaged regimes can be achieved by varying the length of the cutting cycle, increasing or decreasing the volume in periodic harvests, varying the size of openings, favoring different species, or other means. However, large-scale production of multiple benefits may be achieved with greater variation in management regimes that alter vertical and spatial structural features. Mixtures of multiaged and even-aged approaches may be best at meeting these large-scale objectives.

CHAPTER 12

Growth projection in multiaged stands

12.1 Introduction

The ability to project stand growth processes is an important aspect of managing multiaged stands. Some level of predictability is necessary for all management decisions because it reduces uncertainty and assists with risk management. For commodity production, investment decisions may be based on potential production values. For objectives related to non-commodity values, predictability provides short- and long-term forecasting of stand structural features, wildlife habitat, resistance to disturbances, or aesthetics. Regardless of the objective, tools are required to project commodity production and changes in stand structure. The ability to predict changes in stand structure also helps measure performance and provides repeatability in management treatments.

This chapter provides an overview of growth projection tools and their potential uses for projecting multiaged and other complex stand structures. It begins with a discussion of assessing site quality in complex stands, where traditional tools designed for even-aged stand are not suited. Growth projection tools are presented, from the most simple to the most complex. These tools vary in their potential utility for complex stands and also in their data requirements. There have also been many efforts to develop optimal stocking regimes to achieve volume production or economic objectives. Optimization methods rely on growth projection tools, which in turn rely on estimates of site quality.

12.2 Site quality assessment

Forest site quality is important because it determines how rapidly stand development and other changes in stand structure will occur. It is also important for assessing the suitability of species to a site and the capability of the site to support certain levels of density or stocking. Projecting changes in structure and suitability of different types of resource management on forest lands is an important aspect of forest planning and has a long history in forestry dating back several centuries. Reviews of forest site productivity assessments and classifications have been provided by Carmean (1975), Bailey et al. (1978), Daniel et al. (1979), Hägglund (1981), Barnes et al. (1982), Vanclay (1992, 1994), Skovsgaard and Vanclay (2008, 2013), Weiskittel et al. (2011), Bontemps and Bouriaud (2014), and many others.

There are many ways to evaluate site productivity. Foresters traditionally expressed the potential productivity as either total volume or merchantable volume per unit area. This was generally referred to as direct productivity because it measured the variable (wood volume) that was the ultimate objective of management (Jones 1969; Daniel et al. 1979; Vanclay 1992; Skovsgaard and Vanclay 2008). Volume yields or site productivity were recognized as being dependent upon site quality and other factors. Direct measurement of site productivity is difficult, so alternative methods of site productivity assessment are more common. These include phytocentric and geocentric methods (Skovsgaard and Vanclay 2008). Phytocentric methods include site index and analysis of vegetation. Site index is based on the recognition that the height growth of dominant trees in even-aged stands is independent of tree density and is therefore representative of the quality of the site. Vegetation methods are based on the recognition that certain plants are found on, and indicative of, certain environments (Spilsbury and Smith 1947;

Multiaged Silviculture. Kevin L. O'Hara.
© Kevin L. O'Hara 2014. Published 2014 by Oxford University Press.

Cajander 1949). Geocentric methods use soil characteristics, topography, climate, and other variables to estimate site quality (Coile and Schumacher 1953; Hills 1953; Krajina 1965; Steinbrenner 1979).

Site index is the average height of dominant or codominant trees at a specified index age, such as 50 or 100 years (Helms 1998). It is the most common method of site quality estimation. Early forest scientists recognized the relationship between dominant height growth and site productivity and developed volume yield tables for even-aged stands that used a site index as a variable. Site index requires dominant trees that are free of competition, or free-to-grow, through their entire development. This situation is not common in a multiaged stand and nonexistent in stands managed with traditional multiaged systems, where trees generally develop through several periods of suppression (Figure 12.1). However, for methods that rely on openings such as group selection, or methods that use relatively low stockings, suitable free-to-grow trees may be found. Otherwise, conventional site index does not work in multiaged stands.

Vegetation approaches or soil-site relationships may be used in multiaged stands. The presence of existing plants was long ago recognized as indicative of certain site characteristics. However, vegetation is also influenced by management, other human activities, and natural factors such as disturbances.

Indicators therefore have to be reliable in cases with widely different land-use histories. Early uses in forestry included the work of Salisbury and Smith (1947) in the Pacific Northwest, which used the abundance of only a few plants rather than their frequency. Other vegetation approaches included Cajander's (1926, 1949) work in Finland and Daubenmire's work in the western United States (Daubenmire 1952, 1976; Daubenmire and Daubenmire 1968). Cajander used understory plants as indicator species to predict a climax community, and Daubenmire used both understory and overstory plants. These approaches are often termed habitat types or plant associations and have been criticized for their reliance on concepts related to succession and climax vegetation (Cook 1996; O'Hara et al. 1996; Kusbach et al. 2012); nonetheless, they are still in common use.

Geocentric or environmental approaches include Coile and Schumacher's (1953) work relating soil properties to the site index in loblolly and shortleaf pine in the southeastern United States. Coile and Schumacher found that factors such as the depth of the A horizon and water holding capacity of the B horizon were highly correlated with the site index. Steinbrenner (1979) developed a similar system for the Douglas-fir region of the Pacific Northwest that included a range of soil and physiographic variables. Other soil or physiographic systems include those developed by Hills (1953), Hodgkins (1956), Della-Bianca and Olson (1961), and Carmean (1979).

Some systems have attempted to link different types of variables to evaluate site quality. For example, Steinbrenner's (1979) system used both soil descriptions and physiographic variables. In British Columbia, Canada, Krajina (1965, 1969; see Klinka et al. 1989) developed a biogeoclimatic system that integrates vegetation, soils, physiography, and climate into a site classification system. The classification of the vegetation community, or **biogeocoenosis**, is defined much like a Daubenmire (1952, 1976) habitat type in that both are based on climax vegetation but the zonation of soil types limits Krajina's biogeoclimatic ecosystem classification to a single soil series. Ecosystem process models also have the capability to estimate potential or actual forest production at small or large scales. Examples include Forest-BGC (Running and Coughlan 1988), 3-PG (Landsberg and

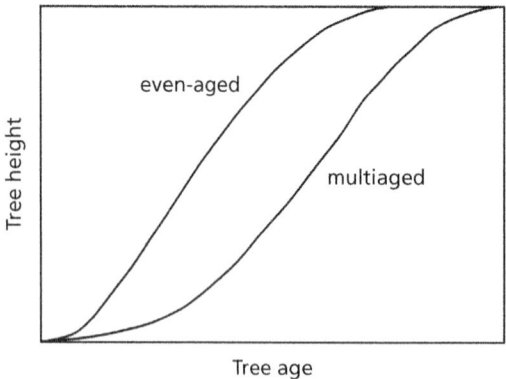

Figure 12.1 Schematic of height growth of a dominant tree in an even-aged stand compared to a tree that becomes dominant in a multiaged stand. The tree in the multiaged stand endures periods of various degrees of suppression before it emerges into the upper canopy, making it unsuitable as a site index tree.

Waring 1997), or CABALA (Battaglia et al. 2004). Milner et al. (1996) used Forest-BGC to quantify potential site productivity as the culmination of mean annual increment on forest sites in the northern Rocky Mountains of North America.

Whereas vegetation and soil/environment site quality assessment approaches are better suited to multiaged stands than site index, these approaches are only developed for a subset of potential forest types where multiaged management may be practiced. However, even where these systems are developed, there may not be a direct link to timber production potential. If productivity estimates are desired, then some systems may have these linkages. For example, Steinbrenner's (1979) system was related to site index, and the habitat type system of Pfister et al. (1977) was related to site classes or groupings of site indices. Rayner's (1992) analysis of different site assessment approaches for karri in southwest Australia found that the vegetation, soils, physiographic, and climatic approaches were not as effective as site index for estimating productivity. Hence, there may be trade-offs associated with predicting volume productivity using methods other than site index. The reliance on site index for predicting volume productivity in these other site quality approaches demonstrates the value of the site index method (Skovsgard and Vanclay 2008).

Estimating the site index of a mixed-species stand may also be difficult. Whereas the intent is to estimate the productive potential of the site, it is well established that species provide different site index estimates on the same site (Doolittle 1958; Olson and Della-Bianca 1959; Carmean 1979; Shoulders and Tiarks 1980). The estimate with one species might indicate a higher or lower productivity than a different species on the same site. These variations may be indicative of differences in potential productivity if the two species were grown in separate monocultures, or suggest that interspecific interactions affected growth, or may simply provide variable estimates of inherent productivity. In both cases, these differences demonstrate that mixed-species stands add another complication to the basic concept of site quality estimation.

There is a long-term recognition of the need for an alternative "site index" for complex stands. In irregular red spruce stands in eastern North America, McLintock and Bickford (1957) used a height-over-diameter relationship as a measure of site quality instead of height over age as a site index. Comparing observed height–diameter relationships for dominant trees to standard curves, like conventional site index, provided an index of productivity. In Australia, Vanclay and Henry (1988) used a similar approach in cypress pine (Figure 12.2), and in subtropical eucalypts (Vanclay 1992). The key in these approaches is using an intermediate diameter within the normal growth range (Figure 12.2), much like the index age used in height-based site index.

In tropical forests in Australia, Vanclay (1992) developed a "growth index" based on diameter increment adjusted for tree size and competition. The method used basal area to represent competition and did not use age or tree height. In temperate forests, coast redwood has a highly irregular height growth pattern, even among dominant trees in even-aged stands. As a measure of productivity, Berrill and O'Hara (2014) found that basal area increment was a better predictor of site productivity than the site index in redwood stands of a range of stand structures. Forms of ecological site classifications (Bailey et al. 1978; Barnes et al. 1982) combine different approaches into a more integrated site evaluation that relies on multiple factors. These approaches provide evaluations for forestland

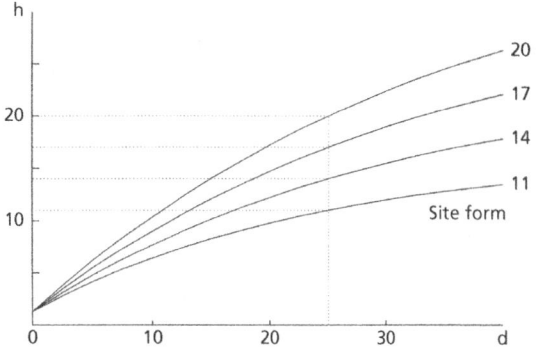

Figure 12.2 Height–diameter curves for multiaged cypress pine in Queensland, Australia. Much like a conventional site index, these curves are fit to existing data, and an index age (25 years in this example) is established. The height (h) and diameter (d) of individual trees would be plotted on the curve to evaluate site productivity or "site form" for a given density and age (from Vanclay and Henry 1988 and Vanclay 1994; with permission from CABI).

suitability for different uses as well as basic productivity, and these approaches are applicable to complex stands. An alternative approach for multiaged stands is to estimate the site index in a nearby and similar even-aged stand and then assume differences are minimal.

Most site quality research has focused on volume productivity in even-aged stands. Nevertheless, the management of complex stands can benefit from this information. This includes information on productivity, information on species suitability, and insights into establishing suitable stocking levels. It could be argued that many complex stands will not be managed to maximize volume production, so site productivity information is not necessary. However, the site productivity information can guide species selections, stocking levels, and help guide treatment intervals. It also is a key variable in most stand growth projection systems. Traditional site index is the most developed procedure for assessing site productivity, but much of the development of site index has been focused on even-aged stands for timber production. Alternative methods that focus on assessing vegetation or environmental variables may provide the best insights into the potential of complex stands to meet management objectives. But at present, these applications are not well developed.

Site quality assessment is an important part of multiple benefit forestry including the management of complex stand structures. Approaches are needed to quantify stand growth and potential productivity for use in growth projection models and provide meaningful comparisons between different sites. Approaches are also needed to provide more qualitative assessments of site potential for other resource uses. Site index is an effective site quality assessment approach for even-aged stand production, but there are other methods that provide alternative information. This is also true for complex structures: methods are needed that can quantify potential timber production and assess site potential for alternative uses.

12.3 Growth projection

The objective of multiaged stand growth prediction is to project future development of the stand and the changes in stand structure. Projections of wood volume production allow the land manager to project future revenue and make decisions on stocking levels and the length of cutting cycles. In other cases, projections may provide information on future stand structures and corresponding information about wildlife habitat, disturbance risks, and the stand component information of large-scale or landscape vegetation patterns. The concepts related to growth prediction of multiaged stands vary depending on the stand structure, or specifically the horizontal and vertical heterogeneity of the stand. Complex forest stands can be viewed as existing on a gradient of spatial patterns ranging from relatively uniform to very heterogeneous. The structural features that distinguish homogeneous from heterogeneous multiaged stands are largely the size and arrangement of openings. Growth prediction concepts vary depending on the size of the openings. Using the terms "small opening" and "large opening," to be consistent with the terminology of Chapter 6, concepts related to small openings created through systems such as single tree selection are presented separately from concepts related to large openings from systems such as group selection.

Multiaged stands managed with the single tree selection system to form regular or uniform stand structures are generally assumed to have negative exponential or reverse-J diameter distributions (see Figure 7.13). Other diameter distributions are possible, as are distributions of other tree dimensions. The distributions change over time but are often assumed to fluctuate from an initial structure to a preharvest structure and then back again. These structures are assumed to have a relatively uniform spatial distribution. Over time, these stands may be managed to a specific target diameter distribution, such as a negative exponential distribution (see Chapter 7). The use of a consistent diameter distribution provides the opportunity to project the changes in this diameter distribution over time. This type of consistency in stand structure over time provides the opportunity to easily project changes.

As opening size increases, stand heterogeneity also increases. This is often an objective of multiaged systems but makes projections of stand development more difficult. A stand with a fully regulated group selection system will essentially have the stand segmented into groups of equal area. If groups are large, traditional even-aged projection tools can be used to project the development of individual groups and then summed for stand growth. For example, even-aged stand yield tables can be used, as yield table ages correspond to age classes in the multiaged stand. Calama et al. (2008) used this approach, adapting even-aged models to multiaged stands of stone pine in Spain. When groups are small or variable in size, achieving equal area per age class becomes more difficult, and there may be significant edge effects. However, even-aged stand projection approaches can be used for these stands, but there must be a greater acceptance of the potential error involved. Alternatively, average stand data for heterogeneous stands can be projected as though it were a more homogeneous stand structure.

12.4 Growth projection tools

There are a variety of systems or tools designed for projecting stand development, and many can be applied in complex stand structures (e.g., Hann and Bare 1979; Vanclay 1995; Peng 2000). They vary in their complexity and their imbedded assumptions but are typically divided into empirical and mechanistic models (Figure 12.3). Many of these tools were designed to model both simple even-aged stands and more complex multiaged stands. Most of these models require a site quality assessment, usually in the form of a site index. Hence difficulties with site quality estimation may limit the tools that are available and affect the accuracy of predictions. Growth prediction tools are presented here from simplest to most complex and from stand-level to individual-tree models per previous classifications (Munro 1974; Burkhart and Brooks 1990). This order of presentation also generally applies to the chronology of development of these tools, as new computing technologies are greatly expanding the capabilities of growth projection. More detailed overviews of growth projection tools have been provided by Avery and Burkhart (2002), Vanclay (1994, 1995), Husch et al. (2003), Peng (2000), Pretzsch (2009), and Weiskittel et al. (2011).

Traditional **yield tables** date back several centuries in Europe and much earlier in China (Vanclay 1994). Yield tables are often organized by age and site index for an individual species. The yield estimates are best in uniform stands. Hence fully stocked, or normal stands, are one type of yield table for even-aged stands. The earliest European yield tables were based on rotation-length data, but tables in North America and elsewhere were based on less complete information and fitted to guide curves (Peng 2000). Another development was the

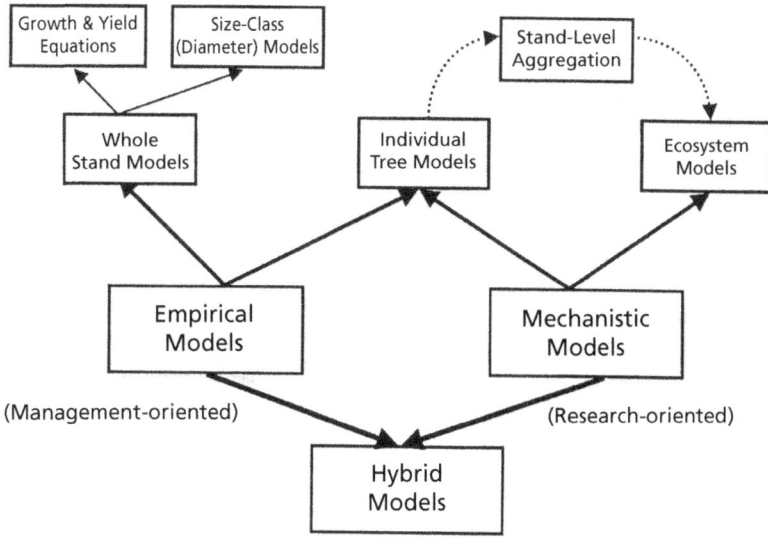

Figure 12.3 Chart showing classification of growth projection tools for multiaged stands (from Peng 2000; reproduced with permission from Elsevier).

variable-density yield table, which presents separate yield tables for a range of stand density. For mixed-species stands or multiaged stands, determining yield is more complicated. Neither the age or site index of even-aged stands will usually be applicable to complex stands. Duerr and Gevorkiantz (1938) suggested finding even-aged groups of trees in multiaged stands and using them to develop yield functions for the entire stand. Others have suggested using the dominant species and then prorating the stand yield estimates by the percentage of the dominant species (MacKinney et al. 1937). Although the yield table approach has served forestry well for decades and even centuries, it has limitations for use with complex stands. These limitations are chiefly the difficulties in expressing stand age and site quality in multiaged stands, and the varying contributions to productivity from different species in mixed-species stands.

Stand **growth and yield equations** are another way to project volume growth in complex stands. These equations project average stand conditions from some initial stand age. They are derived from existing stand inventory data with repeat measurements. The simplest (stand-level) equations do not consider individual trees, and mortality during the estimation period can be modeled separately or is implicitly part of the estimated stand volume growth. A typical form might include an initial volume, a measure of stocking, and a time interval. Equations might be developed for different groups of sites or site may be included as a separate variable in prediction equations (Vanclay 1994). This procedure is crude because it does not generally consider species composition or any tree-level variation related to stand structure. Growth and yield equations are also not particular adept at characterizing management treatments such as thinning. However, the basic concept can be applied to even- or multiaged stands if the only objective is the change in volume or basal area with time. For example, Moser and Hall (1969) and Moser (1972) developed volume growth equations for multiaged hardwood stands in eastern North America.

Whole-stand distribution models project a stand structure—as represented by a size-class distribution—forward over time. Most commonly, diameter size-class distributions are used. These models have been effective at projecting changes in uniform, even-aged stands, particularly plantations where unimodal, normal distributions are common. The size-class distribution is characterized by a mathematical function such as a Weibull or beta function. Existing plot data are used to project changes in the distribution over time based on site quality. For even-aged stands, temporary plots that obtain a stand age can be used to develop these models. Diameter-height relationships, stem taper relationships, and volume equations can be applied to individual size classes to estimate volumes, which are then summed for stand volume. Silvicultural treatments such as thinning can be effectively modeled using changes in the posttreatment size-class distribution.

Whole-stand models can also be used in multiaged stands where a negative exponential size-class distribution (e.g., Figure 12.4) can be represented with a mathematical function, such as an

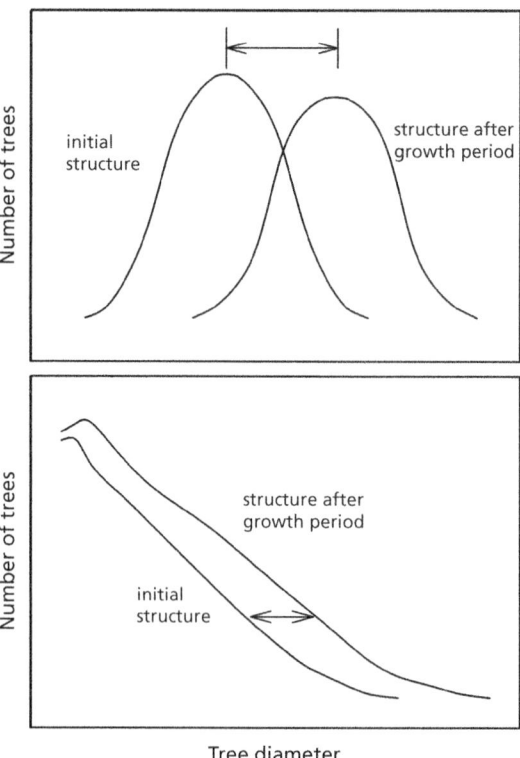

Figure 12.4 Changes in stand structure as represented by diameter distributions for an even-aged stand (above) and a multiaged stand (below). Both diagrams show a stand in an initial stand structure and then after a growth period. Either situation can be modeled with a whole-stand model.

alternate form of a Weibull function. Unlike the even-aged stand, where the stand moves forward, the multiaged size-class distribution is moved forward with growth and back with treatment (Figure 12.4). This may be a simpler modeling problem. Whole-stand modeling can therefore be relatively simple providing that the diameter size-class distribution is also simple. Interval data provided by permanent plots or repeated measurements is needed to project changes because a stand age is not applicable in multiaged stands.

Negative exponential diameter distributions are well suited to whole-stand modeling approaches, and this may be an impetus to manage for a negative exponential distribution: maintaining a uniform and preset size-class distribution improves predictability. Other size-class distributions would be more difficult, particularly if they varied between cutting cycles. Example applications of whole-stand approaches to multiaged stands include ones developed by Ek (1974) and Adams and Ek (1974), who used nonlinear functions to develop stand table projection models in northern hardwoods in North America. In addition, Hyink and Moser (1983) demonstrated how parameters that describe diameter size-class distributions can be predicted in multiaged hardwoods in eastern North America.

Whole-stand models have the capability to project both stand structure and volume yields in multiaged stands. The transformation from evenaged to multiaged stands with these models is relatively simple. However, consistent size-class distributions are necessary, and these models become less successful with more complicated stand structures. Their greatest utility in projecting complex stands will be for relatively simple structures where the primary attribute of the modeling effort is projection of volume growth. However, designing stand structures should be driven by meeting management objectives, not modeling efficiency.

Another form of projection system that is applicable to complex stands uses size classes as the basic unit for projection rather than stands or individual trees. These **size-class models** project the growth of individual size classes and aggregate to the stand level. **Stand table projection** is a form of a size-class model where diameter size classes are advanced through the diameter distribution at a certain rate as a function of tree size, stocking, site quality, and mortality rates. A future stand table could then be developed from a previous table, given growth, mortality, and harvest factors. For a hypothetical 5 cm diameter class, a certain percentage of trees might advance one class, two or more classes, or not at all. Forms of movement ratios or growth-index ratios—developed from inventory records—are used to make these calculations. The stand table projection method has been used in a variety of forest types and management regimes. For example, Korsgaard (1989) developed a simulation system based on stand table projection for mixed forests in Malaysia. **Matrix models** are another form of size-class model (Buongiorno and Michie 1980; Liang and Picard 2013). These models project changes in structure but allow greater efficiency in summarization of transition functions. They have been developed for a number of different forest types including temperate (Hao et al. 2005), boreal (Kolström, 1998), and tropical forests (Mendoza and Setyarso 1986). These models are commonly used in multiaged forests, and although they typically use diameter classes as the state variable, age classes, species, or canopy strata can also be used.

A widely used group of stand projection tools are **individual-tree models**. These models project the growth of individual trees that are aggregated to the stand level. These models are often categorized as either models that do not require spatial locations of trees (distance-independent) and those that do (distance-dependent; Munro 1974; Peng 2000). Although these models are generally designed to guide management, some are more applicable to research (Mohren and Burkhart 1994; Battaglia and Sands 1998; Peng 2000; Groot et al. 2004). In either case, they are important for the analysis and assessment of complex forest structures.

There are a number of classifications of individual-tree growth projection models. There are the distance-dependent/independent classes, the empirical/mechanistic classes, individual-tree models that focus on stands, and others that are focused on gaps. Many of these models were developed to either provide greater predictive potential in complex stands or to gain a better understanding of stand-level processes. Many of these developments can be viewed as moving from describing forest dynamics to explaining forest dynamics (Bossel 1991).

Distance-independent individual-tree models simulate the growth of the individual tree and sum individual-tree results to produce stand-level results. Spatial locations of individual trees are not a requirement of these models. Projections are based on individual trees or by size classes. Stand-level variables that are often required include site index, stand age, and measures of density. Competition is often described with indices that describe density and crown competition (Biging and Dobbertin 1995) based on average stand characteristics. These models generally run on typical stand inventory data. Many of these models are well suited to projecting development of complex stand structures when issues over quantifying site quality and stand age are resolved. Examples include PROGNOSIS/FVS (Stage 1973; Wycoff et al. 1982; Teck et al. 1996), PROGNAUS (Monserud and Sterba 1996), and many others.

Distance-dependent individual-tree models require data on spatial locations of individual trees. These models project individual trees forward, and tree growth is summed to produce stand-level results. Information on tree bole sizes, crown sizes, and spatial patterns among trees can be used to describe competition (Biging and Dobbertin 1992). Mechanistic or process models nearly always fall into this category. These models may attempt to describe photosynthesis, respiration, and nutrient cycling to describe how stands grow and how these processes interact to affect other processes. Examples include the FOREST model (Ek and Monserud 1974), SORTIE model (Coates et al. 2003), SILVA (Pretzsch et al. 2002), and many others.

Mechanistic or process models are capable of simulating scenarios outside the range of empirical data. This is useful for projecting stands of unusual structures or management strategies, or for simulations in changing environments and climate. Distance-dependent individual-tree models are appropriate for projecting changes in complex forest structures, assuming they provide a realistic representation of tree growth across a range of growing conditions (e.g., at/near gap edges, in deep shade, etc.). The primary obstacles to operational use of mechanistic models is the need for stem-mapped inventory data and the common view that these models are best for research purposes. Battaglia and Sands (1998) noted the potential of these models for guiding management decisions in lieu of empirical models. However, the required spatial data may be too expensive for management applications (Vanclay 1995). Distance-dependent individual-tree models have the potential to project complex stand structures and are most capable of all the different stand projection tools at including the intricacies of complex structures.

Gap replacement models are another form of individual-tree, mechanistic model and simulate the development of small patches within stands or larger units of forest (see Section 5.4). The individual tree is usually the simulation unit in these models. Individual trees are aggregated to represent the development of the gap (Shugart 1984; Peng 2000; Bugmann 2001; Schliemann and Bockheim 2011). A single tree tree fall is the usual gap-initiating event. Gap replacement models are generally distance-independent, and most are adaptations of the JABOWA or FOREST models (Botkin et al. 1972; Shugart and West 1977). Gaps are individual patches that do not interact but may be aggregated to represent a stand. Hence any stand-level simulations from gap replacement models are projections of independent patches that model the within-gap processes of growth, regeneration, competition, and mortality; however, these models generally assume no interactions from adjacent patches, including edge effects, on these processes.

Gaps are important structural elements that affect the dynamics of complex forests. Gap replacement models are useful for simulating gap effects and improving our understanding of gap dynamics. These models can simulate a variety of processes such as species changes, habitat development for wildlife, energy exchanges, and others. They have great utility for stand- and ecosystem-level analyses. However, gap replacement models are rarely used for management applications because of the limitations of working at the gap scale and the data requirements for model implementation.

12.5 Optimization tools

Determining the optimal stand structure has been a common objective in multiaged management and the subject of considerable research. **Optimization tools**, such as **linear programming**, are advanced tools within the field of **operations research**. The primary objective function of these optimization models has usually been maximization of a sustainable level of volume production. This is because of the easily quantifiable value of wood volume, and also because maximizing wood volume and value have traditionally been primary drivers in designing multiaged stand structures. Because of the emphasis on volume production, these optimization tools rely on growth projection tools to estimate wood volumes for different scenarios. The scenarios are stocking regimes that present different options for how growing space can be allocated to stand components.

Examples of optimization model applications using stand growth projection models include ones provided by Adams and Ek (1974) with a whole-stand model, Buongiorno and Michie (1980) with a matrix model, and Hasse and Ek (1981) with a distance-dependent model. Many other examples exist (Box 12.1), mostly with northern hardwood forests or ponderosa pine forests in North America (Hall and Bruna 1983; Haight et al. 1985; Haight 1987; Bare and Opalach 1988; Gove and Fairweather 1992; Anderson and Bare 1994, and others). These optimization models include many assumptions, much like studies of relative productivity (see Chapter 13). If the objective is to create or maintain a predetermined stand structure, then there have to be assumptions about species composition, cutting cycle length, and the initial condition or starting point for the analysis.

12.6 Synthesis

Quantification of growth processes for complex stands is essential for their management and for understanding stand- and ecosystem-level processes. Projecting change in multiaged stands has been hindered by limitations in tools to assess site quality, which determines the rate of change in stand structure and volume production. Forest site productivity estimation has largely been based on dominant height growth, a concept with limited application in multiaged stands. Additional research in this area is needed.

Historically, growth projections relied on empirical approaches using large amounts of data to develop reasonably accurate estimates of stand growth and production. However, they are limited by available data and are not robust when accommodating unusual management strategies or changing environmental conditions. These empirical approaches also perform best in even-aged, single-species stands and are less effective in more complex stands. Although these tools have the capability to project stands with a variety of stand structures, many assume the negative exponential diameter distribution as a standard. Many of these stand-level projection tools were developed in North American northern hardwood stands, where plentiful regeneration results in negative exponential diameter distributions.

Developments in individual-tree models provide tools for both management and research. These models range from the empirically based to models based on mechanistic processes in forest ecosystems. The capability of these models—particularly the distance-dependent models—to represent complex processes and interactions between stand elements makes them well suited for modeling complex stands. However, requirements for spatial data for distance-dependent models and the general model complexity of individual-tree models may limit their usage.

Empirical approaches are limited to the past conditions from which the data they are based were derived. This generally limits projections to previous experiences where data exists. Mechanistic models offer the potential to look beyond existing data to explore alternative stand structures or the effects of climate change (Kimmins 1990; Bossel 1991). Peng (2000) described the weaknesses of empirical models as strengths for mechanistic models, and vice versa. These criteria included applicability to management or research, the number of parameters, and other criteria. Hence hybrid approaches that combine process and empirical models (e.g., Kimmins 1990) may offer opportunities to link the strengths of both approaches into useful stand projection systems.

Optimization tools have been used to find optimal stocking regimes for volume or economic production in multiaged stands. Much of the impetus for greater use of complex stand structures is to meet resource demands that are not as easily quantified as wood volume or economic production.

Box 12.1 Optimizing transformation

Operations research in forestry has attempted to identify optimal stand structures for multiaged stands. Because these studies focus on an optimal structure, they attempt to meet a relatively static condition. Another management objective is transformation of an even-aged to a multiaged stand (see Chapter 11). Rojo and Orios (2005) developed a decision support system to maximize net present revenue among options for transforming even-aged maritime pine to multiaged structures in northwestern Spain. The decision support system used a transition matrix growth model and nonlinear programming.

The transformation begins from an even-aged stand and moves toward a target multiaged structure previously identified by Orios et al. (2004). Thinnings are used to reduce density and regenerate new age classes, although trees less than 15 cm diameter were not modeled (Figure 12.5). The decision support system solutions were sensitive to the interest rate used. Higher interest rates led to solutions with heavier cutting in early thinnings whereas with an interest rate of zero the thinnings were late in the transformation process. The analysis used a target structure represented by a basal area of 17 m^2/ha. However, the system found that there were several different stand structures which yield similar benefits. The decision support system was also capable of varying the initial stand structure, the target stand structure, the thinning treatments, and the economic constraints to test alternative transformation systems.

Analyses to develop optimal stand structures or management regimes can include multiaged as well as traditional even-aged applications. As with other studies (Buongiorno 2001; Hanewinkel 2001; Knoke and Plusczyk 2001), Rojo and Orois (2005) found the transformation process to be economically viable. This is apparently due, in large part, to the earlier harvests associated with transformation regimes that are usually attempting to regenerate new age classes earlier than when many even-aged stands would be generating income (Knoke and Plusczyk 2001). Although many non-commodity values are difficult to quantify, similar tools can assess the ability of a transformation regime to achieve objectives other than optimizing a stand structure or maximizing returns. A key component of all these analyses, regardless of the objective, is a stand projection tool.

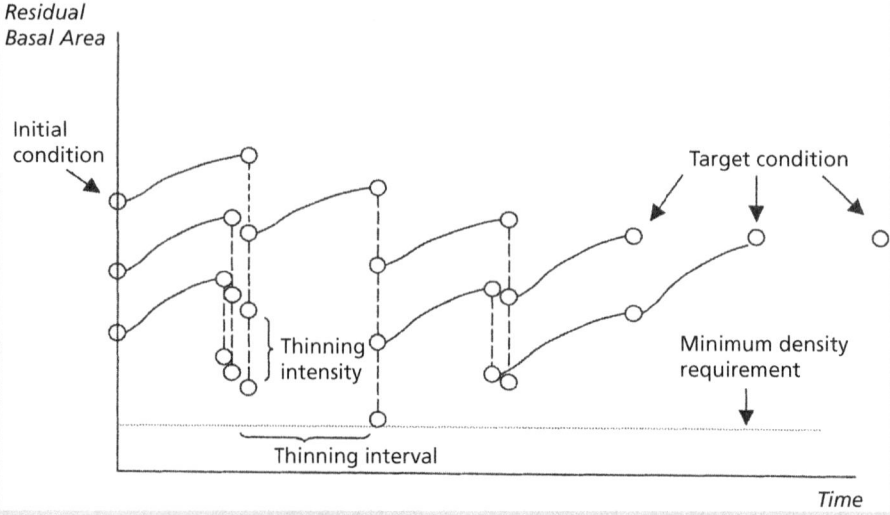

Figure 12.5 Potential solutions from an initial condition of an even-aged stand to the target condition of a multiaged stand. Various thinnings of different intensity or intervals between thinnings were compared as alternate pathways to achieve the target stand structure (from Rojo and Orois 2005; reproduced with permission from Elsevier).

Instead, the same factors motivating the use of mechanistic modeling approaches for unique stand structures or to anticipate effects of climate change are requiring increasingly complex tools. These tools have a role in assessing forest management options. However, on many of the forest lands where managing for complex stand structures will be a dominant objective, managers may be making on-the-ground decisions based on the presence or absence of structural features such as snags, multiple species, or unique spatial patterns. The desire to retain or create these features at the individual-stand level will create unique management decisions at each stand that cannot be easily quantified, modeled, or optimized on an individual-stand basis.

CHAPTER 13

Volume and economic production of multiaged stands

13.1 Introduction

The relative productivity of multiaged stands is an old question that dominates many discussions of alternative silvicultural systems. Forest managers are often concerned about the effects of multiaged stand management strategies on volume production and economic returns. Forest scientists are often interested in the causal mechanisms that affect production in multiaged stand structures. The decision of which system to implement involves many factors that have little to do with volume production or even economic production. There has also been a tendency to compare production of multiaged stands with even-aged stands. This is a natural comparison that benefits our understanding of multiaged stands, since there is much more information available on even-aged stands. There have been a number of useful interesting theories on this topic but relatively little data to support any conclusions. For example, Bourne (1951) hypothesized that there was an advantage to multistrata stands because resources could be shared with more efficient utilization of resources. A similar "advanced growth effect," where a new age class of trees that started as advanced regeneration provided a production advantage, was described by Smith et al. (1997). On the other hand, Assmann (1970) viewed the slow growth of understory trees as evidence of reduced production in multiaged stands. These are examples of the disparate interpretations on this issue.

O'Hara and Nagel (2006) described the volume growth comparison as two faceted, with one aspect being the differences related to stand structure and the other related to differences in management

operations. Multiaged silviculture–and even-aged silviculture–involves both the resultant stand structure and the operations needed to produce it. Both of these considerations will affect production, and both are integral to the question of comparing diverse management regimes. Although stand structures and the forest operations that create them may appear to be separate issues, they are closely interrelated in managed stands and demonstrate the general difficulty of assessing production. Economic comparisons are based on either empirical or modeled growth projections, with assumptions related to specific operations and their timing.

There are many ways to measure production and productivity of forests, and all are important in different ways. For multiaged stands, nearly all previous analyses have focused on stem volume production. These include both empirical studies as well as modeling both volume production and economic returns. This chapter presents a number of facets of volume and economic production in multiaged stands, including basic production relationships, results from empirical studies, modeling studies, and implications for management.

13.2 Stand-level volume production relationships

Production in multiaged stands is governed by the same limitations/constraints as in even-aged stands. Although production of multiaged stands has been studied far less than even-aged plantations or natural stands, we can draw on this knowledge of even-aged stand production to make inferences regarding more complex structures. Common terms that are applicable to both structures include **gross primary production** (GPP), which is the total assimilation of energy and nutrients by a stand, and **net primary production** (NPP), which is gross primary production minus the production used in respiration. Generally, foresters are interested in NPP, as it is more representative of what the stand can produce. However the relationship between GPP and NPP is important in order to understand stand-level productivity.

Cannell (1989) reviewed the functional basis of wood production of even-aged stands. The gross production and net production of even-aged stands are often represented as in Figure 13.1. Gross production is approximated by the total leaf biomass or leaf area index (LAI) of the stand because stand-level photosynthesis is largely a function of the total amount of foliage. LAI is usually assumed to develop rapidly after stand initiation until reaching a maximum, after which it may be stable or decline. The plateau represents an upper limit of LAI—and therefore gross production—that is related to the inherent productivity of a site. Although a plateau is often depicted as the shape of this relationship as a stand ages, the relationship between LAI and stand age has also been represented with a declining relationship (Figure 13.2; Vose et al. 1994). In either case, there are limits on maximum LAI related to site productivity.

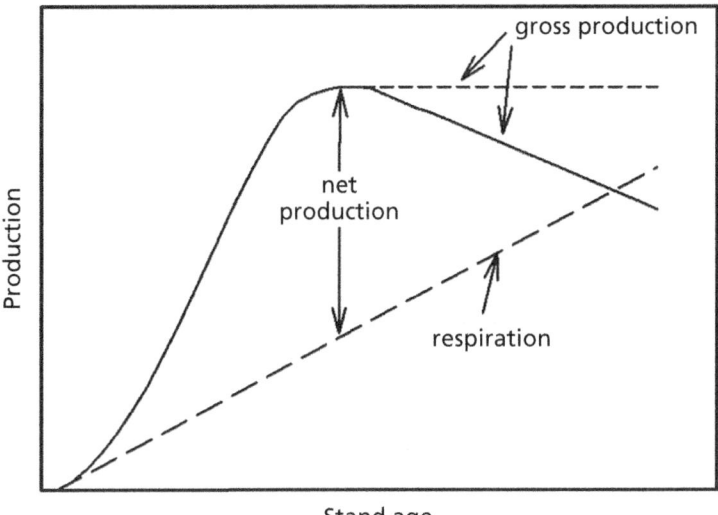

Figure 13.1 Gross and net primary production relationships for idealized even-aged stands (after Kira and Shidei 1967 and Möller et al. 1954). In this diagram, gross production is shown following two hypothetical pathways: one where the leaf area index remains constant and another where it declines after reaching a maximum. Respiration increases and will exceed gross production with either hypothetical pathway. Net production is the difference between gross production and respiration.

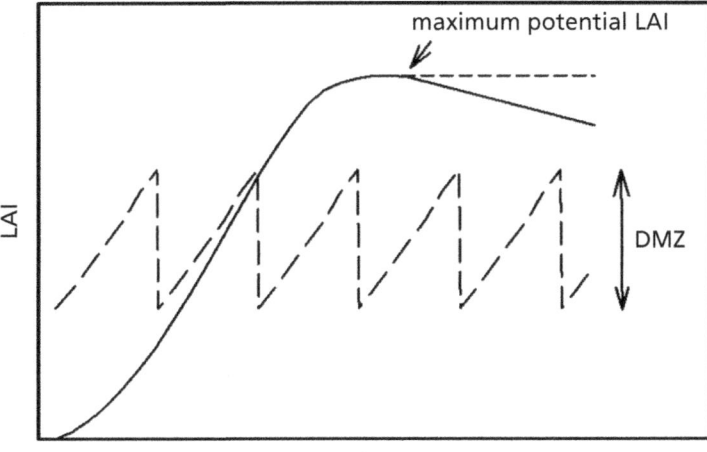

Figure 13.2 Density management zone (DMZ) for a hypothetical managed multiaged stand overlaid on a diagram of potential leaf area index (LAI) for an even-aged stand. The managed multiaged stand exists below the maximum LAI to avoid mortality or stand health problems, or to increase individual tree sizes. Managed even-aged stands are also usually managed well below the maximum potential LAI. Note the two even-aged LAI trajectories correspond to the two gross production hypotheses in Figure 13.1.

Respiration is the net carbon loss from leaves and other living tissues in a plant or in a stand. Leaf respiration is proportional to total leaf biomass or LAI and therefore follows a similar trend as LAI (Figure 13.1). Respiration of other tissues, such as stem and branch wood, roots, twigs, etc., increases as the amount of these tissues increases during stand development. The total respiration of an even-aged stand therefore increases as a stand ages and is often depicted as eventually reaching a level comparable to gross production. In those stand structures–which are often described as old growth, "climax," or simply old forests—net production may approach zero (Figure 13.1).

For foresters, it is usually stem volume production that is the production variable of interest. Above-ground biomass is also becoming an increasingly important consideration because of whole-tree harvesting for energy production. Additionally, the interest in carbon sequestration considers the potential for stored carbon both above- and below-ground in forest systems. For all of these objectives, stands will usually be managed to maintain less than maximum LAI to reduce susceptibility to insects and pathogens and to increase average tree vigor. A resulting LAI trend for even-aged stands over time is shown in Figure 7.5. This managed stand regime also shows the effect of a series of thinning treatments over time. Whereas natural stands would typically approach the maximum for LAI and follow the production relationships shown in Figure 13.1, managed stands are likely to follow similar patterns but at less than maximum LAI (Figure 13.2). This reduces mortality, and susceptibility to insects and pathogens.

Similar relationships govern GPP and NPP in multiaged stands. However, unlike even-aged stands, there is no beginning or end to multiaged stand development. As a result, the production relationships do not build from zero as represented in Figure 13.1. Instead, under a natural disturbance regime where a disturbance such as fire or windthrow periodically occurs, LAI would climb to a maximum level and then fall with each disturbance, which would be followed by a rebuilding period (Figure 13.3). Under an idealized set of conditions, a managed stand would fluctuate through a series of cutting cycles. During these cycles, LAI increases and drops from periodic harvests and thinnings (Figure 13.3). The difference in these two multiaged patterns is the managed stand would be maintained at less than maximum LAI and would have similar cutting cycle lengths over time, whereas the unmanaged stand would probably have higher LAI with disturbances occurring at variable intervals and severities. Leaf respiration follows a similar pattern as LAI in both managed and unmanaged stands. Other forms of respiration would correspond to the amount of living wood biomass on the site and would fluctuate with each cutting cycle.

Management objectives are generally concerned with production of stemwood volume or biomass.

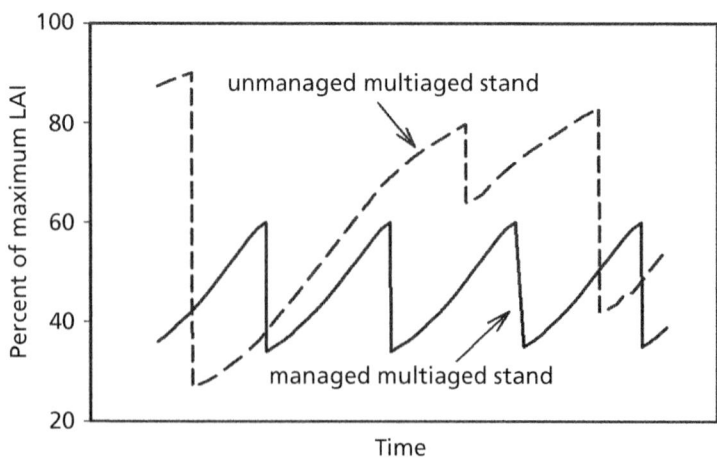

Figure 13.3 Leaf area index for managed and natural multiaged stands. The managed stand fluctuates within a density management zone but the natural stand may approach the maximum, where periodic disturbance may lower it to variable levels.

These objectives are usually achieved when net production—or the difference between gross production and total respiration—is maximized. Measures of growth efficiency or growing space efficiency can be used to express these relationships as the ratio of volume increment to stand LAI (Section 7.2.1; Waring 1983). An efficient stand is therefore one where respiration is minimized. Although these measures of efficiency will often correspond to measures of stand production, they may also vary considerably. For example, a stand with a LAI that is much lower than the maximum potential LAI might be very efficient, but its production might be low because GPP is low. Stand production is therefore a function of GPP and respiration, and management objectives indirectly affect these processes through the design of the stand structure.

13.3 Volume production of even-aged and multiaged stands

Stand volume production varies tremendously because of differences in species compositions, site productivity, management intensity, and other factors. The variation associated with stand structure—from multiple age classes or canopy strata—simply adds another layer of complexity. Nevertheless, given the current management trend toward greater stand complexity, the volume production of multiaged stands is an important consideration.

The common international measure for stand-level volume production data is typically the unit-area measure m^3 ha^{-1} $year^{-1}$. For even-aged stands, the mean annual increment (MAI; total increment divided by stand age) of a mature stand is the best measure because it includes all phases of stand growth from the period of stand establishment through the period of rapid volume accumulation. For multiaged stands, the average volume increment over a full cutting cycle or series of cutting cycles would be a comparable measure. This could also be expressed as the **average annual yield** (AAY; Guldin and Baker 1988). The values in Table 13.1 provide examples of the vast range in potential volume production of forest stands. Considerably higher production levels have been reported in plantations and other even-aged stands (Evans

Table 13.1 Comparisons of published studies showing production rates for both even-aged and multiaged (or even-sized and all-sized) forests.

Species	Location	Management/ Structure	Production (m^3/ha/year)	Source
Conifer	Finland	All-sized	5.2	Lähde et al. 1994a
Conifer–broadleaved		All-sized	5.5	
Conifer		Even-sized	5.4	
Conifer–broadleaved		Even-sized	5.7	
Scots pine	Finland	All-sized	3.8	Lähde et al. 1994b
Conifer–broadleaved		All-sized	4.1	
Scots pine		Even-sized	3.6	
Conifer–broadleaved		Even-sized	3.5	
Ponderosa pine	Oregon, US	Even-aged	4.8	O'Hara 1996
		Multiaged	4.9	
Ponderosa pine	Montana, US	Even-aged	5.0	O'Hara 1996
		Multiaged	5.4	
Lodgepole pine	Montana, US	Even-aged	7.9	Kollenberg and O'Hara 1999
		Multiaged	9.0	
Mixed fir–spruce–beech	Central Europe	Even-aged	10.3	Assmann 1970 (Table 150)
		"Selection"	8.7	
Mixed fir–spruce–beech	Switzerland	Even-aged	10.2	Köhl and Baldauf 2012
		"Uneven-aged"	8.5	
		"Selection"	6.8	

and Turnbull 2004; Jokela et al. 2010; Jones and O'Hara 2012). This is largely because the fastest-growing species are grown in intensively managed even-aged plantations. These plantations also often include species planted outside their native range. Rarely have multiaged stands received such intensive management, and it is uncommon for them to include nonnative tree species.

13.3.1 Empirical comparisons

The relative production of stands with complex age structures versus even-aged stands has been a frequent question in forestry. Although the question is simple, answering the question with empirical field trials is very difficult (Hanewinkel 2002; O'Hara and Nagel 2006). A comparison of even-aged MAI vs. periodic increment or AAY would therefore provide an indication of which management regime was more productive. For a valid comparison, there would have to be long-term plots of both systems that have been managed with a consistent silvicultural regime. Comparable species composition and site quality would be required for the two systems. There are also difficult decisions over what other management treatments are necessary. For example, would regeneration be planted or natural? Would thinning treatments be used? How would equivalent adjustments in density through thinning be performed in the two systems? What rotation length and cutting cycle length would be optimum and comparable for the two management regimes? Holding these variables "constant" over such diverse management regimes would be difficult even under the most ideal conditions. Finally, sufficient sample sizes are needed for meaningful statistical comparisons (O'Hara and Nagel 2006).

Most attempts to compare even-aged and multiaged production have been empirical studies. Some of these date back into the nineteenth century, indicating a long-term interest in resolving this question. Assmann (1970) summarized comparative studies to date in central Europe and concluded that even-aged stands were more productive. However, his interpretation was contrary to the conclusions of some of the studies he cited. For example, Assmann reworked the results of Leibundgut (1945), Mitscherlich (1952), and others to show greater production in even-aged stands.

Hanewinkel (2002) also summarized the central European literature and concluded that nearly every study found greater production in multiaged stands. Laiho et al. (2011) reached similar conclusions from summarizing previous work in Finland.

In North America, Smith and DeBald (1978) discussed previous work from mostly broadleaved stands from the eastern United States and concluded that uneven-aged stand structures had lower volume production than even-aged structures. In mixtures of shortleaf and loblolly pine in the southeastern United States, Cain and Shelton (2001) compared a clearcut, seed-tree regime with an uneven-aged regime and found no differences in production over a 53-year study period. In a separate 36-year trial, an even-aged regime produced more cubic volume than a comparable uneven-aged regime. However, when volume production was measured in board feet—a common measure of wood utilization in the United States—uneven-aged stands were more productive (Guldin and Baker 1988). Thus, units of measurement as well as the standards for volume measurement can have a major effect on interpretations regarding production.

Another approach is through large data sets such as forest survey or national inventory data that include both even-aged and multiaged management regimes. The difference in stand volume from one measurement to the next provides a measure of production. These comparisons also require the same assumptions as smaller scale empirical studies, including comparable site qualities, levels of occupied growing space, and species composition. For a valid comparison, these data must also include a representative sampling of even-aged stand ages and stands at different points in the multiaged cutting cycle. Using inventory data from Finland, Lähde et al. (1994a, b) compared even-sized and uneven-sized mixed stands and stands dominated by Scots pine and found greater productivity in the uneven-sized stands. Although these characterizations of size structure were probably correlated with age structure, the important variable of age structure was not measured in these inventories. Long and Shaw (2010) completed a similar analysis using broad-scale inventory data with ponderosa pine in western North America. Other large-scale inventory data include similar drawbacks related to characterizing age structure; however, they can provide useful comparisons of management approaches that are in widespread use in a particular region.

13.3.2 Functional comparisons

Another approach to the question of production would be a scaled-down examination of differences in resource use or other functional differences between even-aged and multiaged stand structures. O'Hara and Nagel (2006) summarized a series of studies on ponderosa pine to evaluate functional productivity issues. They found no differences in leaf nitrogen or photosynthetic capacity, but multiaged stands had less water stress during dry summer months. The ratio of stem volume increment to leaf area was used as a measure of stand-level efficiency. By comparing even-aged and multiaged stands, these analyses indicated which stand structure used LAI—a measure of growing space occupancy—more efficiently. There were few differences in growing space efficiency (ratio of stem growth to LAI; see Chapter 7) for the two stand structures, with multiaged stands being slightly more efficient. However, a similar comparison with lodgepole pine found that the even-aged stands were more efficient (Kollenberg and O'Hara 1999). Differences in these studies may have been the result of sampling but may also represent differences in how different species perform in different stand structures.

Individual tree efficiency comparisons yield other interesting trends. Older age classes in ponderosa pine show increases in both rate of volume increment and growing space efficiency (O'Hara 1996). Although a greater increment of age classes in upper canopy positions is generally expected, the greater growing space efficiency implied that upper canopy ponderosa pine trees were growing faster, and growing faster per unit of occupied growing space, than lower canopy trees (Figure 13.4). This indicates that the organization of trees in a stand—or the stand structure—can have a large influence on productivity. This was demonstrated by O'Hara (1996) by combining different numbers of trees in four-age class stands, with productivity varying by as much as 380% while keeping total growing space occupancy constant.

Whereas the ponderosa pine example showed greater growing space efficiency in older age

Figure 13.4 Efficiency of age classes in multiaged stands. O'Hara (1996) found older age classes were most efficient in ponderosa pine in the western United States. Seymour and Kennefic (2002) found the opposite pattern in eastern hemlock/red spruce stands in the northeastern United States.

classes, this pattern may vary by species. For example, in eastern hemlock and red spruce, Seymour and Kenefic (2002) found a declining pattern of growing space efficiency with increasing age when crown size was held constant. Differences in growing space efficiency between age classes, strata, or crown classes imply that these components could be rearranged in combinations, or stand structures, that affect stand production for ponderosa pine. However, the reduced efficiency in older eastern hemlock and red spruce trees indicates that the stand structures with the greatest volume production would be those with fewer older trees, in contrast to the results with ponderosa pine (O'Hara 1996). If an ideal multiaged stand structure existed, it would probably be specific to certain forest types.

Theoretical views related to productivity include those of Bourne (1951), who speculated that the overlapping generations of age classes in multiaged stands provided unique efficiencies. Kenk and Guenhe (2001) advanced the same idea for mixed forests in central Europe. Assmann (1970) thought the vertical layering of selection forests reduced efficiency. He speculated that trees had wide crowns in the overstory and were inefficient in the understory. This may be true in some forest types but not in others. Variable growth rates or efficiencies of different age classes could create compensatory effects where poor growth rates in one age class would be compensated by increased growth in other age classes. For example, Seymour and Kenefic (2002) found greater efficiency in younger age classes, and O'Hara (1996) found greater efficiency in older age classes (Figure 13.4). In either case, the greater contribution to growth by the more efficient stand components may compensate for the less efficient stand components.

Following a harvest treatment in a multiaged stand, residual trees may exhibit a variety of responses, and there may be interactions with different components of the residual stand. Compensatory effects may include changes in growth rates or changes in mortality rates. This may be most apparent following establishment of group openings which regenerate a new age class of trees, but may also have an effect on the residual trees surrounding the opening. For example, York and Battles (2008) reported increased radial increment in residual trees surrounding group openings in mixed-conifer forests in the Sierra Nevada of California, US. Brodie and DeBell (2013) found increased volume increment in overstory trees in two-aged Pacific Northwest conifer stands. These results suggested that the temporary loss in production from the group as it regenerated was at least partially compensated by the growth of surrounding or overstory trees. Yamashita et al. (2006) found complex interactions between height growth of planted sugi in group openings and surrounding trees in Japan. Interactions in the opening varied depending on location. In beech forests in New Zealand, Wiser et al. (2005) and Hurst et al. (2012) note that compensatory effects could include positive or negative changes in growth, mortality, or even crown architecture, which could have strong effects on natural regeneration. These compensatory effects are difficult to measure, as they affect stands in a variety of different ways. As a result, they complicate growth comparisons between multiaged and even-aged structures, because effects include so many different parameters besides diameter or height increment.

13.3.3 Modeling comparisons

There is generally a shortage of data for empirical comparisons of volume production between stand structures. As a result, stand growth simulation models have been used to generate stand production data for comparison between stand structures. These production data can then be used for optimization studies and economic comparisons. One of the first was the work of Hasse and Ek (1981) with northern hardwoods in North America. They found comparable production of total stem volume but greater production for merchantable cubic volume in uneven-aged stands. Haight and Monserud (1990) found comparable production between even-aged and uneven-aged mixed-conifer stands in western North America. Tahvonen et al. (2010) found greater production in even-aged stands in Finland. Results from these studies are mixed in part because species and forest types are expected to exhibit variable responses to management. The management regimes included in the models will also vary. Finally, growth projection models can assume a variety of forms (see Chapter 12) and include a variety of assumptions on how stands grow and respond to treatments. Not surprisingly, there can be differences in the empirical results and modeled results, such as between the greater multiaged productivity in Finland from empirical work (Läiho et al. 2011) and the model projections (Tahvonen et al. 2010) for the same forest types. Although there are other studies that have modeled relative production, many of these have been part of economic comparisons that are discussed in the next section.

13.4 Economic productivity

Many decisions over silvicultural methods are based on economic comparisons rather than comparisons of volume production. Economic comparisons require estimates of volume production, target stand structures, and operational costs. They do not, in general, factor in other ecosystem services. Virtually all of these analyses have used models to project development of even-aged and multiaged stands, and these models then become the basis for the economic comparisons of the management regimes. Hanewinkel (2002) noted that basing comparisons on models requires that both the multiaged and even-aged models represent the goals of the decision maker. Analyses of operations generally indicate greater costs with harvesting treatments in more complex stand structures (Kluender et al. 1998). This is due to the greater difficulty in removing trees or logs in the limited spaces between residual trees. These harvest treatments also result in greater potential for worker injury (Fjeld and Granhus 1998), which may indirectly affect costs. Another key differentiating feature of the two types of systems is the revenue stream, which is generally a series of payments that are smaller and more frequent for the multiaged system, and larger and less frequent for the even-aged system.

Assessing the economic potential of different management regimes is often developed as an optimization problem where the net present value or land expectation value is maximized for an infinite series of cutting cycles for multiaged stands or rotations for even-aged stands. Results are highly sensitive to the inputs (e.g., costs, growth rates, mortality functions, etc.) and to the objective functions used. Results indicated a range of responses. For example, studies that indicated multiaged stands were more productive include one by Haight (1987) for ponderosa pine in the southwestern United States, whereas others, such as studies by Hanewinkel (2002) in central Europe and Tahvonen (2011) with Norway spruce in Finland, indicated that even-aged stands were more productive. Many of these analyses did not make comparisons with even-aged stands but were based on analyses of multiaged stands with varying assumptions. These include studies by Adams and Ek (1974), Chang (1981, 1990), Hall (1983), Haight et al. (1985), Bare and Opalach (1988), Buongiorno et al. (1994), Hanewinkel (2002), Tahvonen (2011), and many others.

By and large, these economic studies are not as novel for their assessments of the relative economic productivity of different management regimes as they are for the analytical procedures used in the analyses. These analyses may inform management decisions, but the inherent assumptions necessary to employ these empirical models make them insufficient to base large-scale changes in fundamental management approaches. There may also be a disconnect between what is identified as economically optimal and what is sustainable (López-Torres et al. 2013).

Although there are exceptions (Buongiorno 2001; Knoke and Plusczyk 2001; Hanewinkel and Pretzsch 2000; Tarp et al. 2005), many studies have compared the established even-aged stand to the multiaged stand without the consideration of the costs of transformation. Most studies noted the importance of the starting position for the growth projection process (Haight and Monserud 1990). A large flux of income from a harvest treatment soon after the start of the analysis provides a heavy positive influence on the returns, much like a delayed harvest will have a much smaller effect due to discounting of cash flows.

Results often indicate a preferred solution under some assumptions and a different solution under others. For example, Orois et al. (2004) found the best solution for maritime pine in Spain depended on site index, with even-aged stands being superior on better sites. Likewise, Tahvonen (2011) found multiaged systems were preferable for Norway spruce in Finland when interest rates were higher. Kuuluvainen et al. (2012) provide a summary of similar studies in Fennoscandia and their various results.

13.5 Synthesis

Comparing the relative productivity of multiaged and even-aged stands is a complex problem burdened with a multitude of assumptions and variables. The principle objective in pursuing these alternative management regimes is generally to achieve two different stand structures. However, achieving the different stand structures requires a different set of management operations, which affects subsequent production. Hence, separating the outcome of different stand structures from the means of achieving these outcomes is problematic in itself. In addition, controlling such variables as species composition, growing space occupancy, timing of operations, and management intensity complicates any comparison.

Many studies report greater productivity in even-aged stands, and many others report greater productivity in multiaged stands. Some differences are likely related to species compositions, measurement units, or study methodology. In other cases, real differences probably do exist. However, a universal truth that one management regime is more productive than another is implausible. Instead, there may be ecosystems where one management regime is more productive than another, and other ecosystems where the opposite is true. In most cases, differences in production will probably be small and relatively insignificant in relation to the effects of operational differences in these management regimes. Because these operational differences may lead to different cost structures, economic comparisons may indicate advantages to one regime over another beyond those based purely on production.

From a purely functional comparison, either system will receive similar inputs of light and precipitation and will utilize the soil in similar ways. Inherent productivity would therefore be about the same regardless of structure. Ultimately, management efficiency may be the variable that provides advantages to one system or another. The simpler management of even-aged systems may therefore be a dominant factor.

Likewise, there may be other resource values, such as wildlife habitat or esthetic values, that dominate the decision process rather than simply production or economic returns. This is particularly true for complex stand structures, where the motivation for their implementation may not depend on economics or volume production, but instead will be based on the production of other ecosystem services.

CHAPTER 14

Genetics and multiaged silviculture

14.1 Introduction

Many silvicultural practices have inherent genetic effects that need to be understood and assessed when designing silvicultural operations (Konnert et al. 2007). Multiaged silviculture practices are no exception and have often been cited as potentially having more detrimental genetic effects than even-aged silviculture (Gibbs 1978; Howe 1989). Ironically, multiaged systems are labeled as **dysgenic**, or having negative genetic effects, while they are also perceived as more natural in other contexts. This contradiction is likely because multiaged systems can be highly variable depending on the species involved, their reproductive processes, spacing of sexually mature trees, cutting cycle lengths, and other factors. However, another contributing factor is the traditional view of what is involved in multiaged—or more specifically uneven-aged—silviculture and what constitutes a multiaged stand. Concerns related to genetics have arisen from two areas: 1) stands developing from multiaged systems may have reproductive limitations that hinder regeneration and reduce genetic diversity; and 2) the practice of removing larger trees from a stand is assumed to be dysgenic. Dysgenic harvesting practices in silviculture are one possible effect of **high grading**, or **creaming**, which refer to practices where the most valuable trees are removed. High grading often has genetic, economic, and stand health implications (Helms 1998). This chapter explores multiaged systems and their implications for genetic diversity and population viability.

14.2 Genetic conservation and silviculture

Many forestry activities involve preferences for, or selection of, trees to increase vegetative growth or to meet some other objective. When these preferences affect subsequent reproduction, they are a form of **artificial selection** that directs the stand toward trees that exhibit the preferred characteristics. **Natural selection** is generally assumed to maximize total reproductive fitness in the form of fertility and seed viability (Savolainen and Kärkkäinen 1992). This difference between artificial and natural selection has implications for genetic diversity in forest populations. The concern for multiaged regeneration methods, and silvicultural regeneration methods, is that a reduction in number of breeding-age parents may create a localized population bottleneck resulting in the loss of genetic diversity. For example, regenerating a stand from only a few parents is a form of bottleneck because only a small number of individuals may pass their genes to offspring. This may lead to the loss of some alleles or a form of localized genetic drift. **Genetic drift** is the random loss of a gene variant or allele that may occur in small populations. If a regeneration method was used in many stands over a large area and favored some traits either intentionally or not, the effects on large-scale genetic diversity could be significant.

Genetic diversity is important for forestry because: 1) it affects vulnerability to pests and short- or long-term climatic fluctuations; 2) genetic variants are important for future adaptability or breeding; and 3) potential reductions in stability of entire forest ecosystems may result if genetic diversity is reduced (Ledig 1986). Forest gene conservation is therefore very important for maintaining productivity and adaptability to change (Hughes et al. 2008). The effect of human activities on forest gene pools has been evident for several centuries. Most notably, preferential cutting of the straightest trees has affected the tree form characteristics of the residual trees. Examples of this include the selective cutting of Lebanon cedar in the Middle East, pitch pine in eastern North America, and mahogany in Central America (Makkonen 1967; Styles 1972; Ledig and Fryer 1974). If human selections can inadvertently alter genetic traits, there is great potential to also affect "unseen" traits or to simply reduce genetic diversity.

The importance of favoring desirable phenotypes is well recognized in modern silviculture (Smith et al. 1997; Nyland 2002). Foresters generally consider the perpetuation of desirable phenotypes when designing regeneration methods, thinning methods, and in seed collection/production systems. Foresters are encouraged to leave desirable trees to serve as parents for the next generation or to collect seed from parents that are desirable phenotypes. The goal has been to maintain or upgrade the gene pool with each generation of trees. Breeding-based tree improvement programs represent a higher level of tree selection. Selection in tree improvement programs is often geared toward economic traits or tree health but also generally includes maintenance of a broad genetic base for long-term breeding and gene preservation (Zobel and Talbert 1984; White et al. 2007).

A cornerstone of ecosystem management and related paradigms is using natural processes to guide management. For silviculture, these processes include natural disturbances as a guide for regeneration treatments. Natural disturbance processes are highly variable from one forest ecosystem to the next. They also vary over time. In some ecosystems, a stand-replacement disturbance regime may be dominant, whereas in another, smaller openings may be more common. Even-aged and multiaged stands can result from these two extremes of disturbance regimes. Variations in timing, severity, or frequency of disturbances interact to create a large range in post-disturbance stand structures. If silvicultural practices resemble these natural processes, then it is likely that native genetic variation is maintained.

The **historical range of variability** (or historical range and variability; Keane et al. 2009) is a useful concept for defining a reference set of conditions or benchmarks for management. It defines a set of ecological and evolutionary conditions—including natural disturbance regimes—for a given area that help provide the basis for sustaining biological diversity (Morgan et al. 1994; Landres et al. 1999). Hence, maintaining systems within a range of natural variability provides for the conditions necessary to maintain genetic diversity. The general assumption is that by maintaining stands within a range of preexisting conditions, the preexisting diversity will be maintained. However, this process is dependent on a climate that fluctuates within a range of conditions representative of the reference period. In a period of rapid climate change, these reference conditions may provide little insight into ecosystem reactions to disturbances, and a more thoughtful approach may be needed. For example, Millar et al. (2007) proposed adaptive strategies that included maintenance of system resistance, resilience, and management to assist development of ecosystems in transition stages. We can anticipate that natural environments may experience some predictable changes, and we can attempt to direct stands on trajectories toward those anticipated changes. This becomes a **future range of variability** that we can manage for, rather than the historic range of variability of past conditions (Thompson et al. 2009).

A critical component of genetic conservation in forests is that silviculture should either resemble natural disturbance processes or provide for other mechanisms to supplement genetic diversity. For example, artificial regeneration is a common means of supplementing genetic diversity. Planted seedlings generally involve transfer of genetic material that may increase local genetic diversity (Adams et al. 1998) but not necessarily adaptedness (see Schoppa and Gregorius 2001).

At larger scales, natural disturbance patterns may be used to guide organization of stands on a landscape. For example, the size of regeneration units, or their proximity to each other, might be based

Figure 14.1 Burned stand in California showing low numbers of surviving trees of reproductive age.

in some way on patterns of windthrow or fire. This has genetic implications, because large-scale stand-replacement disturbances tend to increase distances between surviving trees (Figure 14.1). For example, the 1910 fires in the inland Northwest of North America burned an estimated 1.2 million hectares and included many contiguous areas of stand-replacement disturbance (Arno and Allison-Bunnell 2002; Egan 2009).

Alternatively, there may be a tendency to assume some natural systems result from only very light disturbances. For example, natural forest management (also known as NFM; see Bawa and Seidler 1998) attempts to manage forests to achieve sustainability and preserve biotic integrity by using light infrequent harvests. This system is primarily used in tropical moist forests through the use of some form of selection systems. Although it might be a "light" treatment, the genetic implications of this system remain unexplored, and the population sizes and the structure of genetic variation maintained in those systems could affect the viability of these systems (National Research Council 1991). However, achieving genetic conservation goals through emulating natural disturbances requires more than simply applying light treatments to stands. The context of these disturbances and their effect on regeneration and genetic conservation is also critical.

14.3 Reproductive processes

Reproductive processes are highly variable in multiaged stands and affect genetic variability. Both pollination and seed dispersal are important because they are the primary dispersal processes of genetic information. Pollination dynamics vary by species. For example, some species are wind-pollinated, whereas others are pollinated by insects or animals. Additionally, a species that occurs with great abundance in a stand may be exposed to higher levels of wind-blown pollen than a species with low abundance (Koski 1970). Wind-blown pollen can also spread over great distances under mild wind conditions. Hence, for wind-pollinated trees that make up a large segment of a stand, a high proportion of pollen may come from other trees from the same stand or surrounding stands. In insect-pollinated angiosperms, gene flow may also be substantial, but within-stand pollination may primarily be between the nearest flowering trees (Adams 1992; Di-Giovanni et al. 1995; White et al. 2007). For tropical strangler figs, Nason et al. (1998) found insect pollination over distances of 14 km. Hence, the potential for gene flow in both wind- and insect-pollinated trees is strong, with the possible exception of less-common trees.

Pollen is dispersed in the greatest volume near the parent tree but with a long tail that may be measured in kilometers (Figure 14.2; Bohrerova et al. 2009). If sexually mature trees are spatially dispersed, the probability of selfing may increase. **Selfing** occurs when plants self-pollinate and results in greater numbers of homozygous loci and greater expression of deleterious recessive genes, both of which generally lead to poor seed formation and germination, abnormal growth, and poor competitiveness (Zobel and Talbert 1984). Fortunately, most trees are almost completely outcrossed, with a few exceptions (Mitton 1992). Many species have adaptations to avoid selfing, such as timing of male and female flowering or the position of male and female flowers in a tree crown. Others, such as species in the ash, juniper, and cottonwood families, are dioecious and have separate male and female trees. Still others are "self-incompatible" or seed embryo development is aborted due to the expression of homozygous lethal genes (Seavey and Bawa 1986; Ledig 1986). Inbreeding presents a variety of potential problems to plants as well. **Inbreeding** results in reduced genetic variation, high levels of relatedness, and a decline in general fitness or vigor. With a greater proportion of homozygous gene combinations, there is a greater potential for the expression of undesirable traits or deleterious genes. However, inbreeding and relatedness do occur naturally and are found in natural stands. For example, neighborhood patterns of related trees may exist in a stand where trees in other neighborhoods may be only distantly related (Coles and Fowler 1976; Tigerstedt et al. 1982). Mixed-species stand structures have greater average pollination distances because breeding trees are more likely to be dispersed (Figure 14.3). This is most pronounced in many tropical forests, where tree-species diversity is high, but is also true in many temperate forests as well. As a result, there may be greater potential for inbreeding for a given species in a mixed stand than for the same species in a pure stand, independent of stand age structure.

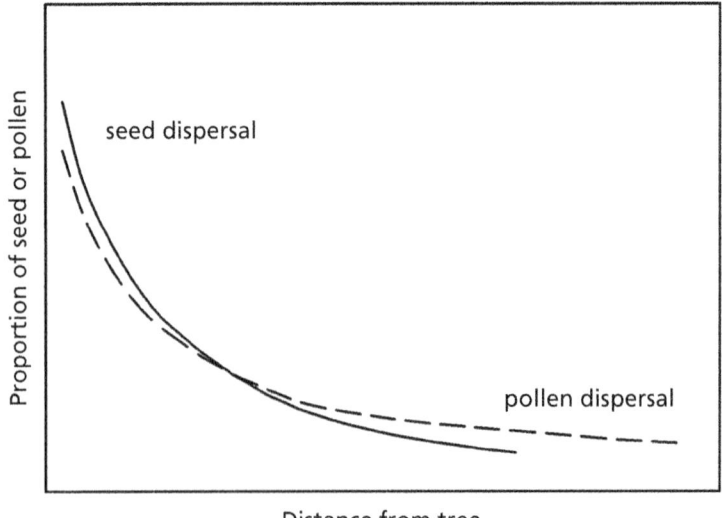

Figure 14.2 Diagram of dispersal distances of wind-blown pollen and seed from a temperate conifer. Although both relationships form similar shapes, the pollen dispersal relationship has a much longer tail than seed dispersal does.

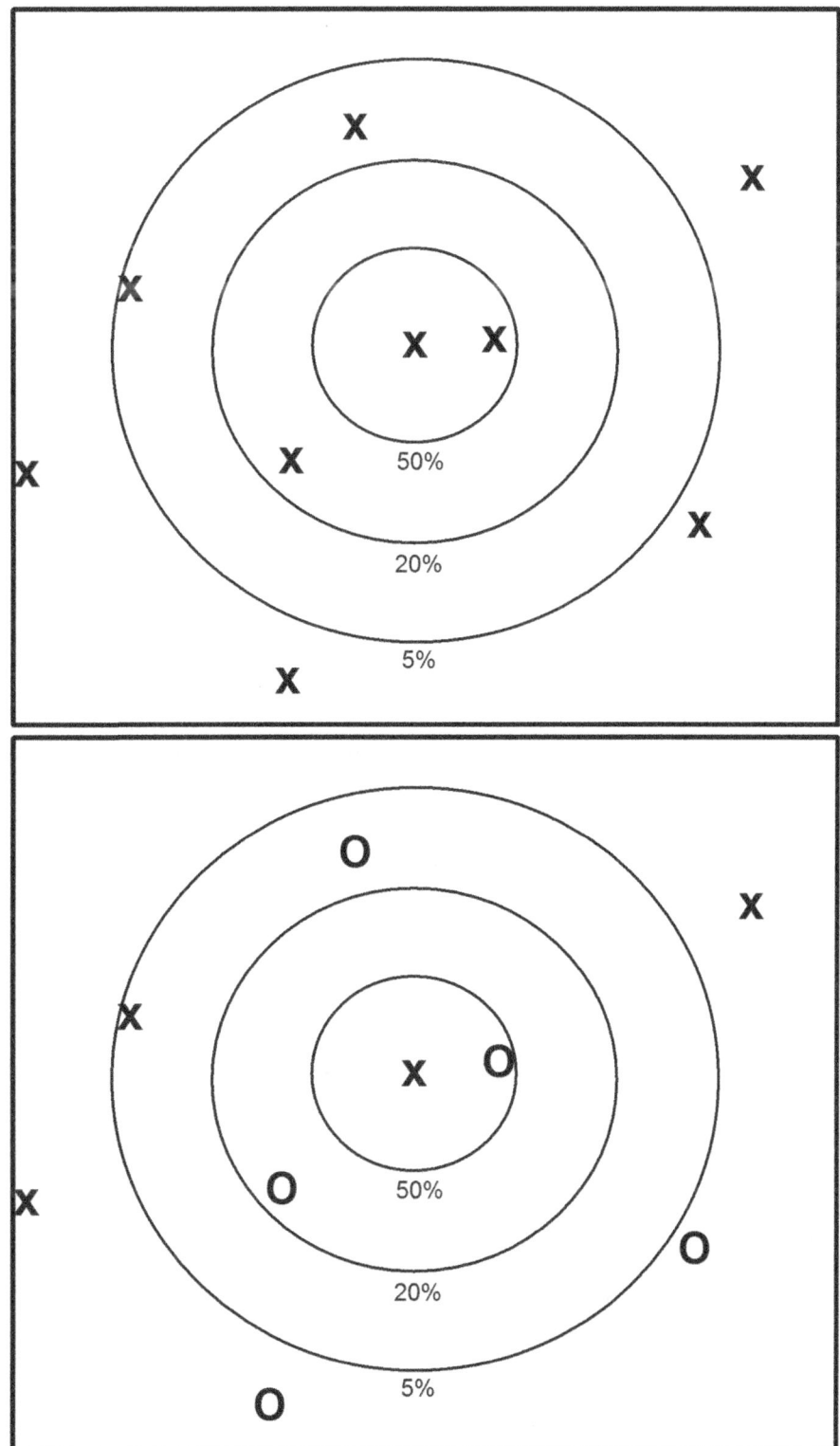

Figure 14.3 Diagram of spatial mating patterns for two hypothetical stands. The three concentric circles represent the concentration of the pollen cloud originating from the center tree. In the upper diagram, which consists of an overstory of one species, four trees (represented by X) would be within the 5% band. In the lower diagram, which includes two species (X and O), only one X tree occurs within the 5% band. Hence, with equal density, the mixed stand has a lower probability of outcrossing than the single-species stand.

Seed dispersal is also highly variable, ranging from wind dispersion to animal or water dispersion. Regardless of dispersal mechanism, dispersal generally follows a pattern of rapid decline with distance from source (Figure 14.2). For example, for coast Douglas-fir, most seed was dispersed within 60 m of the source, with comparatively little or no dispersal beyond 300 m (Isaac 1930). Seed dispersion may be strongly affected by stand structure in some species: seed fall would range further from the mother tree in more open conditions and less far in the more complex structures of multiaged stands.

The presence of advance regeneration or a seed bank has genetic implications in multiaged stands. Both advance regeneration and stored seed essentially extend the regeneration period from immediately after a harvest treatment (or disturbance) to anytime during the cutting cycle. Hence, the regeneration process may be continuous, even though the cutting cycle may be long. Finally, many systems involve vegetative reproduction such as stump sprouts, and the potential life spans of these individuals may extend for many cutting cycles. As a result, the contribution of vegetative reproduction to the genetic pool may be very small over a cutting cycle but more significant over the life span of the individual tree. Similarly, stored seed may dramatically increase the potential parents beyond those present at the beginning of a cutting cycle.

Although the regeneration process in multiaged stands is most likely to occur shortly after a disturbance or harvest treatment, it can occur at any time during the cutting cycle. As a result, variations in timing of fertility can lead to different matings over time. In single-species stands, there is probably little concern over inbreeding and reduced genetic diversity. Likewise, there may be fewer concerns over genetic diversity of species that regenerate vegetatively, form advance regeneration, or develop significant seed banks. As the number of species increase, however, so does the concern over genetic diversity. Increases in the number of species translate directly into greater average distances between species if density is constant and spatial patterns are assumed to be random. The presence of dioecious species has similar effects, since twice as many individuals are needed to provide a comparable number of breeding individuals as with a similar monoecious species.

A unique feature of multiaged stands is the possible presence of generational patterns of mating, where trees in individual age classes may be more closely related to each other than to trees in other age classes. This may be an advantage for the multiaged stand because it could reduce high levels of relatedness at the stand level. It could also be viewed as greater total genetic variation within the stand (Roberds and Conkle 1984) although a variable pattern of age-related sexual maturity may form a barrier to complete gene dispersion across age classes.

There are therefore offsetting processes occurring in a breeding-age stand in any stand structure: the presence of fewer breeding individuals results in a greater potential for inbreeding (Farris and Mitton 1984; Krakowski and El-Kassaby 2003). Alternatively, silvicultural treatments that disperse breeding-age trees may reduce family structures (Marquardt et al. 2007). There are probably critical threshold densities that provide sufficient numbers of parents to maintain genetic diversity over time. These will vary by species due to pollination mechanisms, pollen dispersal, periodicity of seed production, and even seed dispersal patterns. Potential problem areas are those stands with high species diversity, such as the humid tropics, or containing species with unsynchronized patterns of flowering. An additional factor is that nonsexual forms of reproduction, or strategies which expand the opportunity or time period for breeding, such as advance regeneration or dormant seeds, can reduce inbreeding potential.

14.4 Genetics and natural disturbances

Natural disturbance patterns are an important aspect of understanding regeneration processes. These disturbances are a primary mechanism for the natural regeneration of trees and occur over a range of frequencies and severities. The typical scenario for the development of a multiaged stand is a high frequency and low severity disturbance regime, whereas even-aged stands typically result from more severe but less frequent disturbances (Oliver and Larson 1996; also see Chapter 3). But many combinations of severity and frequency are possible, and this range of disturbances produces a diversity of stand structures. Disturbances also affect patterns of reproduction in forest trees and the genetic structure of the ensuing tree populations.

Much of silviculture can be described as an attempt to manage competition to favor certain species. This is often achieved by mimicking the effects of natural disturbances to favor the species that might have evolved under those disturbance regimes. Examples include the overstory release of advance regeneration and its resemblance to windthrow or the resemblance of the clearcut to a stand-replacement fire. In both examples, the harvest treatment results in large-scale removal of wood that is not characteristic of the corresponding natural disturbance, but both create an environment for regeneration that is similar to that created in a natural disturbance: the overstory removal might favor more shade-tolerant species, whereas the clearcut may favor fast-growing, shade-intolerant species.

Silviculture can produce stand-structure changes with genetic effects that resemble those from natural disturbances. Silviculture can also produce effects that do not resemble those from natural disturbances. Human effects that are detrimental to genetic qualities of future generations are termed dysgenic, whereas those that are positive are **eugenic**. It is important to note that every silvicultural treatment need not be eugenic or have a positive effect on stand-level gene structure. To achieve sustainability, residual stand structures only need to fall within an acceptable range of variability of natural disturbance effects. If silviculture also has a eugenic effect beyond that from the disturbance process, then this provides an additional means of meeting some management objectives. To achieve sustainability, residual stand structures have generally been assumed to only need to fall within the range of variability of natural disturbance effects. The standard for many modern ecosystem management treatments is the historical range of variability (Landres et al. 1999). However, in the context of global climate change, we may have little guidance from past conditions to guide future structures (Millar et al. 2007). Instead, management objectives to favor unique species mixtures, enhance genetic diversity, or assist with species migrations may be necessary as we strive for achieving anticipated future ranges in variation. Multiaged silviculture can be an important strategy in promoting these objectives.

14.5 Genetics and multiaged silviculture

Emerging studies using genetic markers provide new insights into how silviculture affects genetic diversity following silvicultural treatment. These can guide the development of silvicultural treatments in multiage stands, which have unique stand dynamics compared to even-aged stands.

14.5.1 Genetic implications of multiaged stand dynamics

From a tree-genetics perspective, a primary difference between even-aged and multiaged systems is the overlapping of generations in multiaged stands. At any given time, several age classes or cohorts will be present in a multiaged system (Figure 14.4). The even-aged system will typically decline in density during stand development because of mortality or artificial thinning, and surviving trees will generally reach sexual maturity at the same time. In the multiaged system, density will also decline during each cutting cycle, but only trees in the older age classes will be sexually mature. The number of sexually mature trees will therefore be lower in multiaged stands than in comparable even-aged stands. Over a cutting cycle, the number of trees in a multiaged stand will rise initially and then perhaps reach an asymptote later (Figure 14.5). This is because the growing space that is available initially after a harvest treatment may not be available later in the cutting cycle. The number of sexually mature or breeding-age trees will generally increase during a cutting cycle (Figure 14.5). But the number of sexually mature trees will always be lowest at the beginning of the cutting cycle, when growing space is most available.

Because sexually mature trees occur at lower densities in multiaged stands, they may also be more spatially dispersed than in even-aged stands. Depending on the spatial pattern in a stand, they may be aggregated or dispersed. For example, a traditional group-selection stand with large openings would have no sexually mature trees inside recent group openings but would likely have mature trees in surrounding areas of the matrix. Spatial patterns of mature trees would therefore be aggregated in group-selection stands, but opening size may

Figure 14.4 Multiaged ponderosa pine stand in Black Hills of South Dakota, showing several age classes.

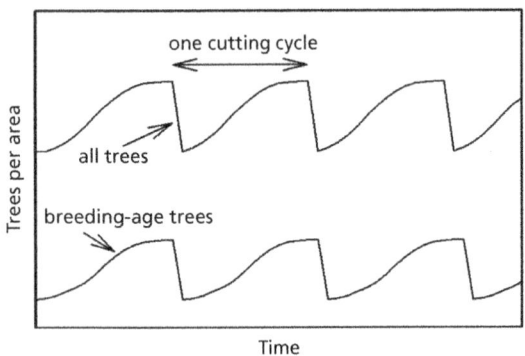

Figure 14.5 Total number of trees and number of breeding-age trees varies from a high and a low during the cutting cycle of a hypothetical multiaged stand.

14.5.2 Regeneration methods and genetic diversity

A central variable in assessing the genetic effects of any silvicultural method that relies on natural regeneration is the number of sexually mature or breeding-age trees left after treatment (Figure 14.3). Traditional silvicultural regeneration methods that rely on natural regeneration vary considerably in the number of sexually mature trees left after treatment. These numbers can vary by forest type, between species, or within a species. Generally, a seed-tree method retains the smallest number of these trees, often in the 5–30 trees/ha range, or a **selection ratio** of approximately 1:20 trees (Daniel et al. 1979; Nyland 2002). The shelterwood method generally leaves a higher number, with a selection ratio of approximately 1:4. Both the seed-tree and shelterwood methods are comparable to a selection treatment because although selection methods typically leave more trees, not all are sexually mature. However, multiaged systems vary widely in

exceed the normal range of seed dispersal. For systems with small openings, such as with single tree selection, mature trees would be more randomly dispersed, with greater distances between trees for pollen distribution but shorter distances for seed dispersal (Figure 14.2).

numbers of residual trees and are further complicated by species mixtures.

Our understanding of the genetic structure of tree populations has increased dramatically in recent decades. Much of this advancement is the result of the development of genetic markers that allow the estimation of genetic diversity in populations and answer other questions related to tree breeding. Isozyme markers are one type of molecular marker that has been used for studying genetic variation within and between populations. Although isozymes, or allozymes, have largely been replaced with DNA-based markers, they remain an important and cost-effective tool for studies of variation in forest genetics. DNA-based markers provide a virtually limitless number of markers for identifying genotypes or mapping a genome. Although DNA markers are more expensive and difficult to apply (Neale et al. 1992), they have become the more common tool. Both types of markers have been used to study genetic diversity in forest populations following silvicultural treatment including selection systems.

In coast Douglas-fir in the Pacific Northwest of North America, a shelterwood that removed approximately 85% of parent trees had no effect on allelic frequency at 10 allozyme loci (Neale 1985). This study compared genetic materials from an uncut stand with material from shelterwood leave trees, seed from shelterwood trees, and the three- to five-year-old regeneration, indicating similar selection pressures on all life stages. Additionally, very little evidence of inbreeding was found even with the small number of parent trees (15 and 35/ha at the two sites studied).

A Scots pine seed-tree system in Sweden with 122 residual trees/ha produced similar results from allozyme comparisons of parent trees and regeneration (Yazdani et al. 1985). However, significant differences were found when these results were compared to seed embryos, indicating high levels of homozygotes and some inbreeding at the seed-level. The apparent elimination of these genotypes by age 10–20 indicated they were being selected against. In Norway spruce, Finkeldey (1995) found high rates of outcrossing in seed from seed trees. Likewise, in European beech shelterwood stands, plantations, and stands receiving limited management, there were no significant differences in diversity (Buiteveld et al. 2007). Interestingly, a slightly higher inbreeding coefficient was observed for the limited-management stands than for highly managed stands.

In Japanese beech, only minor differences in genetic variability were found between unharvested areas and those resulting from a seed-tree method, and these differences were expected to disappear with self-thinning (Takahashi et al. 2000). In silvertop ash in Australia, no significant differences in genetic variability were found between seed-tree treatments that left only 10% of the overstory and uncut controls (Glaubitz et al. 2003a). Although this analysis showed no change in allelic richness or heterozygosity, it indicated a loss of some rare alleles. A similar study with another eucalypt, yertchuk, indicated a loss in diversity with a seed-tree treatment and also reported lost alleles (Glaubitz et al. 2003b). In eastern white pine, Marquardt et al. (2007) found that stands managed with a variety of methods, including shelterwood treatments, had less inbreeding than mature trees in similar unharvested stands; they suggested that this result was due to a reduction in the genetic spatial structure of mature trees and that this reduction had been caused by the harvests.

Several studies have compared variants of selection regeneration methods to unmanaged stands and stands treated with other regeneration methods. In coast Douglas-fir, a group-selection system with 0.20 ha groups was compared to a shelterwood with 15–30 trees/ha and an untreated area (Adams et al. 1998). All treatments also received some planting of seedlings from local seed sources. Populations of natural seedlings were similar in all treatments based on analyses of 17 allozyme loci. Some rare alleles that were probably deleterious were removed when cutting smaller trees in shelterwoods. Retention of a range of tree sizes in shelterwoods was recommended to avoid this potential problem. It should be noted that leaving a range of tree sizes is a normal practice in single tree selection systems. As expected, greater diversity was found in artificial regeneration than in natural regeneration.

Another Pacific Northwest study compared shelterwood treatment, variable retention, large group cuts, and a clearcut with unmanaged mixed western hemlock and Pacific silver fir stands (El-Kassaby et al. 2003). The variable retention was the most severe treatment and left 25 windfirm trees/ha. However, only the shelterwood treatment had a negative effect on genetic diversity, and only on western hemlock, as Pacific silver fir was unaffected. This result indicated how different species could have different responses to the same disturbances. The shelterwood in this study retained a variety of crown classes, unlike traditional shelterwood treatments, in which only upper crown classes are retained. With silver fir, only slight differences in genetic variation were found between adult trees and regeneration in Plenter forests in Germany (Hussendörfer and Konnert 2000a, 2000b). Other studies have reported similar results with selection systems in European conifers and broadleaved trees (Müller 1990; Starke et al. 1996; Hosius et al. 2006).

In tropical forests, intraspecies densities are often low because of high species diversity, and many species are pollinated by insects or animals. For the *Shorea* species Honda beraliya, higher rates of selfing were found following "selective" logging that included only 2 mature trees/ha compared to 10 mature trees/ha in an undisturbed forest (Murawski et al. 1994). This could lead to inbreeding depression; but as the authors acknowledged, additional natural selection of offspring may subsequently increase heterozygosity. Pautasso (2009) reported that no previous studies found detrimental effects of "selective" logging in other tropical species, but other modeling studies suggested long-term effects of these practices.

Several temperate studies have at least proposed that homozygosity may decline with stand development (Yadzani et al. 1985; Takahashi et al. 2000). Many of these studies of genetic diversity after regeneration treatments are comparisons of parent and regeneration trees. A better measure of inbreeding caused by different regeneration methods may be changes in gene frequencies at the end of a rotation or cutting cycle rather than at the beginning. However, these comparisons would require long-term study sites with carefully maintained records of treatments and treatment effects.

14.5.3 Silviculture strategies for gene conservation

Savolainen and Kärkkäinen (1992) noted that natural selection maximizes total fitness, whereas traditional forest management seeks to maximize vegetative growth. There are many justifications for multiaged forestry, and only one of them is maximizing wood production. Nevertheless, a genetic conservation objective need not focus on attempting to mimic natural mating processes. Fitness can be improved through artificial selection, which can also advance other objectives such as improved growth, form, wood quality, or disease resistance. Genetic conservation will therefore attempt to maintain or enhance fitness while advancing other economic or ecological objectives.

Multiaged systems may resemble natural process for unmanaged forests in some situations. In other cases, these multiaged systems will be substantially different from natural processes. Schaberg et al. (2008) used a modeling strategy to assess effects of silviculture on eastern hemlock and eastern white pine. They concluded silviculture could have could have either negative or positive effects depending on the selection criteria used. Keys to long-term genetic sustainability are to develop silvicultural methods with selection criteria that either emulate these natural processes, develop practices which can enhance fitness and long-term viability, or use some combination of both approaches. There are several silvicultural strategies for either approach.

A principal concern with multiaged stands is that reduced population densities may affect gene flow, resulting in spatially clumped genotypes, inbreeding, increased selfing, and reduced fertility. Reduced genetic diversity and adaptability could ultimately result in weaker population structures and decreased adaptive potentials for the species (Finkeldey 2001). Compared to even-aged systems, multiaged silviculture reduces the density of sexually mature trees. However, the number of breeding-age trees in multiaged systems may be more comparable to those obtained with traditional seed-tree or shelterwood methods. The regeneration period is increased to be somewhat continuous in the multiaged stand. One strategy to reduce inbreeding is to design stocking regimes that maintain greater numbers of large trees relative to small trees or maintain more than one age class of breeding-age trees (see Chapter 7). Many stocking prescriptions assume large numbers of small or replacement trees are necessary. For example, the reverse-J diameter distribution defines an exponentially increasing number of trees with decreasing diameter. If some of the growing space allocated to small trees is shifted to larger trees, there will be greater numbers of sexually mature trees and reduced chances of inbreeding.

Short cutting cycle lengths have been implicated as a potential cause of inbreeding (Finkeldey and Ziehe 2004). This reasoning may apply primarily to forests with many species, such as tropical forests. Cutting cycle lengths are proportional to cutting severity, with longer cutting cycles corresponding to heavier harvest treatments (see Chapter 7). Longer cutting cycles may have other benefits by reducing numbers of harvest entries and providing more open conditions for regeneration.

Assessments of genetic effects of silvicultural systems are largely based on stand-level comparisons. But larger scale patterns of stand structures also affect the dynamics of gene transfer (Figure 14.6). At larger scales, such as multiple stands or landscapes, a forest of multiaged stands may have advantages for pollen and seed dispersal. By forming a relatively homogeneous forest of multiaged stands, sexually mature trees would be present in every stand thereby providing additional means of gene exchange across stand boundaries. Conversely, the regulated even-aged forest that consisted of stands of a range of ages may include many stands that were not sexually mature. The potential for inter-stand spread of pollen and seed may therefore be greater in multiaged stands.

Tree planting and thinning treatments are often assumed to be operations exclusive to even-aged stands, but multiaged systems and either treatment are not mutually exclusive. Artificial regeneration offers an opportunity to introduce new genotypes. Planting can thereby enhance genetic diversity as well as modifying species composition in multiaged stands. Planting can also focus on gaps including group openings, or attempt to get understory trees established before the overstory is removed.

Thinning even-aged stands has long been recognized as an early or midrotation tool for removing undesirable phenotypes and improving genetic quality of stands. Like artificial regeneration, thinning can also modify species composition. Thinning younger age classes is an important aspect of stocking control in multiaged stands (see Chapter 9). Each cutting cycle represents an opportunity to

Figure 14.6 Landscape patterns showing many broadleaf and conifer stands in Alberta, Canada.

adjust the numbers of trees in younger age classes when overstory trees are removed. This thinning can include cutting of commercial-sized trees or respacing/precommercial thinning of smaller trees. Although this selection is based on phenotypic characteristics, it has the potential to remove undesirable genotypes and improve the genetic quality of a stand. As with low thinning in even-aged stands, multiaged thinning that favors superior genotypes has the potential to accelerate stand development.

14.6 Synthesis

It is difficult to make generalizations about the genetic effects of management in any ecological system because of the complexity and variability in the ecological system and the species present. Nearly all silvicultural activities, including multiaged systems, have genetic effects. For multiaged systems, one concern is that the small number of breeding individuals will cause a dysgenic effect. Multiaged systems have multiple generations of trees present that overlap in age and sexual maturity. This results in fewer breeding-age individuals at any given time than a comparable population of end-of-rotation, even-aged trees of breeding age. This potential problem is compounded in mixed-species stands, where numbers of breeding-age individuals of a species are further reduced (Figure 14.3). Multiaged systems compensate, at least in part, by having more frequent reproduction periods than even-aged systems. The presence of advance regeneration, vegetative reproduction, or seedlings originating from buried seeds represent other mechanisms that may reduce potential problems from small numbers of breeding trees. Additionally, the number of breeding trees in multiaged systems may be no smaller than in seed-tree or shelterwood regeneration methods.

A second concern with multiaged systems is the assumption that the process of removing larger trees will be dysgenic. A related point of confusion has been the application of the similar terms "selective" and "selection" to the silviculture of multiaged stands. "Selection" is used to describe both a thinning method and single tree and group methods to regenerate multiaged stands. "Selective" cutting is a more general term used to describe any form of partial cutting and is often synonymous with high grading and its dysgenic effects (Helms 1998). There is a long history of dysgenic harvesting in forestry, and much has occurred under the name of "selective cutting" (Ledig 1986). The common confusion of "selection" with "selective" has provided the unintentional association between selection and dysgenic practices. In contrast to even-aged stands, larger trees in multiaged stands will typically be older and of a different age class than smaller trees. The removal of these larger, older trees is not dysgenic. Instead, it is the final step in a process of tree development from regeneration to maturity, to reproduction, and final removal.

Although previous studies provide no clear evidence of genetic problems associated with multiaged systems, there is reason for concern in some situations. Potential problems include systems with high species diversity or systems where preferences for large numbers of sexually immature trees reduce numbers of sexually mature trees. In their work with tropical forests, Finkeldey and Ziehe (2004) concluded that the most severe threats to genetic variation were selection systems with short cutting cycles. However, in temperate forests, where tree-species diversity is lower, the threats of inbreeding and reduced diversity are much less severe. At either extreme, silviculture can reduce these threats.

There is a strong conceptual basis for potential dysgenic selection problems with multiaged silvicultural systems because there is a history of high grading under the guise of selection silviculture. This has occurred under names such as selective cutting, diameter-limit cutting, "plunder" or Plenter cutting, and others. However, there is a shortage of data indicating genetic problems. This may be due to: 1) the need for stands with a consistent and long-term management history for analysis; 2) potential ameliorating effects from surrounding stands; and 3) the relatively recent emergence of techniques for genetic studies. The dynamics of gene transfer in forests may be sufficiently resilient, given the many factors that interact to affect genetic diversity, that only the most dysgenic treatments produce measurable effects. Alternatively, this may indicate that the potential threats from multiaged silvicultural systems are overstated.

Good silviculture is first step in gene conservation (Ledig 1986), and multiaged silviculture can be a part of this. Multiaged silviculture can also be a viable strategy in an era of climate change. These structures provide some resilience through the presence of multiple age classes that is not provided in even-aged stands (O'Hara and Ramage 2013). This silviculture must recognize the role of natural disturbances in affecting reproductive and gene transfer dynamics and that a system may work well in one forest type but not in another. Numbers of breeding-age trees will have a strong influence on genetic diversity, and those forests with the greatest species diversity may require the greatest care in assuring sufficient numbers of parent trees. Artificial regeneration, thinning, and favoring desirable phenotypes in cutting decisions will provide similar benefits as with even-aged stands. Finally, the associations between selection systems and dysgenic practices are poorly founded, and dysgenic effects are avoidable with sound silvicultural practices.

CHAPTER 15

Multiaged structures and stand health

15.1 Introduction

Stand structure affects resistance to disturbances such as insects, pathogens, and wind damage. **Resistance** is defined as the ability of a stand to avoid or prevent disturbance impacts (Larsen 1995; National Academy of Sciences 2000; Millar et al. 2007). Stand structure may have a direct effect on resistance, such as with interactions between structure and fire behavior, or it may have a more indirect effect by improving tree vigor and thereby increasing resistance to an insect or pathogen. Stand structure also affects the ability to recover following disturbance events and thereby has implications for stand- and ecosystem-level **resilience** or **elasticity** (Larsen 1995; National Academy of Sciences 2000; Brang 2001; Millar et al. 2007). Managing for complex structures therefore has implications for stand health and the capacity of stands to meet other objectives both pre- and post-disturbance. Anticipated changes in climate make stand-level health issues more important and also more complicated over multiple scales. Climate change is essentially an ecosystem stressor that can combine with other stressors to exacerbate many forest health issues. For example, a warming climate may have contributed to the massive expansion of the range of mountain pine beetle in western Canada (de la Giroday et al. 2012) or greater incidence of fire in Mediterranean Europe (Lindner et al. 2010).

This chapter focuses on stand-level relationships, although many of these disturbances also affect forests at much larger scales. Many forests may be specifically managed as protection forests to reduce effects of fire, snow damage, or for some other protective function. Multiaged stands are important options for protection forests (O'Hara 2006).

The objective in this chapter is to discuss general interactions between complex stands and their ability to resist and respond to disturbance. It does not attempt to provide a comprehensive overview of interactions between stand structure and disturbance, nor does it attempt to present all types of disturbances.

15.2 Wind and other storm damage

Wind is a common forest disturbance agent on trees and forests around the world. Other weather-related disturbance events that have similar effects include snow and ice damage (see Chapter 3). Most of our knowledge of storm damage is related to wind damage. There are complex interactions between wind and stand structure that apparently make complex structures more resistant to wind damage. Stand structure affects wind damage either through ameliorating wind stresses within a stand or by developing trees that are more wind resistant. Interactions between stand structure and wind damage, and post-disturbance resilience to wind damage are discussed here. All types of wind damage—stem breakage, uprooting, and loss of branches or foliage—are included, but generally not differentiated.

Mitchell (2013) described trees as "self-designing structures" whose form is the result of available resources and stress loads on the tree. Trees with the resources or carbon to allocate to xylem production can increase the strength of stems, roots, and branches, whereas low vigor trees do not have these resources. Although allocation of carbon to stems and branches is a normal process, it is accentuated for those structural elements where stresses occur. Hence, a tree exposed to wind stress will become more wind resistant than a similar tree without those stresses. Stands therefore have the potential to become more stable or resistant to wind damage when they have sufficient resources, on a per tree basis, to adjust strength of branches, stems, or roots.

There is also potential for greater stand resilience to wind damage when stands have multiple age classes. Many studies have described a greater predisposition to wind damage in simpler stand structures in a wide range of forest types, either through observation or simulations (Ruth and Yoder 1953; Everham and Brokaw 1996; Mason 2002; Mitchell 2013). If responses to wind stress result in trees that are preconditioned or acclimatized to wind and therefore more resistant, then what might cause trees in complex structures to have greater resistance? Perhaps the complex structure results in upper canopy trees that develop resistance because of their emergent canopy position. Perhaps larger trees shield smaller trees from damaging winds. Or perhaps the complex structure allows a greater degree of wind penetration into the canopy, thereby resulting in greater resistance in all trees.

Height-to-diameter ratios (H:D) are common measures of tree stability because they represent the cumulative effects of carbon allocation on tree form (Mitchell 2000). Lower ratios are correlated with greater stability because the typical response of vigorous trees under stress is to allocate greater carbon to diameter or girth instead of height. Less vigorous, or more suppressed trees, allocate a lower proportion of carbon to diameter and have higher H:D ratios. Likewise, trees grown at narrow spacings will have higher H:D ratios than trees grown at wider spacings. Threshold H:D ratios for stability vary by species and range from approximately 70 to 100 when both dimensions are measured in equal units (Table 15.1). However, tree stability is a complex process, and trees with lower H:D ratios may also have larger crowns, which provide a confounding effect (Albrecht et al. 2012).

In even-aged stands grown at high densities, trees become increasingly slender with age. Collectively, they may approach or exceed stability thresholds represented by H:D ratios but are supported and sheltered by the surrounding trees of similar size. Accordingly, entire even-aged stands may become vulnerable and, indeed, entire stands can be windthrown simultaneously (Mason 2002; Gardiner et al. 2005, 2008; Albrecht et al. 2012; Mitchell 2013). Several authors have reported lower H:D ratios in multiaged stands, as compared to even-aged stands, in central Europe (Kenk and Guehne 2001; Mason 2002), and these observations imply a greater inherent stability in more complex stands, based on the sum of individual trees. An irregular, complex stand structure has a more varied canopy surface, with upper canopy trees being more exposed, and possibly a more varied wind environment below. This may provide the minor ongoing stresses that result in greater stability in more complex stand structures. Additionally, multiaged forests typically do not have the abrupt edges that are found in even-aged stands and where windthrow is more common (Figure 15.1).

Table 15.1 Examples of threshold height-to-diameter (H:D) ratios for selected trees species. Height and diameter (at 1.3 m) are measured in equal units.

Species	H:D threshold	Reference
Western larch	80	Wonn and O'Hara 2001
Coast Douglas-fir	85	Wilson and Oliver 2000
Interior Douglas-fir	80	Wonn and O'Hara 2001
Lodgepole pine	80	Wonn and O'Hara 2001
Ponderosa pine	80	Wonn and O'Hara 2001
Radiata pine	74	Cremer et al. 1982
Scots pine	80	Peltola et al. 1997
Norway spruce	80	Peltola et al. 1997

Stand structure interacts with disturbances in affecting how stands respond after a disturbance event. However, recovery is strongly influenced by the characteristics of the disturbance event: the extent and magnitude of the disturbance, and the types of effects on trees (e.g., uprooted trees, broken stems, or branch loss). The presence of multiple generations of trees of different sizes and susceptibilities to damage gives multiaged stands a greater resiliency to storm damage. For example, the advance regeneration beneath the canopy provides a replacement source for the next stand after overstory windthrow. In a stand with many age classes where only the oldest is affected, the post-disturbance stand would maintain many functions and processes.

Figure 15.1 Wind-damaged Sitka spruce along the cutting edge of an even-aged stand in Scotland.

15.3 Insects, pathogens, and invasives

There are many insects and pathogens that affect forests, and many of these have effects that are related to stand structure. Invasive insects and pathogens are also grave threats to many forests, as are invasive plants. The threats from invasives are also highly variable. As a result, it is not possible to provide broad generalizations about the health of complex stands as related to insects or pathogens. Instead, this section focuses on several specific threats and discusses how they interact with stand structure and thus may affect management of complex stands. More complete overviews of forest insect and pathogen problems have been provided by Edmonds et al. (2000), Lundquist and Hamelin (2005), Castello and Teale (2011), Ciesla (2011), Wylie and Speight (2012), and Gonthier and Nicolotti (2013). Insects and pathogens may also interact with fire or wind to exacerbate stand-health problems (Box 15.1).

15.3.1 Insects

Insects have a wide variety of interactions with stand structure. Some insects are more successful in multistoried stands of susceptible species. One example from western North America is the western spruce budworm, which hatches in the tops of trees and moves downward as it develops. Multistoried canopies favor the development of this insect and so typically receive greater damage (Carlson and Wulf 1989; Hummel and Agee 2003). Complex stands structures may be advantageous in some situations. For example, in western North America, the mountain pine beetle attacks a variety of pine species, including lodgepole, ponderosa, and whitebark pines. This insect is most effective in stands of low vigor trees where resistances are low. The mountain pine beetle generally attacks trees greater than approximately 20 cm in diameter (Fettig et al. 2007). Stand structures consisting of many low vigor trees larger than this threshold

Box 15.1 High grading and stand health in western North America

The history of forest management in many western North America forest types began with initial removals of the most valuable species (Hessburg and Agee 2003). The initial harvesting often removed the most commercially valuable, or high-grade, trees. In some cases, surviving trees and regeneration developed into new structures. In other cases, the sites were intentionally burned following harvest in a stand replacement disturbance that resulted in an even-aged stand. Fire suppression often allowed a suite of species to thrive that were tolerant of shade but intolerant of fire.

The preharvest age structures of these forests were often complex multiaged stands formed by variable fire frequencies and consisting of species capable of withstanding these fire regimes. Some of these forest types formed "savannah" or "open old forest" structures (O'Hara et al. 1996) where site distances were large and tree densities were low (Figure 15.2). A feature of these stands was the highly variable diameter distributions, which often had more trees in larger than smaller size classes (Weaver 1943; Cooper 1960; Agee 1993; Covington and Moore 1994; Arno et al. 1995, 1997).

In the postharvest stands that experienced this pattern of treatment, species composition shifted to more shade-tolerant species that were also more prone to insect and pathogen problems. The suppression of fire contributed to high stockings and accumulating fuel loads. This resulted in stand-health problems on a massive scale: the stands became prone to large-scale insect and pathogen problems (Sampson and Adams 1994; Clark and Sampson 1995; Stine et al. 2014), and complete changes in fire regime from low- and mixed-severity fires to severe fires on massive scales, or megafires (Attiwill and Binkley 2013).

Many of the present forests are multiaged, or thought to be multiaged, because they have formed stratified mixtures of species. They also exist at relatively high stocking levels (Figure 15.3). A common management approach in the past has been to promote these structures by continuing to

Figure 15.2 Old ponderosa pine stand in western Montana, US, with large, widely spaced trees and open understory in a savannah or open old forest stand structure. The background shows a denser stand with invasion of understory trees.

> **Box 15.1** *Continued*

remove more valuable species and encouraging these stands to form negative exponential diameter distributions. This management has often been under the guise of selection silviculture. This created further insect and pathogen problems and exacerbated fuels problems. These problems continue today (Stine et al. 2014).

The western North America pattern of high grading is not unusual in how initial harvests were conducted, as forests were exploited around the world. But the large-scale species conversion and "densification" of these forests is more unique. It is evidence of a poor characterization of age structure and a poor understanding of stand dynamics. This has contributed to a variety of stand-health issues that exist at scales that make them difficult to manage. These stand-health issues and association with "selection" cutting have contributed to negative opinions of multiaged silviculture and contributed to the cyclic shift to even-aged silviculture in recent decades.

Figure 15.3 Dense stand of conifers in the Blue Mountains of eastern Oregon, US.

are most susceptible. Hence, the worst damage occurs in dense, even-aged stands of susceptible species. Multiaged stands typically include trees that are both above and below the threshold providing some resistance (Fettig et al. 2014). Additionally, the overstory trees in multiaged stands often exist at low densities and are of higher vigor than overstory trees in even-aged stands (O'Hara 1996). A mixed-species structure will also be beneficial, as it reduces the ability of beetles to move from tree to tree.

These examples demonstrate just two of many types of interactions between insects and stand structure. They demonstrate the difficulty in generalizing about interactions between insect damage and stand size or age structure. Tree vigor, stand structure, and tree size are important with some insects but not others. As a result, there are a wide range of possible interactions. However, the host specificity of most insects results in mixed-species stands having greater resistance to infestations. Whereas increased species diversity is generally beneficial, the effects of complex age structures are highly specific to individual insects.

15.3.2 Pathogens

A highly varied suite of pathogens affect forests, but the primary effects of stand structure are through species composition and tree vigor. Among root diseases, various species of *Armillaria* are ubiquitous around the world in boreal, temperate, and tropical forests. This genus includes species that cause bole decay or tree mortality (Shaw and Kile 1991). These fungi attack broadleaves, conifers, and shrubs, as well as agricultural crops. Resistance to this pathogen is related to tree vigor and variations in susceptibility among species. As plantations usually consist of a single species, they are often prone to armillaria. The advantages of complex stand structures are therefore related to species composition. Establishing stands of resistant species or thinning to favor resistant species is effective for reducing armillaria incidence. Adjustments in total stocking in any structure will improve tree vigor and resistance. However, there are no known inherent advantages or disadvantages related to multiaged structures with armillaria or other root diseases. Any structure can be directed towards certain species compositions or treated to enhance tree vigor.

Dwarf mistletoes are a group of plants that parasitize several conifer species in North America as well as Asia, Africa, and South America. They receive most of their nutritional needs from the host. Effects on the host are a reduction in vigor, growth, and a variety of deformities such as dead tops or formation of brooms. Dwarf mistletoes reproduce by ejecting seeds that may land on and infect a susceptible host. Seeds may also be spread by small mammals or birds. Spread is most effective when the seed can be projected from a large tree to a smaller tree (Mathiasen 1996; Hawksworth and Wiens 1996). A multistoried structure with sufficient space between tree crowns to maximize seed dispersal represents an ideal structure for mistletoe spread (Figure 15.4). Multiaged stands are therefore more susceptible to dwarf mistletoe than simpler structures. However, dwarf mistletoes are host specific, and a mixed-species structure is an effective means of reducing spread.

Like forest insect problems, there are few generalities about the interactions between pathogens and stand structure other than the beneficial aspects of greater tree species diversity. Complexity in stand structure provides few advantages as compared to simple stand structures with a host-specific pathogen. Treatments to enhance tree vigor may be beneficial with some pathogens regardless of stand structure.

15.3.3 Invasives

A variety of invasive plants, insects, and microorganisms threaten the health and ecological integrity of forest stands and larger ecosystems (Pejchar and Mooney 2009, Britton and Liebhold 2013). Ecosystem function and structure can be substantially changed when a nonnative organism becomes established. In the most visible cases, non-natives modify ecosystems and negatively affect production of ecosystem services (Vicente et al. 2013). However, many non-natives become innocuous in the communities they invade.

Management can be effective in reducing the effects, or spread, of invasives: however, eliminating an established invasive is generally not possible unless the invasion is recognized very quickly. Even when eradication is impossible, management efforts to limit effects are most effective when treatments are early in the invasion process (Waring and O'Hara 2005, Liebhold 2012). Stands with greater complexity probably have greater resistance to invasives, particularly when native species have a variety of life strategies or occupy variable niches. This provides a broader range of competition for invasive plants and greater resilience for ecosystem functioning after the invasion. Multiaged stands

Figure 15.4 Western larch overstory trees in eastern Washington State, US, infected with dwarf mistletoe. The position of these trees, well spaced above regeneration, provides ideal conditions for the spread of mistletoe to understory western larch.

may also benefit from fewer edges and the continuous cover that may limit spread of shade intolerant invasive plants.

15.4 Fire

Fire is a dominant disturbance agent in many forest types. Although fire is often viewed as a negative factor, it has several important ecosystem functions including rapid recycling of various types of forest fuels, reducing stocking, and stimulating regeneration. Natural fires maintain multiaged structures when they periodically occur at low severities and result in, at most, only partial mortality (Agee 1993). Stand structure determines the amount, type, and organization of fuels and interacts with fire behavior. Fuel characteristics can be controlled by management in any type of silvicultural system. The primary difference between even-aged and multiaged systems is the more varied vertical and horizontal arrangement of fuels in multiaged stands and the effects of this arrangement on fire behavior. The amount and type of fuels will therefore vary with the silvicultural system, but can be controlled with management.

There are two apparent advantages related to stand structure and fire that favor even-aged stands. One is the resistance to fire when even-aged stands reach a development stage where crowns are separated from surface fuels. The vertical arrangement of fuels in multiaged stands is generally more continuous than for even-aged stands (Figure 15.5). The small trees near the ground in many multiaged stands provide a direct connection between surface fuels and crown fuels and thus form **fuel ladders** that allow fire to move upward into the canopy. The fuel ladders that are formed when stands have multiple strata are generally assumed to make them prone to fire. However, if both even-aged and multiaged stands are assumed to have equal total leaf area and production, then fuel density must be lower in the multiaged stand, imparting some advantages.

Figure 15.5 Schematic diagram of vertical fuel profiles in hypothetical even-aged (top) and multiaged stands (bottom).

The second advantage is that even-aged stands with crowns separated from surface fuels are more suitable for prescribed burning. The crown separation that provides resistance therefore provides options for management in the form of fuel reductions that impart even greater resistance for even-aged stands. Whereas this is conceptually correct, it tends to oversimplify the potential disadvantages of multiaged stands or presumes a continuous recruitment of trees in multiaged stands. Although even-aged stands have periods where prescribed fire may be a useful fuel treatment, burning is not suitable during early stand development. Many forest types in fire-prone ecosystems with high fire frequencies form multiaged stands with the natural fire regime. This indicates multiaged structures and prescribed burning can coexist. These stands often have relatively low stocking levels where fuels are discontinuous both vertically and horizontally (Figure 15.2). Many species in these forests have adaptations to survive fire, such as thick bark, postfire sprouting ability, or insulated buds. Additionally, regeneration events are episodic and may coincide with fire events. Prescribed burning therefore has potential as a fuels treatment in either structure, although it may be more flexible in even-aged stands.

Resiliency to fire may be in the form of rapid tree establishment or the maintenance of functionality in the post-disturbance structure. Multiaged stands with surviving trees may have the advantage of regeneration sources being present immediately postfire and thereby providing for a more rapid recovery.

The presence of surviving trees, which may be more likely in complex structures, also provides opportunities to maintain ecosystem functions.

15.5 A dynamic climate

Global climate change is expected to fundamentally change forest ecosystems in a variety of ways. Additionally, increasing atmospheric CO_2 will also directly affect plant communities, including forests, independent of climatic changes. Projections indicate a generally warmer climate and a drier climate in many places. Locations that do not experience reduced precipitation may still be more prone to water stress because of warmer temperatures. Changes in atmospheric chemistry may enhance forest productivity or, in localized areas, induce stress. Hence, even direct climate change effects are difficult to predict (Reyer et al. 2013).

One anticipated effect of climate change is an intensification of disturbance regimes in the form of greater frequency and severity of disturbances. This includes storm-related disturbances and also an increase in the incidence and severity of insect and pathogen activities. Warmer and drier conditions are likely to also increase fire problems. Combinations of increases stresses from climate change and other anthropogenic effects, such as densification of many forests, may compound these problems (Raffa et al. 2008). Greater disturbance incidence is already apparent in some regions such as Europe, where increases in disturbance frequency and severity in recent decades have been reported (Seidl et al. 2011).

Insects and pathogens will interact with climate change in complex ways that contribute to a variety of temporal patterns of mortality (Allen et al. 2010). Grulke (2011) described an environment–host–pathogen triangle, where climate change affects both the host and the pathogen in ways that may make the host more vulnerable. Changes in climate may result in gradual increases in pathogen or insect activity that may not be immediately apparent. Hence, the development of large-scale infections or infestations may be delayed or asynchronous from climate changes. Likewise, there may be significant delays between management efforts to reduce insects and pathogens and any apparent decline in these agents (Seidl et al. 2009). It is also important to separate climatic effects from those vulnerabilities related to stand age and tree sizes. Thorpe and Daniels (2012) examined long-term mortality patterns in conifer forests in Alberta, Canada. They found increasing mortality over their 50-year study period but attributed it to stand age and density, not climate change.

Climatic shifts will not be consistent in all regions, and some regions will experience increased productivity and possibly expanded management options. For example, boreal forests in northern Europe are expected to have greater productivity and will be capable of supporting greater species diversity. Mediterranean forests may see a transition to lower stocking levels as well as treatments that encourage or establish drought-resistant species (Linder et al. 2010). In general, the most anticipated effects are greater stresses on trees, greater incidence of insect and pathogen problems, and intensification of disturbance regimes.

15.6 Management options

There are options to reduce the potential threat or the severity of disturbance effects. Increasing tree species diversity is likely the most effective means to reduce insect and pathogen threats. Many insects and pathogens have coevolved with their hosts and have strong host dependencies. Mixed-species forests therefore have greater resistance and resiliency to respond to insects or pathogens that affect only a few species. Enhancing tree vigor through stocking control is another means to improve resistance. Although increased vigor in trees does not reduce susceptibility to all insects and pathogens, it can help with many, such as some defoliating insects and many bark beetles (Fettig et al 2007). Hence, maintaining trees with high vigor, as well as maintaining species diversity, can improve stand health in the form of greater resistance and resilience.

Prescribed burning represents a viable management option in many multiaged forests. Smaller, but more frequent, windows of opportunity for prescribed burning exist in multiaged as compared to even-aged stands. Although this may be viewed as a disadvantage for multiaged stands, it does not preclude prescribed burning. Multiaged stands with open structures of fire-resistant species are the best candidates. Controlling stocking as well as species composition are keys to successful use of fire in managing multiaged stands.

Climate change will provide additional stresses on forests. These stresses will combine with the legacies of past management that have simplified age structures or favored single-species stands over broad areas to exacerbate many forest health issues (Seidl et al. 2011). Management responses to these threats should be to encourage species diversity as well as assist the dispersal of species to more appropriate climates. Management may also encourage nonnative species that are better adapted to new sites. Increased planting of coast Douglas-fir in central Europe is an example of expanding the range of this nonnative species because of its greater tolerance to drought than many native species and its ability to grow in multiaged structures (Schütz and Pommerening 2013). Identifying drought-tolerant species and varieties is therefore an ongoing research issue (Reyer et al. 2010).

Although presented here as stand-health issues, these issues are often large-scale problems that involve many stands over large landscapes or regions. These problems should be addressed at large scales, but making changes in landscapes, or even regions, will involve stand-level manipulations as the primary scale for intervention (O'Hara and Nagel 2013). Management will also need to anticipate future conditions and plan for future ranges of variation rather than those related to historic conditions (Duncan et al. 2010; Stephens et al. 2010; Stine et al. 2014).

Another approach is a "disturbance integration" strategy (O'Hara and Ramage 2013) that promotes emulation of disturbance effects in the design of multiaged systems. These systems can be flexible to accommodate variable cutting cycle lengths and treatments that vary in their severity, while also managing to encourage high tree vigor and tree species diversity. Ongoing changes in climate and uncertainty associated with any management approaches will require a greater emphasis on adaptive strategies that, in turn, rely on models (Yousefpour et al. 2012). The future will also involve trade-offs between providing ecosystem services and developing adaptive capacities in our forests (Seidl and Lexer 2013). Because these strategies potentially lack the consistency over time that is characteristic of many silvicultural systems, there will be inconsistent production of ecosystem services, including timber, over time. These systems may be more resistant and resilient to disturbances, so that when disturbances do occur, they represent a relatively minor perturbation in the long-term management of the stand. Integrated management strategies that attempt to anticipate disturbance effects and reduce susceptibilities are an important means of improving stand health.

Salvage cutting refers to the removal of dead trees or trees damaged or dying due to some sort of disturbance (Helms 1998). These treatments have traditionally had a strong economic component, as trees were removed to capture value that otherwise would be lost. However, salvage treatments should be designed to create desirable stand structures based on what is retained—including dead or dying trees—rather than what is removed (O'Hara and Ramage 2013). These treatments can be used to promote nontimber ecosystem services, including resistance and resilience to disturbances. Various forms of multiaged stands are viable candidates for salvage treatments.

Finally, these recommendations for potential management treatments are highly generalized and deliberately vague. Management to develop complex forests using multiaged systems must consider the risks of using these systems in the specific environments under consideration. The risks and uncertainties on a site-by-site basis are highly variable, as are the legacies of historical disturbance events, management activities, and other human interventions. Hence the management activities on any site should be designed specifically to meet stand-level conditions within the landscape context.

15.7 Synthesis

Complex stands in general, and multiaged stands in particular, provide some advantages as well as disadvantages over simpler stand structures related to maintaining stand resistance to disturbance events. The primary resistance inherent to multiaged stands is the diversity of tree ages and sizes that have variable susceptibilities to disturbance. The distinction between complex and multiaged becomes important for stand-health issues because complexity in species composition provides resistance to forest pathogens and insects. However, complexity in age structure provides advantages in limited cases and disadvantages in others. This is the case for insect, pathogen, and fire disturbances, for which specific advantages and disadvantages depend on the disturbance type and the forest structure. Additional advantages related to stand structure include the resistance of trees of high vigor. For storm-related disturbances, such as wind, the available evidence suggests complex structures of multiaged stands provide greater resistance.

Complex multiaged stands may have greater resilience than simpler stand structures because the multiple age classes have variable susceptibilities and offer multiple pathways for recovery after disturbance. Post-disturbance, multiaged stands, because of their range of tree sizes and ages, will also potentially retain many pre-disturbance functions. Variable tree sizes and ages also enable a wide range in response mechanisms, as do different species. As a result, complex structures may exhibit resiliencies not related to resistance that provide additional advantages over simpler stand structures (O'Hara and Ramage 2013).

Climate change confounds these problems by increasing disturbance risk and uncertainty. Potential shifts in climate averages and extremes will interact with disturbances to create new disturbance regimes that are less predictable. Potential reductions in tree vigor will also affect disturbance regimes and post-disturbance responses. Multiaged silviculture may offer ways to create stands that are more resistant to disturbance and more capable of responding post-disturbance in a more uncertain climate. A "disturbance integration" strategy (O'Hara and Ramage 2013) offers the potential to create more resistant and resilient forests, but may be most applicable to publicly owned forests, where wood production is a secondary objective. These changes will challenge our abilities to anticipate change in environmental situations and to assess the trade-offs associated with increased risk and uncertainty.

CHAPTER 16

Social justifications for multiaged silviculture

16.1 Introduction

Many justifications for managing complex stands are based on ecological objectives such as providing wildlife habitat, emulating natural disturbance processes, or providing for stand structural diversity. However, successful forest management also depends on a level of public understanding and acceptance, particularly on public lands. Some forest management activities, such as managing for aesthetics or scenic beauty, are intended to provide attractive stands or landscapes. A related consideration is maintaining some level of naturalness in both unmanaged and managed forests. Forests are dynamic, and naturalness is a fuzzy concept with multiple potential methods of interpretation. Perceptions of what is natural vary with context: people will judge a stand as natural depending on their background, the location of the stand, and its relationship to other stands. Naturalness is, in many ways, a social construct. Multiaged silviculture provides an option for meeting objectives related to both scenic quality and naturalness. Multiaged stands may also have some negative aspects, such as greater management costs and possibly greater education or experience requirements for foresters. This is consistent with other trade-offs in natural resources management. This chapter discusses some of the social aspects of silviculture. It does not attempt to cover all of the social aspects of forestry but is limited to discussing social justifications for the use of multiaged systems.

16.2 Aesthetics

The esthetic qualities of silvicultural treatment are probably the single greatest factor influencing public opinion of these activities. Many studies have shown a strong preference against clearcut treatments and in favor of treatments that leave residual stand structure (Benson et al. 1985; Ribe 1989, 1999, 2009; Magill 1992; Johnson et al. 1994; Brunson 1996; Egan et al. 1997; Pâquet and Bélanger 1997; Karjalainan and Komulainan 1999; Shelby et al. 2003; Burchfield et al. 2003; Tonnes et al. 2004; Bradley and Kearney 2007; Ford et al. 2009a, b; Kearney and Bradley 2011; Ribe et al. 2013). The bulk of this work is from the Pacific Northwest of North America, but it also includes work in other regions of the world. Bliss (2000) described the opposition to clearcutting as widespread throughout the world and relatively constant from one region to another. A common theme in this research is how deeply ingrained this opposition is and how it is the result of factors beyond just esthetics (Bliss 1990; Shindler et al. 2002; Ribe 2006). Bliss (2000) described it as a deep opposition rooted in multiple factors, including an association with **deforestation**, or the permanent removal of forest cover (Helms 1998). Public perceptions of silvicultural activities generally improve as stands recover and develop posttreatment. For example, Shelby et al. (2003) measured a steady convergence of perception of different types of silvicultural treatments with time.

Favorable preferences for forest structures are also generally consistent across regions and forest

Figure 16.1 European beech stand with open understory and large trees.

types. Stands with large trees, moderate densities, shrub or ground cover, and random or dispersed spatial patterns of trees are preferred (Figure 16.1). Ribe (2009) used the DEMO studies' variable retention treatments in Oregon and Washington, US, (see Aubry et al. 2009) to examine preferences. He reported preferences for stocking levels of these mature stands at 110–155 m^2/ha of basal area and 700–900 trees/ha. In contrast, preferences in Tasmanian eucalyptus stands were for aggregated patterns of large trees (Ribe et al. 2013). These preferences do not always coincide with what is natural: Ribe (2009) reported a preference for stands with low levels of woody debris on the forest floor. Preferences also vary by background or experience of the public: stakeholder groups more accustomed to forestry activities generally have higher preferences for silvicultural activities than stakeholders that are less accustomed to forestry activities (Bliss 2000; Bradley and Kearney 2007).

Scenic beauty estimation studies (see Daniel and Boster 1976) that evaluate the aesthetics of silvicultural treatments rarely separate the stand structure from the processes that create it. The appearance of the structure is evaluated, and the silviculture is blamed. Whereas clearcutting can easily be labeled as the means for creating clearcuts, complex stand structures can often result from multiple developmental pathways and different silvicultural treatments. An aesthetic evaluation of a complex stand structure is therefore not easily attached to a silvicultural treatment, particularly when the structure might also result from natural processes. Structures resulting from natural disturbances may be more acceptable than if the same structure resulted from silviculture (Brunson 1993). Any silvicultural treatment may be applied in a sound and careful manner, or it can be poorly applied. A silvicultural treatment may therefore be unfairly credited or discredited if a label is not applied carefully. Likewise, there is often flexibility within any silvicultural treatment to enhance or detract from aesthetics.

Multiaged and other complex stand structures are not explicitly identified in many of these surveys of visual preferences of stand structures or silvicultural treatments. Many studies have focused on comparisons of clearcutting systems with an assortment of available alternatives. These alternatives often included thinning treatments, various levels of retention, or, occasionally, selection treatments. Single tree selection treatments are not common in the coastal regions of the Pacific Northwest, where many of these studies took place. However, studies demonstrate a strong preference for residual stand structures with significant retention of large trees. This overriding preference for large tree retention implies that a multiaged silvicultural strategy—that develops an attractive residual stand structure in a sustainable way to regenerate new age classes of desirable trees—will be acceptable to the public. It is noteworthy that the emphasis is not on a particular form of silviculture or a specific silvicultural system. Instead, the preponderance of information on public preferences for silvicultural treatment suggests flexible means to retain large trees and regenerate new trees will be perceived in a more favorable way compared to traditional even-aged systems.

The basis of visual preferences is the conditioning and context of the viewer (Shindler et al. 2002). The strong preferences against silviculture that uses clearcut systems are associated with concerns about ecological effects and motives of foresters (Bliss 2000; Shindler et al. 2002). There is both an effect of previous knowledge and also attitudes toward forestry (Kearny and Bradley 2011). Whereas visual effects are the most apparent results of silvicultural treatments, particularly regeneration methods, visual preferences may represent historical views of regional silvicultural practices, local environmental conditions, or long-held views on what is natural. Societal values that affect visual preferences and how natural resources are viewed also are dynamic; they change with time (Hull and Buhyoff 1986; Axesson and Angelstam 2011) and between regions (Angelstam et al. 2011; Ribe et al. 2013). These values therefore have a spatial, temporal, and a social context (Brunson 1996; Shindler et al. 2002; Clausen and Schroeder 2004).

16.3 Naturalness

Esthetic appeal is an important consideration for posttreatment stand structures, but there are also human preferences for treatments that are assumed to be more natural or that achieve a greater degree of naturalness. **Naturalness** is defined (following Reif and Walentowski 2008) as a condition with minimal human interference. This opens a complicated situation where human influence is ignored but where human indifference may itself affect the naturalness of ecosystems. Hence the term "natural" reflects an older interpretation of nature that predates our present understanding of disturbance effects, ecological stability, and equilibrium ecosystem conditions (Pickett and White 1985; Oliver and Larson 1990; Botkin 1990; Sprugel 1991). Hull et al. (2001) made three observations regarding the way humans, primarily in North America, view nature. First, there is an assumed existence of a "balance of nature" whereby our natural environments are best left alone without human interventions. Hence, a proportion of people assume our natural ecosystems, especially our forests, are superior if left unmanaged. The second observation was that humans recognize that a range of naturalness exists. This is different from the concepts of a historical or future range of variability in ecosystem states. Instead, Hull et al. (2001) referred to the recognition that different natural states existed in different ecosystems. This distinction is important because it indicates recognition of spatial variation in naturalness, not necessarily of temporal variation. Third, naturalness is a social construct that results from the diversity of cultures, values, and beliefs of people. Perceptions of what is natural will therefore vary over time, between regions, and from different segments of the population.

The context of naturalness also has a strong social component. It is not surprising that studies of forest management approaches reveal preferences for natural conditions, (e.g., Burchfield et al. 2003). Shindler et al. (2002) indicated "natural" is a desirable condition in our culture. Perceptions of natural and naturalness are highly variable and very much affected by context (Figure 16.2). Manipulations by humans, such as through silviculture, are generally viewed as neither natural nor desirable. Naturalness

Figure 16.2 An even-aged stand of coast redwood in New Zealand. Although this stand of a nonnative conifer is artificial, it is revered because of its spiritual and esthetic values.

can be viewed as something that varies with space and time. The public has a view of naturalness that may not coincide with that of scientists or foresters. And it may not coincide with any traditional definition of the term "natural" (Shindler et al. 2002). Naturalness is highly valued in western North America where landscapes are viewed as less manipulated than many other places. For multiaged silviculture, these landscapes, and the stands within them, are examples for how stand structures can be designed. In many other parts of the world, fewer examples of "natural" stands or landscapes exist and there is a recognition that stand structures can be designed to fit other models, even if they are more artificial. The Plenter structure in central Europe is often cited as an example of close-to-nature silviculture, yet Schütz (2001b) noted the unnaturalness of this system. Thus cultural differences, and the conditioning that accompanies them, can lead to different interpretations of naturalness.

Management that emulates natural disturbance is a common theme of contemporary silviculture (see Chapter 3) and a means to achieve a greater level of naturalness. However, it has only been in last half century that ecologists have embraced the importance of disturbances in affecting ecological systems (O'Hara and Ramage 2013). Although describing silviculture as disturbance based goes back at least to Hawley (1921), public perceptions of disturbance emulation are mixed at best. In a case study of attitudes to forest management in western Oregon, US, Shindler and Mallon (2009) found the public skeptical about disturbance-based management. A majority of responses to their survey indicated support for the idea that following nature was preferable to human intervention. These responses may indicate unfamiliarity with concepts related to disturbance emulation, or simply a dual skepticism about both management and emulating something—disturbance—that has traditionally had a negative connotation.

Views of naturalness also rarely recognize the dynamism of natural ecosystems. Rather than viewing ecosystems as dynamic entities that are always moving to a different state, they are generally regarded as relatively static. Magnuson (1990) referred to the inability to recognize long-term change as the "invisible present." If humans are unable to foresee changes in stand structures, for example, then it is difficult to accept treatments that result in dramatic changes. If the anticipated recovery of an ecosystem following treatment cannot be visualized, then there will be a reluctance to accept those treatments. Hence, treatments that result in less dramatic changes in structure, such as with multiaged silviculture, will be seen as more desirable than treatments that result in more drastic changes, regardless of their relative naturalness.

If the management that results in the smallest change is assumed to be most desirable, then it follows that no management may be best. This provides some explanation for why people might be reluctant to accept any management. However, it is an indication that people do not recognize that change is inevitable in any system and that a "do nothing" or no management option has consequences that may be very undesirable.

A fairly recent trend is to provide new names for silvicultural treatments that generally occur within the range of traditional silvicultural regeneration methods described by Troup (1928) and Matthews (1989). The development of an alternative nomenclature or terminology for multiaged silviculture can be viewed as attempts to be more descriptive or to present them in a way that makes them more tolerable (Table 1.2). Some of these terms—such as "nature-based," "close-to-nature," or "near-natural"—also describe new philosophical approaches to management and particularly toward approaches that achieve a higher level of naturalness. They generally are variations of single tree selection or other forms of multiaged silviculture. These new terms can be seen as a reaction to the strong public opinion against forestry, and particularly even-aged silviculture. They are a social reaction rather than a new development in silviculture.

This leads to the question: can naturalness be achieved using silviculture, or is this a contradiction in terms? And does "close-to-nature silviculture" or "near-natural silviculture" have any meaning if silviculture and naturalness are incompatible? Close-to-nature silviculture has become a common term, particularly in Europe (Schütz 1999). It describes management approaches that attempt to work within the productive capability of the forest using multiaged systems and natural regeneration. However, do close-to-nature systems achieve naturalness? The concept of **hemeroby** is used to evaluate naturalness of ecosystems, with higher levels of hemeroby corresponding to less anthropogenic effects (Hill et al. 2002). The naturalness of a site might therefore correspond to the overall condition of a site, but the hemeroby describes only the portion related to human influences. The use of hemeroby, or some similar scale, allows us to evaluate the level of naturalness rather than to view it as a black-and-white type of issue. It allows us to separate the anthropogenic effects on ecological systems from other factors. Çolak et al. (2003) and Reif and Walentowski (2008) discussed strategies to enhance naturalness with silviculture. In their analyses, clearcut systems were viewed as achieving very unnatural results (low hemeroby) and multiaged systems as achieving more natural structures (high hemeroby). Thus close-to-nature or near-natural or nature-based approaches, with their use of multiaged silviculture, are generally viewed as means of achieving a higher level of naturalness. This resembles Leopold's (1949) land ethic that something is "right when it tends to preserve the integrity, stability and beauty of the biotic community."

16.4 Social acceptability of multiaged silviculture

Kimmins (1993) defined "green religion" as a faith-based belief system with many strongly held beliefs about silviculture. There is also little pretense that these beliefs are based in science. One belief is that clearcutting is always harmful, planted forests are not forest ecosystems, and the only timber harvesting should be through single tree selection (Kimmins 1993). These beliefs influence public policy and can contribute to cyclic patterns in the use of forestry practices.

Cyclic trends in the popularity and use of silvicultural practices, particularly from even-aged to multiaged, have been apparent for several centuries (see Chapter 2; O'Hara 2002, Pommerening and Murphy 2004). These cycles are the result of complex social patterns that interact with science and systems of governance. They have been evident in the frequency of use of different silviculture systems and in government policy which may, for example, discourage multiaged silviculture in one country (e.g., Finland or Japan) and encourage it in another (e.g., Switzerland). Cycles are also the result of developments in science. The great expansion of even-aged forestry in the latter half of the twentieth century was, in part, a result of the great advancements in the science of forest establishment, genetics, and growth and yield. Even-aged plantations were productive and predictable.

Multiaged stands involve relatively frequent harvest treatments. These frequent harvests represent an advantage to many owners of small forest properties that want a more continuous income stream and the flexibility to harvest when market conditions are best or when income is needed. For managers of larger forest properties, the frequent stand-level harvests may result in the same volume production from a greater number of stands. This may present operational difficulties such as needing more harvest crews at any given time or the need for more supervision. Complex silvicultural systems also require greater managerial skill and possibly more and better-trained personnel for tree marking. Hence there are long-term socioeconomic costs and benefits associated with which silvicultural systems are used.

16.5 Naturalness and the "easy way out"

Michael Pollan (1998) wrote that "when you don't trust yourself to make wise decisions about the land, letting nature decide the matter is an appealingly straightforward approach." This captures a common approach to land management from the public: with a poor understanding of forest dynamics and management, and a low level of trust in managers, why would anybody want to disrupt what are perceived to be natural systems? However, if some level of wood production necessitates some level of management, then it is a logical approach to practice a form of minimal management. Multiaged silviculture is often, therefore, the management method of choice. It represents a conservative approach for managing a resource that often requires time commitments that exceed typical human lifespans.

Education of the public is often thought to be a means of improving understanding of forestry and forest practices. Studies of public perceptions have found that prefacing choices with background information improves assessments of scenic beauty or naturalness (Bliss 2000; Bradley and Kearney 2007). However, education has its limits, and achieving major swings in public opinion on forestry issues is at best a long-term challenge. There is an underlying theme to education arguments that foresters are misunderstood and would be more appreciated if they were only understood better (Bliss 2000). A more successful approach may be to recognize the limits of educating the public and, along with education, attempt to integrate public concerns into management strategies. This involves addressing the lack of public trust in natural resources management (Shindler et al. 2002). It will likely require compromises between what is optimal from a scientific basis, and demands from the public. Multiaged silviculture may be a means of providing for public concerns over aesthetics and naturalness while also providing other ecosystem services.

16.6 Synthesis

It has been argued that forestry practices without societal acceptance will fail (Clawson 1975; Romm 1994; Shindler et al. 2002). The failure will not be due to ecological science or its application. It will be due to assuming the ecological science should take precedence over societal concerns, regardless of how poorly founded in science these concerns may be.

There is a strong negative reaction against even-aged forestry in many regions. This reaction is not based on science. Instead, it is the result of strong negative perceptions rooted in aesthetics and negative associations related to deforestation and forest degradation (Bliss 2000). Even-aged systems, particularly clearcutting, are esthetically unappealing and viewed as unnatural. Clearcutting is also viewed as environmentally destructive, even in places where stand-replacement natural disturbances may have dominated.

Multiaged silvicultural approaches have a number of advantages over approaches that result in simpler stand structures. They provide a higher level of esthetic quality and are perceived as being more natural. Demands for changes in forestry are not necessarily demands for multiaged or complex stands. Instead they are calls for naturalness, given the limitations of what public perceptions of what naturalness means. Multiaged stands are well suited to meet these social justifications but not necessarily because of ecological reasons. Multiaged stands and complex mixtures satisfy these perceptions and also avoid the sudden changes in structure that many people find offensive. For many landowners, multiaged systems provide a more constant flow of income or fuelwood. Alternatively, multiaged systems are inherently more difficult, and possibly more expensive, to manage. Hence, as with any decision regarding natural resources, there are trade-offs, where some resource values or services are enhanced and others are not. The relative constancy in stand structure that provides more consistent revenue to the landowner or scenic beauty to the viewer is a form of "social resilience" for multiaged stands. This resilience is similar to the ecological resilience of multiaged stands to natural disturbances because it also results in greater level of resilience to the economic and visual aberrations associated with even-aged systems.

Epilogue

Multiaged stands are many things. They result from combinations of diverse disturbances affecting stands at different intervals and on variable sites. These events provide multiaged stand structures that serve as models for an endless array of potential silvicultural treatments. The highly varied treatments that encompass multiaged silviculture include the majority of contemporary silvicultural options. These options include the silviculture to create complex, mixed-species stands with many age classes, the relatively simple shelterwood with reserve trees, and all the structural combinations and permutations that exist between these extremes.

In 1972, Smith wrote that "silviculture fitted to demonstrable realities of nature and human need will call forth the evolution of methods of treatment more varied than our wildest present imagination can encompass." In the decades since this quote, a combination of socioeconomic and ecological factors have resulted in greater and more complex demands on forests. These demands include an expanding range of ecosystem services which are supported by much greater ecological knowledge to guide management. However, the evolution of a more varied suite of methods of treatment to meet these demands has been slow. Forestry is a long-term endeavor, and management decisions are typically conservative. Foresters often have a tendency to adopt a narrow view of management that focuses on singular objectives. This may be because the foundation of forestry is often tied to agriculture, where the production of a single crop on an annual basis often dominates decision making. Or perhaps this narrow view is related to the assumption that forestry is either intensive or extensive, focused on timber or not. There is a fertile middle ground where we can achieve many benefits from multiaged stands, including the production of wood products. However, the silviculture available to implement multiaged stands has not yet approached the variability Smith envisioned.

A central premise of this book is that the rules that have guided various forms of multiaged silviculture in the past are primarily human constructs, not ecological rules. These rules were the result of early attempts to emulate nature or meet relatively simple societal demands such as providing fuelwood, timber, or habitat for game species. They were developed in an earlier time with less ecological understanding than we have at present. We should embrace our new understanding and develop new guidelines that are inherently flexible so they can be modified as needed to accommodate variable stand structures or site conditions. Multiaged silviculture is not a "start-over" operation with a beginning and an end; instead, it is an ongoing process that works within the constraints of the site and the structural elements that are present. With our current understanding of stand dynamics and disturbance emulation, forest managers should feel empowered to expand the range of possible multiaged stand structures. If silviculture is going to produce this array of stand structures, it needs to be unencumbered from the silviculture of the past, which was often too regimented, too inflexible, and too narrowly focused on only a small number of options.

When we attempt to emulate natural processes with silviculture, concepts such as the historical range of variability and historical disturbance regimes are useful. They provide information on disturbance frequencies and other ecosystem characteristics. However, they represent a window to the past, and only include the ecological variation, not the variation in societal demands. A future range of variability might include the integration of social factors and the ecological knowledge from the past (Thompson et al. 2009). These are guiding principles for creating forests that meet human needs and provide ecosystem services within broad ecological constraints. However, in today's forestry, constraints on creativity often limit alternatives to a very narrow set of stand structures or silvicultural systems and hinder our ability to meet diverse objectives. Instead, we should embrace the diversity of potential stand structures that exist and recognize that these represent opportunities, not constraints. Irregular stand structures may be more difficult to model and project forward. They may also vary from stand to stand and be more difficult to manage. These challenges should be justifications for greater flexibility and creativity in management.

The development of new silvicultural regimes has been hindered by a variety of misconceptions about multiaged stands. These misconceptions have, at different times, suggested that multiaged stands are less productive, have greater stand health problems, or that multiaged regimes lead to degrading the genetic quality of forests. While each of these is poorly founded in science, they have affected silviculture decision making. There is also a perception that multiaged silviculture is more difficult than even-aged silviculture. Multiaged silviculture is, of course, highly variable and can be intensively or extensively applied. Even-aged silviculture is also highly variable. Either type of system can rely on site manipulation, planting, vegetation control, and short or long thinning intervals or cutting cycles. Multiaged silviculture requires a different set of skills and should involve more variable prescriptions on a stand-by-stand basis. It is more of an art form. This artful approach to silviculture may be more difficult, but the challenges facing forestry are also difficult, or even wicked, problems (i.e., Allen and Gould 1986).

As we attempt to organize forests to maintain ecosystems and their services, we must make compromises between management efficiency, ecosystem integrity, and social values. We must also recognize that management based on the emulation of natural processes may have a limited suite of available management options. Complicating our attempts at emulation is a changing climate and the need to anticipate future ranges of variation in natural processes and social demands that are outside historic norms. Multiaged silviculture is uniquely suited to address the challenges of compromising between disturbance emulation and meeting social objectives in a changing climate (O'Hara and Ramage 2013).

Freeing Silviculture

Given these factors, the best way to manage forests in a changing climate, over great site variation and with a range of existing structural variation, is to assume each stand will be different and require different management. This may require novel stand structures, in much the same manner as novel ecosystems (e.g., Seastedt et al. 2008) are outside the historic range of variability. It also requires that we accommodate existing site and stand structural conditions. Also required is a novel silviculture that works without the impediments of the regimented approaches of the past and is flexible enough to accommodate the variation that exists in space and time. The widespread application of multiaged silviculture will require, in some cases, a freer silviculture that is fully adaptable to developing individual prescriptions. It will require some level of artfulness, or the ability to interpret site factors and develop viable prescriptions that meet objectives.

Between the extremes of a highly regimented silviculture and an artful silviculture lie the approaches that blend the art of silviculture with quantitative systemization of some aspects of prescription design and implementation. For multiaged stands, this systemization might include ranges of acceptable stocking or allocations to different canopy strata. It should include on-the-ground flexibility to accommodate existing structural features or site conditions. The key point is that these approaches, which fall between both extremes, offer potential benefits as well as drawbacks. The best solutions will be somewhat repeatable over time and space, while maintaining the freedom and flexibility to accommodate existing structural features and site conditions. Although repeatability in application of silvicultural prescriptions is a sound goal, it is only necessary to a certain extent, beyond which some variation is desirable. What is needed is the general guidance that comes from broad-scale management directives and planning, with an increased level of freedom and flexibility for on-the-ground decision making and implementation. The era of Smith's "methods of treatment more varied than our wildest present imagination" is upon us, but we need to move forward with flexibility to achieve what we can imagine.

Multiaged stands are many things, and they should be. Our silvicultural capabilities allow us to do many things to a forest, although they may not all be economical or socially acceptable. Within the constraints of economics and social acceptance, we can manipulate a forest in many ways. These treatments may go beyond historic ranges of ecological conditions or processes, or what is considered "natural." Multiaged stands can be managed to achieve objectives in ways that can exceed the way they are achieved without management. In this sense, multiaged systems can be part of a management that may not necessarily be close to nature, but can build on our understandings to provide ecosystems and ecosystem services in ways that nature cannot. The best solutions will be those that meet a combination of objectives, and some of these solutions may create stand structures that are outside ecological ranges of variability. Multiaged stands are many things and we have the ability, through the artful application of silviculture, to shape them to meet diverse objectives.

List of Species

alpine ash	*Eucalyptus delegatensis*	European ash	*Fraxinus excelsior*
American beech	*Fagus grandifolia*	European beech	*Fagus sylvatica*
armillaria root disease	*Armillaria* spp.	fig, strangler	*Ficus* spp.
ash, alpine	*Eucalyptus delegatensis*	fir, Balsam	*Abies balsamea*
ash, common or European	*Fraxinus excelsior*	fir, grand	*Abies grandis*
ash, silvertop	*Eucalyptus sieberi*	fir, Pacific silver	*Abies amabilis*
ash, white	*Fraxinus americana*	fir, noble	*Abies procera*
aspen species	*Populus* spp.	fir, silver	*Abies alba*
aspen, quaking	*Populus tremuloides*	fir, subalpine	*Abies lasiocarpa*
balsam fir	*Abies balsamea*	fir, white	*Abies concolor*
beech, American	*Fagus grandifolia*	forest tent caterpillar	*Malacosoma disstria*
beech, European	*Fagus sylvatica*	garlic, wild	*Allium ursinum*
beech, Japanese	*Fagus crenata*	giant sequoia	*Sequoiadendron gigantea*
birch, paper	*Betula papyrifera*	grand fir	*Abies grandis*
birch, yellow	*Betula alleghaniensis*	hemlock, eastern	*Tsuga canadensis*
black cherry	*Prunus serotina*	hemlock, Himalayan	*Tsuga dumosa*
black locust	*Robinia pseudoacacia*	hemlock, western	*Tsuga heterophylla*
black oak	*Quercus velutina*	hiba	*Thujopsis dolabrata*
blackjack oak	*Quercus marilandica*	Himalayan hemlock	*Tsuga dumosa*
blueblossom	*Ceanothus thyrsiflorus*	hinoki	*Chamaecyparis obtusa*
Calabrian pine	*Pinus nigra* ssp. *laricio*	Honda beraliya	*Shorea megistophylla*
cedar, Lebanon	*Cedrus libani*	incense-cedar	*Calocedrus decurrens*
chestnut blight	*Cryphonectria parasitica*	interior Douglas-fir	*Pseudotsuga menziesii* var. *glauca*
cherry, black	*Prunus serotina*		
coast Douglas-fir	*Pseudotsuga menziesii* var. *menziesii*	interior spruce	*Picea glauca* × *Picea engelmannii*
coast redwood	*Sequoia sempervirens*	jack pine	*Pinus banksiana*
common ash	*Fraxinus excelsior*	Japanese beech	*Fagus crenata*
cottonwood species	*Populus* spp.	karri	*Eucalyptus diversicolor*
cypress pine	*Callitris* spp.	laminated root rot	*Phellinus weirii*
deerbrush	*Ceanothus integerrimus*	larch species	*Larix* spp.
Douglas-fir, coast	*Pseudotsuga menziesii* var. *menziesii*	larch, Sikkim	*Larix griffithiana*
		larch, western	*Larix occidentalis*
Douglas-fir, interior	*Pseudotsuga menziesii* var. *glauca*	Lebanon cedar	*Cedrus libani*
		lime	*Tilia* spp.
dwarf mistletoe	*Arceuthobium* spp.	loblolly pine	*Pinus taeda*
eastern hemlock	*Tsuga canadensis*	locust, black	*Robinia pseudoacacia*
eastern redcedar	*Juniperus virginiana*	lodgepole pine	*Pinus contorta*
eastern spruce budworm	*Choristoneura fumiferana*	longleaf pine	*Pinus palustris*
eastern white pine	*Pinus strobus*	maple, red	*Acer rubrum*
Engelmann spruce	*Picea engelmannii*	maple, striped	*Acer pensylvanicum*

maple, sycamore	*Acer pseudoplatanus*	redwood, coast	*Sequoia sempervirens*
maple, sugar	*Acer saccharum*	scarlet oak	*Quercus coccinea*
maritime pine	*Pinus pinaster*	Scots pine	*Pinus sylvestris*
mahogany	*Swietenia mahagoni*	sequoia, giant	*Sequoiadendron gigantea*
mountain pine beetle	*Dendroctonus ponderosae*	sessile oak	*Quercus petraea*
manzanita	*Arctostaphylos* spp.	shoestring root disease	*Armillaria* spp.
noble fir	*Abies procera*	shortleaf pine	*Pinus echinata*
northern red oak	*Quercus rubra*	Sikkim larch	*Larix griffithiana*
Norway spruce	*Picea abies*	silver fir	*Abies alba*
oak, black	*Quercus velutina*	silvertop ash	*Eucalyptus sieberi*
oak, blackjack	*Quercus marilandica*	Sitka spruce	*Picea sitchensis*
oak, northern red	*Quercus rubra*	spruce species	*Picea* spp.
oak, post	*Quercus stellata*	spruce budworm, eastern	*Choristoneura fumiferana*
oak, scarlet	*Quercus coccinea*	spruce budworm, western	*Choristoneura occidentalis*
oak, sessile	*Quercus petraea*	spruce, Engelmann	*Picea engelmannii*
oak, water	*Quercus nigra*	spruce, interior	*Picea glauca* × *Picea engelmannii*
oak, white	*Quercus alba*		
Pacific silver fir	*Abies amabilis*	spruce, Norway	*Picea abies*
paper birch	*Betula papyrifera*	spruce, red	*Picea rubens*
phytophthora	*Phytophthora* spp.	spruce, Sitka	*Picea sitchensis*
pine, Calabrian	*Pinus nigra* ssp. *laricio*	spruce, white	*Picea glauca*
pine, cypress	*Callitris* spp.	strangler fig	*Ficus* spp.
pine, eastern white	*Pinus strobus*	stone pine	*Pinus pinea*
pine, jack	*Pinus banksiana*	striped maple	*Acer pensylvanicum*
pine, loblolly	*Pinus taeda*	subalpine fir	*Abies lasiocarpa*
pine, lodgepole	*Pinus contorta*	sugar maple	*Acer saccharum*
pine, longleaf	*Pinus palustris*	sugar pine	*Pinus lambertiana*
pine, maritime	*Pinus pinaster*	sugi	*Cryptomeria japonica*
pine, pitch	*Pinus rigida*	sycamore maple	*Acer pseudoplatanus*
pine, ponderosa	*Pinus ponderosa*	tanoak	*Notholithocarpus densiflorus*
pine, radiata	*Pinus radiata*	water oak	*Quercus nigra*
pine, Scots	*Pinus sylvestris*	western hemlock	*Tsuga heterophylla*
pine, shortleaf	*Pinus echinata*	western larch	*Larix occidentalis*
pine, stone	*Pinus pinea*	western redcedar	*Thuja plicata*
pine, sugar	*Pinus lambertiana*	western spruce budworm	*Choristoneura occidentalis*
pine, whitebark	*Pinus albicaulis*	western white pine	*Pinus monticola*
pitch pine	*Pinus rigida*	white ash	*Fraxinus americana*
ponderosa pine	*Pinus ponderosa*	white fir	*Abies concolor*
post oak	*Quercus stellata*	white oak	*Quercus alba*
quaking aspen	*Populus tremuloides*	white spruce	*Picea glauca*
radiata pine	*Pinus radiata*	whitebark pine	*Pinus albicaulis*
redcedar, eastern	*Juniperus virginiana*	wild garlic	*Allium ursinum*
redcedar, western	*Thuja plicata*	yellow-cedar	*Callitropsis nootkatensis*
red maple	*Acer rubrum*	yellow birch	*Betula alleghaniensis*
red oaks	*Quercus* spp. section *Lobatae*	yellow-poplar	*Liriodendron tulipifera*
red spruce	*Picea rubens*	yertchuk	*Eucalyptus consideniana*

References

Abrams, M.D. and Kubiske, M.E. (1990). Leaf structural characteristics of 31 hardwood and conifer tree species in central Wisconsin: influence of light regime and shade tolerance rank. *Forest Ecology and Management*, **31**(4), 245–253.

Acker, S.A., Zenner, E.K., and Emmingham, W.H. (1998). Structure and yield of two-aged stands on the Willamette National Forest, Oregon: implications for green tree retention. *Canadian Journal of Forest Research*, **28**(5), 749–758.

Adams, D.M. and Ek, A.R. (1974). Optimizing the management of uneven-aged forest stands. *Canadian Journal of Forest Research*, **4**(3), 274–287.

Adams, W.T. (1992). Gene dispersal within forest tree populations. *New Forests*, **6**, 217–240.

Adams, W.T., Zuo J.H., Shimizu, J.Y., and Tappeiner, J.C. (1998). Impact of alternative regeneration methods on genetic diversity in coastal Douglas-fir. *Forest Science*, **44**(3), 390–396.

Agee, J.K. (1993). *Fire Ecology of Pacific Northwest Forests*. Island Press, Washington, DC.

Albrecht, A., Hanewinkel, M., Bauhus, J., and Kohnle, U. (2012). How does silviculture affect storm damage in forests of south-western Germany? Results from empirical modeling based on long-term observations. *European Journal of Forest Research*, **131**(1), 229–247.

Alexander, R.R. and Edminster, C.B. (1977). *Regulation and control of cut under uneven-aged management*. USDA Forest Service, Rocky Mountain Forest and Range Experiment Station. Research Paper RM-RP-182.

Allen, C.D., Macalady, A.K., Chenchouni, H., Bachelet, D., McDowell, N., Vennetier, M., Kitzberger, T., Rigling, A., Breshears, D.D., Hogg, E.H., Gonzalez, P., Fensham, R., Zhang, Z., Castro, J., Demidova, N., Lim, J., Allard, G., Running, S.W., Semerci, A., and Cobb, N. (2010). A global overview of drought and heat-induced tree mortality reveals emerging climate change risks for forests. *Forest Ecology and Management*, **259**(4), 660–684.

Allen, G.M. and Gould, E.M. (1986). Complexity, wickedness, and public forests. *Journal of Forestry*, **84**(4), 20–23.

Ammer, C. and Mosandl, R. (2007). Which grow better under the canopy of Norway spruce—planted or sown seedlings of European beech? *Forestry*, **80**(4), 385–395.

Anderson, D.J. and Bare, B.B. (1994). A dynamic programming algorithm for optimization of uneven-aged forest stands. *Canadian Journal of Forest Research*, **24**(9), 1758–1765.

Angelstam, P., Axelsson, R., Elbakidze, M., Laestadius, L., Lazdinis, M., Nordberg, M., Patru-Stupariu, I., and Smith, M. (2011). Knowledge production and learning for sustainable forest management on the ground: pan-European landscapes as a time machine. *Forestry*, **84**(5), 581–596.

Angelstam, P.K. (1998). Maintaining and restoring biodiversity in European boreal forests by developing natural disturbance regimes. *Journal of Vegetation Science*, **9**(4), 593–602.

Antos, J.A., Parish, R., and Conley, K. (2000). Age structure and growth of the tree-seedling bank in subalpine spruce-fir forests of south-central British Columbia. *American Midland Naturalist*, **143**(2), 342–354.

Arbogast, C., Jr. (1957). *Marking guides for northern hardwoods under the selection system*. USDA Forest Service, Lake States Forest Experiment Station. Station Paper 56.

Arno, S.F. and Allison-Bunnell, S. (2002). *Flames in Our Forests: Disaster or Renewal?* Island Press, Washington, DC.

Arno, S.F., Scott, J.H., and Hartwell, M.G. (1995). *Age-class structure of old growth ponderosa pine/Douglas-fir stands and its relationship to fire history*. USDA Forest Service, Intermountain Research Station. Research Paper INT-RP-481.

Arno, S.F., Smith, H.Y., and Krebs, M.A. (1997). *Old growth ponderosa pine and western larch stand structures: influences of pre-1900 fires and fire exclusion*. USDA Forest Service, Intermountain Research Station. Research Paper INT-RP-495.

Assmann, E. (1970). *The Principles of Forest Yield Study*. Pergamon Press, Oxford.

Attiwill, P.M. (1994). The disturbance of forest ecosystems: the ecological basis for conservation management. *Forest Ecology and Management*, **63**(2–3), 247–300.

Attiwill, P. and Binkley, D. (2013). Exploring the mega-fire reality: a "Forest Ecology and Management" conference. *Forest Ecology and Management*, **294**, 1–3.

Atwood, C.J., Fox, T.R., and Loftis, D.L. (2009). Effects of alternative silviculture on stump sprouting in the southern Appalachians. *Forest Ecology and Management*, **257**(4), 1305–1313.

Aubin, I., Garbe, C.M., Colombo, S., Drever, C.R., McKenney, D.W., Messier, C., Pedlar, J., Saner, M.A., Venier, L., Wellstead, A.M., Winder, R., Witten, E., and Ste-Marie, C. (2011). Why we disagree about assisted migration: ethical implications of a key debate regarding the future of Canada's forests. *Forestry Chronicle*, **87**(6), 755–765.

Aubry, K.B., Halpern, C.B., and Peterson, C.E. (2009). Variable-retention harvests in the Pacific Northwest: a review of short-term findings from the DEMO study. *Forest Ecology and Management*, **258**(4), 398–408.

Avery, E.A. and Burkhart, H.E. (2002). *Forest Measurements*, 5th edn. McGraw-Hill, Inc., Boston.

Axelsson, R. and Angelstam, P. (2011). Uneven-aged forest management in boreal Sweden: local forestry stakeholders' perceptions of different sustainability dimensions. *Forestry*, **84**(5), 567–579.

Bagnaresi, U., Giannini, R., Grassi, G., Minotta, G., Paffetti, D., Prato, E.P., and Placidi, A.M.P. (2002). Stand structure and biodiversity in mixed, uneven-aged coniferous forests in the eastern Alps. *Forestry*, **75**(4), 357–364.

Bailey, R.G., Pfister, R.D., and Henderson, J.A. (1978). Nature of land and resource classification—a review. *Journal of Forestry*, **76**(10), 650–655.

Baker, J.B., Cain, M.D., Guldin, J.M., Murphy, P.A., and Shelton, M.G. (1996). *Uneven-aged silviculture for the loblolly and shortleaf pine forest cover types*. USDA Forest Service, Southern Research Station. General Technical Report SO-GTR-118.

Baker, J.B. and Shelton, M.G. (1998). Rehabilitation of understocked loblolly-shortleaf pine stands: I. Recently cutover natural stands. *Southern Journal of Applied Forestry*, **22**(1), 35–40.

Balandier, P., Collet, C., Miller, J.H., Reynolds, P.E., and Zedaker, S.M. (2006). Designing forest vegetation management strategies based on the mechanisms and dynamics of crop tree competition by neighbouring vegetation. *Forestry*, **79**(1), 3–27.

Baldocchi, D. and Collineau, S. (1994). The physical nature of solar radiation in heterogeneous canopies: spatial and temporal attributes. In M.M. Caldwell and R.W. Pearcy, eds. *Physiological Ecology; Exploitation of Environmental Heterogeneity by Plants: Ecophysiological Processes Above- and Belowground*, pp. 21–71. Academic Press, San Diego.

Bare, B.B. and Opalach, D. (1988). Determining investment-efficient diameter distributions for uneven-aged northern hardwoods. *Forest Science*, **34**(1), 243–249.

Barnes, B.V., Pregitzer, K.S., Spies, T.A., and Spooner, V.H. (1982). Ecological forest site classification. *Journal of Forestry*, **80**(8), 493–498.

Barrett, S.W., Arno, S.F., and Key, C.H. (1991). Fire regimes of western larch-lodgepole pine forests in Glacier National Park, Montana. *Canadian Journal of Forest Research*, **21**(12), 1711–1720.

Barth, A. (1916). Norges skoger med stormskridt mot undergangen [The forests of Norway are heading for their ruin in high speed]. *Tidsskrift for Skogbruk*, **24**, 123–154.

Battaglia, M.A., Mou, P., Palik, B., and Mitchell, R.J. (2002). The effect of spatially variable overstory on the understory light environment of an open-canopied longleaf pine forest. *Canadian Journal of Forest Research*, **32**(11), 1984–1991.

Battaglia, M., Sands, P., White, D., and Mummery, D. (2004). CABALA: a linked carbon, water and nitrogen model of forest growth for silvicultural decision support. *Forest Ecology and Management*, **193**(1–2), 251–282.

Battaglia, M. and Sands, P. (1998). Process-based forest productivity models and their application in forest management. *Forest Ecology and Management*, **102**(1), 13–32.

Bauhus, J. (1996). C and N mineralization in an acid forest soil along a gap-stand gradient. *Soil Biology & Biochemistry*, **28**(7), 923–932.

Bauhus, J., Puettmann, K., and Messier, C. (2009). Silviculture for old-growth attributes. *Forest Ecology and Management*, **258**(4), 525–537.

Bauhus, J., Vor, T., Bartsch, N., and Cowling, A. (2004). The effects of gaps and liming on forest floor decomposition and soil C and N dynamics in a *Fagus sylvatica* forest. *Canadian Journal of Forest Research*, **34**(3), 509–518.

Bawa, K.S. and Seidler, R. (1998). Natural forest management and conservation of biodiversity in tropical forests. *Conservation Biology*, **12**(1), 46–55.

Beaudet, M., Brisson, J., Messier, C., and Gravel, D. (2007). Effect of a major ice storm on understory light conditions in an old-growth *Acer-Fagus* forest: pattern of recovery over seven years. *Forest Ecology and Management*, **242**(2–3), 553–557.

Beaudet, M. and Messier, C. (1998). Growth and morphological responses of yellow birch, sugar maple, and beech seedlings growing under a natural light gradient. *Canadian Journal of Forest Research*, **28**(7), 1007–1015.

Beese, W.J., Dunsworth, B.G., Zielke, K., and Bancroft, B. (2003). Maintaining attributes of old-growth forests in coastal BC through variable retention. *Forestry Chronicle*, **79**(3), 570–578.

Benecke, U. (1996). Ecological silviculture: the application of age-old methods. *New Zealand Forestry*, **41**(2), 27–33.

Benson, R.E., McCool, S.F., and Schlieter, J.A. (1985). *Attaining visual quality objectives in timber harvest areas: landscape architects' evaluation*. USDA Forest Service,

Douglas-fir stands. *Canadian Journal of Forest Research*, **31**(2), 302–312.

Brang, P. (2001). Resistance and elasticity: promising concepts for the management of protection forests in the European Alps. *Forest Ecology and Management*, **145**(1–2), 107–119.

Bravo, F., Osorio, L.F., Pando, V., and Del Peso, C. (2010). Long-term implications of traditional forest regulation methods applied to Maritime pine (*Pinus pinaster* Ait.) forests in central Spain: a century of management plans. *iForest—Biogeosciences and Forestry*, **3** (March 2010), 33–38.

Britton, K.O. and Liebhold, A.M. (2013). One world, many pathogens! *New Phytologist*, **197**(1), 9–10.

Brodie, L.C. and DeBell, D.S. (2013). Residual densities affect growth of overstory trees and planted Douglas-fir, western hemlock, and western redcedar: Results from the first decade. *Western Journal of Applied Forestry*, **28**(3), 121–127.

Brokaw, N. and Busing, R.T. (2000). Niche versus chance and tree diversity in forest gaps. *Trends in Ecology & Evolution*, **15**(5), 183–188.

Brokaw, N.V.L. (1982). The definition of treefall gap and its effect on measures of forest dynamics. *Biotropica*, **14**(2), 158–160.

Brokaw, N.V.L. (1985a). Treefalls, regrowth and community structure in tropical forests. In S.T.A. Pickett and P.S. White, eds. *The Ecology of Natural Disturbance and Patch Dynamics*, pp. 53–69. Academic Press, Inc., San Diego.

Brokaw, N.V.L. (1985b). Gap-phase regeneration in a tropical forest. *Ecology*, **66**(3), 682–687.

Brouwer, R. (1983). Functional equilibrium: sense or nonsense? *Netherlands Journal of Agricultural Sciences*, **31**, 335–348.

Brunson, M.W. (1993). "Socially acceptable" forestry: what does it imply for ecosystem management? *Western Journal of Applied Forestry*, **8**(4), 116–119.

Brunson, M.W. (1996). Human dimensions in silviculture. In A.W. Ewert, ed. *Natural Resource Management: The Human Dimension*, pp. 91–108. Westview Press, Boulder.

Bugmann, H. (2001). A review of forest gap models. *Climatic Change*, **51**(3–4), 259–305.

Buiteveld, J., Vendramin, G.G., Leonardi, S., Kamer, K., and Geburek, T. (2007). Genetic diversity and differentiation in European beech (*Fagus sylvatica* L.) stands varying in management history. *Forest Ecology and Management*, **247**(1–3), 98–106.

Buongiorno, J. (2001). Quantifying the implications of transformation from even to uneven-aged forest stands. *Forest Ecology and Management*, **151**(1–3), 121–132.

Buongiorno, J., Dahir, S., Lu, H.C., and Lin, C.R. (1994). Tree size diversity and economic returns in uneven-aged forest stands. *Forest Science*, **40**(1), 83–103.

Buongiorno, J. and Michie, B. (1980). A matrix model of uneven-aged forest management. *Forest Science*, **26**(4), 609–625.

Burchfield, J.A., Miller, J.M., Allen, S., Schroeder, R.F., and Miller, T. (2003). *Social implications of alternatives to clearcutting on the Tongass National Forest: an exploratory study of residents' responses to alternative silvicultural treatments at Hanus Bay, Alaska*. USDA Forest Service, Pacific Northwest Research Station. General Technical Report PNW-GTR-575.

Burk, D.A. (1970). *The Clearcut Crisis: Controversy in the Bitterroot*. Jursnick Printing, Great Falls.

Burkhart, H.E. and Brooks, T.M. (1990). Status and future of growth and yield models. In V.J. LaBau and T. Cunia, eds. *Symposium on State-of-the-Art Methodology of Forest Inventory*, pp. 409–414. USDA Forest Service, Pacific Northwest Research Station. General Technical Report PNW-GTR-263.

Burschel, P. and Huss, J. (1987). *Grundriss des Waldbaus*. Parey, Berlin.

Butler, P., Swanston, C.W., Janowiak, M., Parker, L., St. Pierre, M., and Brandt, L. (2012). Chapter 2. Adaptation strategies and approaches. In C. Swanston and M. Janowiak, eds. *Forest adaptation resources: climate change tools and approaches for land managers*, pp. 15–34. USDA Forest Service, Northern Research Station. General Technical Report NRS-GTR-87.

Cain, M.D. and Shelton, M.G. (2001). Natural loblolly and shortleaf pine productivity through 53 years of management under four reproduction methods. *Southern Journal of Applied Forestry*, **24**(1), 6–18.

Cain, M.D. and Shelton, M.G. (2002). Does prescribed burning have a place in regenerating uneven-aged loblolly-shortleaf pine stands? *Southern Journal of Applied Forestry*, **26**(3), 117–123.

Cain, M.D., Wigley, T.B., and Reed, D.J. (1998). Prescribed fire effects on structure in uneven-aged stands of loblolly and shortleaf pines. *Wildlife Society Bulletin*, **26**(2), 209–218.

Cajander, A.K. (1926). The theory of forest types. *Acta Forestalia Fennica*, **29**(3), 1–108.

Cajander, A.K. (1949). Forest types and their significance. *Acta Forestalia Fennica*, **56**(Art. 5), 1–71.

Calama, R., Barbeito, I., Pardos, M., del Río, M., and Montero, G. (2008). Adapting a model for even-aged *Pinus pinea* L. stands to complex multi-aged structures. *Forest Ecology and Management*, **256**(6), 1390–1399.

Cameron, A.D., Mason, W.L., and Malcolm, D.C. (2001). Transformation of plantation forests: papers presented at the IUFRO Conference held in Edinburgh, Scotland, 29 August to 3 September 1999. *Forest Ecology and Management*, **151**(1–3), 1–5.

Canham, C.D. (1988a). Growth and canopy architecture of shade-tolerant trees: response to canopy gaps. *Ecology*, **69**(3), 786–795.

Canham, C.D. (1988b). An index for understory light levels in and around canopy gaps. *Ecology*, **69**(5), 1634–1638.

Canham, C.D. (1989). Different responses to gaps among shade-tolerant tree species. *Ecology*, **70**(3), 548–550.

Canham, C.D., Denslow, J.S., Platt, W.J., Runkle, J.R., Spies, T.A., and White, P.S. (1990). Light regimes beneath closed canopies and tree-fall gaps in temperate and tropical forests. *Canadian Journal of Forest Research*, **20**(5), 620.

Canham, C., Kobe, R., Latty, E., and Chazdon, R. (1999). Interspecific and intraspecific variation in tree seedling survival: effects of allocation to roots versus carbohydrate reserves. *Oecologia*, **121**(1), 1–11.

Cannell, M.G.R. (1989). Physiological basis of wood production: a review. *Scandinavian Journal of Forest Research*, **4**(1–4), 459–490.

Carey, A.B. (2003). Biocomplexity and restoration of biodiversity in temperate coniferous forest: inducing spatial heterogeneity with variable-density thinning. *Forestry*, **76**(2), 127–136.

Carey, A.B. and Curtis, R.O. (1996). Conservation of biodiversity: a useful paradigm for forest ecosystem management. *Wildlife Society Bulletin*, **24**(4), 610–620.

Carlson, C.E. and Wulf, N.W. (1989). *Spruce budworms handbook: silvicultural strategies to reduce stand and forest susceptibility to the western spruce budworm*. USDA Agriculture Handbook No. 676. US Government Printing Office, Washington, DC.

Carmean, W.H. (1975). Forest site quality evaluation in the USA. *Advances in Agronomy*, **27**, 209–269.

Carmean, W.H. (1979). Soil-site factors affecting hardwood regeneration and growth. In H.A. Holt and B.C. Fisher, eds. *Proceedings: regenerating oaks in upland hardwood forests. The 1979 John S. Wright Forestry Conference*, pp. 61–74. Purdue University, West Lafayette.

Caspersen, J. (2006). Elevated mortality of residual trees following single-tree felling in northern hardwood forests. *Canadian Journal of Forest Research*, **36**(5), 1255–1265.

Caspersen, J. and Kobe, R. (2001). Interspecific variation in sapling mortality in relation to growth and soil moisture. *Oikos*, **92**(1), 160–168.

Castello, J.D. and Teale, S.A., eds. (2011), *Forest Health: An Integrated Perspective*. Cambridge University Press, New York.

Chang, S.J. (1981). Determination of the optimal growing stock and cutting cycle for an uneven-aged stand. *Forest Science*, **27**(4), 739–744.

Chang, S.J. (1990). An economic comparison of even-aged and uneven-aged management of southern pines in the mid-South. In C.A. Hickman, ed. *Proceedings of the Southern Forest Economics Workshop on Evaluating Even- and All-Aged Timber Management Options for Southern Forest Lands*, pp. 45–52. USDA Forest Service, Southern Experimental Station. General Technical Report SO-GTR-79.

Chapman, H. (1945). The effect of overhead shade on the survival of loblolly pine seedlings. *Ecology*, **26**(3), 274–282.

Chazdon, R. and Pearcy, R. (1991). The importance of sunflecks for forest understory plants. *Bioscience*, **41**(11), 760–766.

Chen, H.Y.H. (1997). Interspecific responses of planted seedlings to light availability in interior British Columbia: survival, growth, allometric patterns, and specific leaf area. *Canadian Journal of Forest Research*, **27**(9), 1383–1393.

Chen, H.Y.H. and Klinka, K. (1998). Survival, growth, and allometry of planted *Larix occidentalis* seedlings in relation to light availability. *Forest Ecology and Management*, **106**(2–3), 169–179.

Chen, H.Y.H., Klinka, K., and Kayahara, G.J. (1996). Effects of light on growth, crown architecture, and specific leaf area for naturally established *Pinus contorta* var *latifolia* and *Pseudotsuga menziesii* var *glauca* saplings. *Canadian Journal of Forest Research*, **26**(7), 1149–1157.

Chisman, H.H. and Schumacher, F.X. (1940). On the tree-area ratio and certain of its applications. *Journal of Forestry*, **38**(4), 311–317.

Chrimes, D. and Nilson, K. (2005). Overstorey density influence on the height of *Picea abies* regeneration in northern Sweden. *Forestry*, **78**(4), 433–442.

Christensen, N.L. (1988). Succession and natural disturbance: paradigms, problems, and preservation of natural ecosystems. In J.K. Agee and D.R. Johnson, eds. *Ecosystem Management for Parks and Wilderness*, pp. 62–86. University of Washington Press, Seattle.

Churchill, D.J., Larson, A.J., Dahlgreen, M.C., Franklin, J.F., Hessburg, P.F. and Lutz, J.A. (2013). Restoring forest resilience: from reference spatial patterns to silvicultural prescriptions and monitoring. *Forest Ecology and Management*, **291**, 442–457.

Ciesla, W.M. (2011). *Forest Entomology: A Global Perspective*. John Wiley & Sons, Chischester.

Clark, L.R. and Sampson, R.N. (1995). *Forest Ecosystem Health in the Inland West: A Science and Policy Reader*. Forest Policy Center of American Forests, Washington, DC.

Clausen, D.L. and Schroeder, R.F. (2004). *Social acceptability of alternatives to clearcutting: discussion and literature review with emphasis on southeast Alaska*. USDA Forest Service, Pacific Northwest Research Station. General Technical Report PNW-GTR-594.

Claveau, Y., Comeau, P.G., Messier, C., and Kelly, C.P. (2006). Early above- and below-ground responses of subboreal conifer seedlings to various levels of deciduous canopy removal. *Canadian Journal of Forest Research*, **36**(8), 1891–1899.

Clawson, M. (1975). *Forests for Whom and for What?* Resources for the Future, Inc., Washington, DC.

Cleary, B.D., Greaves, R.D., and Hermann, R.K., eds. (1978). *Regenerating Oregon's Forests: A Guide for the Regeneration Forester*. Oregon State University Extension Service, Corvallis.

Clebsch, E. and Busing, R. (1989). Secondary succession, gap dynamics, and community structure in a southern Appalachian cove forest. *Ecology*, **70**(3), 728–735.

Clements, F.E. (1916). *Plant Successsion. An Analysis of the Development of Vegetation*. Carnegie Institute, Washington, DC.

Coates, K.D. (2000). Conifer seedling response to northern temperate forest gaps. *Forest Ecology and Management*, **127**(1–3), 249–269.

Coates, K.D. (2002). Tree recruitment in gaps of various size, clearcuts and undisturbed mixed forest of interior British Columbia, Canada. *Forest Ecology and Management*, **155**(1–3), 387–398.

Coates, K.D. and Burton, P.J. (1997). A gap-based approach for development of silvicultural systems to address ecosystem management objectives. *Forest Ecology and Management*, **99**(3), 337–354.

Coates, K.D. and Burton, P.J. (1999). Growth of planted tree seedlings in response to ambient light levels in northwest interior cedar-hemlock forests of British Columbia. *Canadian Journal of Forest Research*, **29**(9), 1374–1382.

Coates, K.D., Canham, C.D., Beaudet, M., Sachs, D.L., and Messier, C. (2003). Use of a spatially explicit individual-tree model (SORTIE/BC) to explore the implications of patchiness in structurally complex forests. *Forest Ecology and Management*, **186**(1–3), 297–310.

Cochran, P.H., Geist, J.M., Clemens, D.L., Clausnitzer, R.R., and Powell, D.C. (1994). *Suggested stocking levels for forest stands in northeastern Oregon and southeastern Washington*. USDA Forest Service, Pacific Northwest Research Station. Research Note PNW-RN-513.

Coile, T.S. and Schumacher, F.X. (1953). Relation of soil properties to site index of loblolly and shortleaf pines in the piedmont region of the Carolinas, Georgia, and Alabama. *Journal of Forestry*, **51**(10), 739–744.

Çolak, A.H., Rotherham, I.D., and Çalikoglu, M. (2003). Combining "naturalness concepts" with close-to-nature silviculture. *Forstwissenschaftliches Centralblatt*, **122**(6), 421–431.

Coles, J.F. and Fowler, D.P. (1976). Inbreeding in neighboring trees in two white spruce populations. *Silvae Genetica*, **25**(1), 29–34.

Collet, C. and Chenost, C. (2006). Using competition and light estimates to predict diameter and height growth of natural regenerated beech seedlings growing under changing canopy conditions. *Forestry*, **79**(5), 489–502.

Collet, C. and Le Moguedec, G. (2007). Individual seedling mortality as a function of size, growth and competition in natural regenerated beech seedlings. *Forestry*, **80**(4), 359–370.

Comeau, P.G., Wang, J.R., and Letchford, T. (2003). Influences of paper birch competition on growth of understory white spruce and subalpine fir following spacing. *Canadian Journal of Forest Research*, **33**(10), 1962–1973.

Cook, J.E. (1996). Implications of modern successional theory for habitat typing: a review. *Forest Science*, **42**(1), 67–75.

Cooper, C.F. (1960). Changes in vegetation, structure, and growth of southwestern pine forests since white settlement. *Ecological Monographs*, **30**(2), 129–164.

Covington, W.W. and Moore, M.M. (1994). Southwestern ponderosa forest structure: changes since Euro-American settlement. *Journal of Forestry*, **92**(1), 39–47.

Cremer, K.W., Borough, C.J., McKinnell, F.H., and Carter, P.R. (1982). Effects of stocking and thinning on wind damage in plantations. *New Zealand Journal of Forestry Science*, **12**(2), 244–268.

Curtis, R.O. (1970). Stand density measures: an interpretation. *Forest Science*, **16**(4), 403–414.

Curtis, R.O. (1971). Tree area power function and related stand density measures for Douglas-fir. *Forest Science*, **17**(2), 146–159.

Curtis, R.O. (1982). A simple index of stand density for Douglas-fir. *Forest Science*, **28**(1), 92–94.

Curtis, R.O. (1995). *Extended rotations and culmination age of coast Douglas-fir: old studies speak to current issues*. USDA Forest Service, Pacific Northwest Research Station. Research Paper PNW-RP-485.

Curtis, R.O. (1998). "Selective cutting" in Douglas-fir: history revisited. *Journal of Forestry*, **96**(7), 40–46.

Curtis, R.O., Marshall, D.D., and Bell, J.F. (1997). LOGS: a pioneering example of silvicultural research in coast Douglas-fir. *Journal of Forestry*, **95**(7), 19–25.

Cyr, D., Gauthier, S., Bergeron, Y., and Carcaillet, C. (2009). Forest management is driving the eastern North American boreal forest outside its natural range of variability. *Frontiers in Ecology and the Environment*, **7**(10), 519–524.

Dai, X.B. (1996). Influence of light conditions in canopy gaps on forest regeneration: a new gap light index and its application in a boreal forest in east-central Sweden. *Forest Ecology and Management*, **84**(1–3), 187–197.

D'Amato, A.W., Bradford, J.B., Fraver, S., and Palik, B.J. (2011). Forest management for mitigation and adaptation to climate change: insights from long-term silviculture experiments. *Forest Ecology and Management*, **262**, 803–816.

Dana, S.T. and Fairfax, S.K. (1980). *Forest and Range Policy: Its Development in the United States*. McGraw-Hill Book Company, New York.

Daniel, R.C. and Boster, R.S. (1976). *Measuring landscape esthetics: the scenic beauty estimation method*. USDA Forest Service, Rocky Mountain Forest and Range Experiment Station. Research Paper RM-RP-167.

Daniel, T.W., Helms, J.A., and Baker, F.S. (1979). *Principles of Silviculture*, 2nd edn. McGraw-Hill, New York.

Daubenmire, R. and Daubenmire, J.B. (1968). *Forest vegetation of eastern Washington and northern Idaho*. Washington Agriculture Experiment Station. Technical Report 60. Washington State University, Pullman.

Daubenmire, R. (1952). Forest vegetation of northern Idaho and adjacent Washington, and its bearing on concepts of vegetation classification. *Ecological Monographs*, **22**(4), 301–330.

Daubenmire, R. (1976). Use of vegetation in assessing productivity of forest lands. *Botanical Review*, **42**(2), 115–143.

de la Giroday, H.-M.C., Carroll, A.L., and Aukema, B.H. (2012). Breach of the northern Rocky Mountain geoclimatic barrier: initiation of range expansion by the mountain pine beetle. *Journal of Biogeography*, **39**(6), 1112–1123.

de Liocourt, F. (1898). De l'amenagement des sapinieres. *Bulletin Trimestriel, Societe Forestie re de Franche-Comte et Belfort*, juillet 1898, 396–409.

Deal, R.L. and Tappeiner, J.C. (2002). The effects of partial cutting on stand structure and growth of western hemlock—Sitka spruce stands in southeast Alaska. *Forest Ecology and Management*, **159**(3), 173–186.

Deal, R.L., Oliver, C.D., and Bormann, B.T. (1991). Reconstruction of mixed hemlock spruce stands in coastal southeast Alaska. *Canadian Journal of Forest Research*, **21**(5), 643–654.

DeBano, L.F., Neary, D.G., and Pfolliott, P.F. (1998). *Fire Effects on Ecosystems*. John Wiley & Sons, Inc., New York.

Della-Bianca, L. and Olson, D.F. (1961). Soil-site studies in Piedmont hardwood and pine-hardwood upland forests. *Forest Science*, **7**(4), 320–329.

Denslow, J.S. (1980). Gap partitioning among tropical rainforest trees. *Biotropica*, **12**(2), 47–55.

Denslow, J.S. and Guzman, S. (2000). Variation in stand structure, light and seedling abundance across a tropical moist forest chronosequence, Panama. *Journal of Vegetation Science*, **11**(2), 201–212.

Devall, B., ed. (1993). *Clearcut: The Tragedy of Industrial Forestry*. Sierra Club Books, San Francisco.

Dey, D.C. and Jensen, R.G. (2002). Stump sprouting potential of oaks in Missouri Ozark forest managed by even- and uneven-aged silviculture. In S. R. Shifley and J. M. Kabrick, eds. *Proceedings of the Second Missouri Ozark Forest Ecosystem Project Symposium: Post-treatment Results of the Landscape Experiment*, pp. 102–113. USDA Forest Service, North Central Forest Experiment Station. General Technical Report NC-227.

Dey, D.C., Johnson, P.S., and Garrett, H.E. (1996). Modeling the regeneration of oak stands in the Missouri Ozark Highlands. *Canadian Journal of Forest Research*, **26**(4), 573–583.

Dey, D.C., Gardiner, E.S., Kabrick, J.M., Stanturf, J.A., and Jacobs, D.F. (2010). Innovations in afforestation of agricultural bottomlands to restore native forests in the eastern USA. *Scandinavian Journal of Forest Research*, **25**(S8), 31–42.

Diaci, J. (2006). *Nature-based forestry in Central Europe: alternatives to industrial forestry and strict preservation*. Studia Forestalia Slovenica Nr. 126. Biotechnical Faculty, Department of Forestry and Renewable Natural Resources, Ljubljana.

Diaci, J., Adamic, T., and Rozman, A. (2012). Gap recruitment and partitioning in an old-growth beech forest of the Dinaric Mountains: influences of light regime, herb competition and browsing. *Forest Ecology and Management*, **285**, 20–28.

Di-Giovanni, F., Kevan, P.G., and Nasr, M.E. (1995). The variability in settling velocities of some pollen and spores. *Grana*, **34**(1), 39–44.

Donoso, P.J. (2005). Crown index: a canopy balance indicator to assess growth and regeneration in uneven-aged forest stands of the coastal range of Chile. *Forestry*, **78**(4), 337–351.

Doolittle, W.T. (1958). Site index comparisons for several forest species in the Southern Appalachians. *Soil Science Society of America Proceedings*, **22**(5), 455–458.

Drengston, A. and Taylor, D., eds. (1997). *Ecoforestry: The Art and Science of Sustainable Forest Use*. New Society Publishers, Gabriola Island.

Drever, C.R., Peterson, G., Messier, C., Bergeron, Y., and Flannigan, M. (2006). Can forest management based on natural disturbances maintain ecological resilience? *Canadian Journal of Forest Research*, **36**(9), 2285–2299.

Drever, C. and Lertzman, K. (2001). Light-growth responses of coastal Douglas-fir and western redcedar saplings under different regimes of soil moisture and nutrients. *Canadian Journal of Forest Research*, **31**(12), 2124–2133.

Drew, T.J. and Flewelling, J.W. (1977). Some recent Japanese theories of yield-density relationships and their application to Monterey pine plantations. *Forest Science*, **23**(4), 517–534.

Drew, T.J. and Flewelling, J.W. (1979). Stand density management: an alternative approach and its application to Douglas-fir plantations. *Forest Science*, **25**(3), 518–532.

Ducey, M.J. (2009). The ratio of additive and traditional stand density indices. *Western Journal of Applied Forestry*, **24**(1), 5–10.

Ducey, M.J. and Larson, B.C. (2003). There a correct stand density index? An alternate interpretation. *Western Journal of Applied Forestry*, **18**(3), 179–184.

Ducey, M.J. and Valentine, H.T. (2008). Direct sampling for stand density index. *Western Journal of Applied Forestry*, **23**(2), 78–82.

Duerr, W.A. and Gevorkiantz, S.R. (1938). Growth prediction and site determination in uneven-aged timber stands. *Journal of Agricultural Research*, **56**, 81–98.

Duncan, S.L., McComb, B.C., and Johnson, K.N. (2010). Integrating ecological and social ranges of variability in conservation of biodiversity: past, present, and future. *Ecology and Society*, **15**(1), 5.

Duncker, P.S., Barreiro, S.M., Hengeveld, G.M., Lind, T., Mason, W.L., Ambrozy, S., and Spiecker, H. (2012). Classification of forest management approaches: a new conceptual framework and its applicability to european forestry. *Ecology and Society*, **17**(4), 51.

Dunning, D. (1928). A tree classification for the selection forest of the Sierra Nevada. *Journal of Agricultural Research*, **36**, 755–771.

Duryea, M.L. and Dougherty, P.M., eds. (1991). *Forest Regeneration Manual*. Kluwer Academic Publishers, Dordrecht.

Edmonds, R.L., Agee, J.K., and Gara, R.I. (2000). *Forest Health and Protection*. McGraw Hill, Inc., Boston.

Egan, A.F., Rowe, J., Peterson, D., and Philippi, G. (1997). West Virginia Tree Farmers and consulting foresters: a comparison of views on timber harvesting. *Northern Journal of Applied Forestry*, **14**(1), 16–19.

Egan, T. (2009). *The Great Burn*, Houghton Mifflin Harcourt, Boston.

Egler, E.F. (1954). Vegetation science concepts. I. Initial floristic composition, a factor in old-field vegetation development. *Vegetatio*, **35**(6), 95–105.

Ek, A.R. (1974). Nonlinear models for stand table projection in northern hardwood stands. *Canadian Journal of Forest Research*, **14**(1), 23–27.

Ek, A.R. and Monserud, R.A. (1974). *FOREST: a computer model for simulating the growth and reproduction of mixed species forest stands*. School of Natural Resources, University of Wisconsin. Research Paper R2635.

El-Kassaby, Y.A., Dunsworth, B.G., and Krakowski, J. (2003). Genetic evaluation of alternative silvicultural systems in coast montane forests: western hemlock and amabilis fir. *Theoretical and Applied Genetics*, **107**(4), 598–610.

Endler, J. (1993). The color of light in forests and its implications. *Ecological Monographs*, **63**(1), 1–27.

Ernst, R.L. and Knapp, W.H. (1985). *Forest stand density and stocking: concepts terms, and the use of stocking guides*. USDA Forest Service. General Technical Report WO-44.

Evans, J. and Turnbull, J. (2004). *Plantation Forestry in the Tropics*, 3rd edn. Oxford University Press, Oxford.

Everham, E.M. and Brokaw, N.V.L. (1996). Forest damage and recovery from catastrophic wind. *Botanical Review*, **62**(2), 113–185.

Faber, P.J. and Sissingh, G. (1975). Stability of stands to wind. [I] A theoretical approach. [II] The practical viewpoint. *Nederlands Bosbouw Tijdschrift*, **47**(7/8), 179–193.

Fairfax, S. and Achterman, G. (1977). Monongahela controversy and political-process. *Journal of Forestry*, **75**(8), 485–487.

Farris, M.A. and Mitton, J.B. (1984). Population-density, outcrossing rate, and heterozygote superiority in ponderosa pine. *Evolution*, **38**(5), 1151–1154.

Favrichon, V. (1998). Modeling the dynamics and species composition of a tropical mixed-species uneven-aged natural forest: effects of alternative cutting regimes. *Forest Science*, **44**(1), 113–124.

Fenton, N.J., Simard, M., and Bergeron, Y. (2009). Emulating natural disturbances: the role of silviculture in creating even-aged and complex structures in the black spruce boreal forest of eastern North America. *Journal of Forest Research*, **14**(5), 258–267.

Ferguson, D.E. and Adams, D.L. (1980). Response of advance grand fir regeneration to overstory removal in northern Idaho. *Forest Science*, **26**(4), 537–545.

Fernow, B.E. (1911). *A Brief History of Forestry. In Europe, the United States, and Other Countries*, Revised and Enlarged Edition. University Press and Forestry Quarterly, Toronto, Canada, and Cambridge, MA.

Fettig, C.J., Gibson, K.E., Munson, S., and Negrón, J.F. (2014). Cultural practices for prevention and mitigation of mountain pine beetle infestations. *Forest Science*, **60**(3), 450–463.

Fettig, C.J., Klepzig, K.D., Billings, R.F., Munson, A.S., Nebeker, T.E., Negrón, J.F., and Nowak, J.T. (2007). The effectiveness of vegetation management practices for prevention and control of bark beetle infestations in coniferous forests of the western and southern United States. *Forest Ecology and Management*, **238**(1–3), 24–53.

Finkeldey, R. (1995). Homogeneity of pollen allele frequencies of single seed trees in *Picea-Abies* (L) Karst plantations. *Heredity*, **74**(5), 451–463.

Finkeldey, R. (2001). Genetic variation of oaks (*Quercus* spp.) in Switzerland. 2. Genetic structures in "pure" and "mixed" forests of pedunculate oak (*Q robur* L.) and sessile oak (*Q petraea* (Matt.) Liebl.). *Silvae Genetica*, **50**(1), 22–30.

Finkeldey, R. and Ziehe, M. (2004). Genetic implications of silvicultural regimes. *Forest Ecology and Management*, **197**(1–3), 231–244.

Fjeld, D. and Granhus, A. (1998). Injuries after selection harvesting in multi-storied spruce stands—the influence of operating systems and harvest intensity. *Journal of Forest Engineering*, **9**(2), 33–40.

Ford, E.D. (1975). Competition and stand structure in some even-aged plant monocultures. *Journal of Ecology*, **63**(1), 311–333.

Ford, R.M., Williams, K.J.H., Bishop, I.D., and Hickey, J.E. (2009a). Public judgements of the social acceptability of silvicultural alternatives in Tasmanian wet eucalypt forests. *Australian Forestry*, **72**(4), 157–171.

Ford, R.M., Williams, K.J.H., Bishop, I.D., and Hickey, J.E. (2009b). Effects of information on the social acceptability of alternatives to clearfelling in Australian wet eucalypt forests. *Environmental Management*, **44**(6), 1149–1162.

Foster, D.R. (1988). Species and stand response to catastrophic wind in central New England, USA. *Journal of Ecology*, **76**(1), 135–151.

Foster, D.R., Knight, D.H., and Franklin, J.F. (1998). Landscape patterns and legacies resulting from large, infrequent forest disturbances. *Ecosystems*, **1**(6), 497–510.

Franklin, J.F. (1989). Towards a new forestry. *American Forests*, **95**(11/12), 37–44.

Franklin, J.F., Mitchell, R.J., and Palik, B.J. (2007). *Natural disturbance and stand development principles for ecological forestry*. USDA Forest Service, Northern Research Station. General Technical Report NRS-GTR-19.

Franklin, J.F., Spies, T.A., Van Pelt, R., Carey, A.B., Thornburgh, D.A., Berg, D.R., Lindenmayer, D.B., Harmon, M.E., Keeton, W.S., Shaw, D.C., Bible, K., and Chen, J.Q. (2002). Disturbances and structural development of natural forest ecosystems with silvicultural implications, using Douglas-fir forests as an example. *Forest Ecology and Management*, **155**(1–3), 399–423.

Franklin, J.F. and Johnson, K.N. (2011). Societal challenges in understanding and responding to regime shifts in forest landscapes. *Proceedings of the National Academy of Sciences of the United States of America*, **108**(41), 16863–16864.

Franklin, J., Shugart, H., and Harmon, M. (1987). Tree death as an ecological process. *Bioscience*, **37**(8), 550–556.

Frelich, L.E. (2002). *Forest Dynamics and Disturbance Regimes: Studies from Temperate Evergreen—Deciduous Forests*. Cambridge University Press, Cambridge.

Frelich, L.E. and Reich, P.B. (1999). Neighborhood effects, disturbance severity, and community stability in forests. *Ecosystems*, **2**(2), 151–166.

Fries, C., Johansson, O., Pettersson, B., and Simonsson, P. (1997). Silvicultural models to maintain and restore natural stand structures in Swedish boreal forests. *Forest Ecology and Management*, **94**(1–3), 89–103.

Fritz, E.C. (1989). *Clearcutting: A Crime Against Nature*. Rainforest Action Network, Austin.

Fujimori, T. (2001). *Ecological and Silvicultural Strategies for Sustainable Forest Management*. Elsevier, Amsterdam.

Fulé, P.Z., Covington, W.W., and Moore, M.M. (1997). Determining reference conditions for ecosystem management of southwestern ponderosa pine forests. *Ecological Applications*, **7**(3), 895–908.

Gadow, K., Zhang, C.Y., Wehenkel, C., Pommerening, A., Corral-Rivas, J., Korol, M., Myklush, S., Hui, G.Y., Kiviste, A., and Zhao, X.H. (2012). Forest structure and diversity. In T. Pukkala and K. von Gadow, eds. *Managing Forest Ecosystems*, 2nd edn., pp. 29–83. Springer, Dordrecht.

Gamborg, C. and Larsen, J.B. (2003). "Back to nature" - a sustainable future for forestry? *Forest Ecology and Management*, **179**(1–3), 559–571.

Gardiner, B., Marshall, B., Achim, A., Belcher, R., and Wood, C. (2005). The stability of different silvicultural systems: a wind-tunnel investigation. *Forestry*, **78**(5), 471–484.

Gardiner, B., Byrne, K., Hale, S., Kamimura, K., Mitchell, S.J., Peltola, H., and Ruel, J. (2008). A review of mechanistic modelling of wind damage risk to forests. *Forestry*, **81**(3), 447–463.

Gardiner, E. and Helmig, L. (1997). Development of water oak stump sprouts under a partial overstory. *New Forests*, **14**(1), 55–62.

Garfitt, J.E. (1995). *Natural Management of Woods: Continuous Cover Forestry*. Research Studies Press, Ltd., Taunton.

Geldenhuys, C.J. (2010). Managing forest complexity through application of disturbance-recovery knowledge in development of silvicultural systems and ecological rehabilitation in natural forest systems in Africa. *Journal of Forest Research*, **15**(1), 3–13.

Gersonde, R., Battles, J.J., and O'Hara, K.L. (2004). Characterizing the light environment in Sierra Nevada mixed-conifer forests using a spatially explicit light model. *Canadian Journal of Forest Research*, **34**(6), 1332–1342.

Gersonde, R.F. and O'Hara, K.L. (2005). Comparative tree growth efficiency in Sierra Nevada mixed-conifer forests. *Forest Ecology and Management*, **219**(1), 95–108.

Gibbs, C.B. (1978). Uneven-aged silviculture and management? Even-aged silviculture and management? Definitions and differences. In *Uneven-aged Silviculture & Management in the United States*, pp. 18–24. USDA Forest Service. General Technical Report WO-24.

Gingrich, S.F. (1967). Measuring and evaluating stocking and stand density in upland hardwood forests of the central states. *Forest Science*, **13**(1), 38–53.

Givnish, T. (1988). Adaptation to sun and shade: a whole-plant perspective. *Australian Journal of Plant Physiology*, **15**(1–2), 63–92.

Glaubitz, J., Murrell, J., and Moran, G. (2003a). Effects of native forest regeneration practices on genetic diversity in *Eucalyptus consideniana*. *Theoretical and Applied Genetics*, **107**(3), 422–431.

Glaubitz, J., Wu, H., and Moran, G. (2003b). Impacts of silviculture on genetic diversity in the native forest species *Eucalyptus sieberi*. *Conservation Genetics*, **4**(3), 275–287.

Glöde, D. (2002). Survival and growth of *Picea abies* regeneration after shelterwood removal with single- and double-grip harvester systems. *Scandinavian Journal of Forest Research*, **17**(5), 417–426.

Goff, F.G. and West, D. (1975). Canopy-understory interaction effects on forest population structure. *Forest Science*, **21**(2), 98–107.

Golser, M. and Hasenauer, H. (1997). Predicting juvenile tree height growth in uneven-aged mixed species stands in Austria. *Forest Ecology and Management*, **97**(2), 133–146.

Gonthier, P. and Nicolotti, G., eds. (2013). *Infectious Forest Diseases*. CABI Publications, Wallingford.

Goodburn, J.M. and Lorimer, C.G. (1999). Population structure in old-growth and managed norther hardwoods: an examination of the balanced diameter distribution concept. *Forest Ecology and Management*, **118**(1–3), 11–29.

Gove, J.H. and Fairweather, S.E. (1992). Optimizing the management of uneven-aged forest stands: a stochastic approach. *Forest Science*, **38**(3), 623–640.

Graham, R.T. and Jain, T.B. (2005). Application of free selection in mixed forests of the inland northwestern United States. *Forest Ecology and Management*, **209**(1–2), 131–145.

Grassi, G. and Bagnaresi, U. (2001). Foliar morphological and physiological plasticity in *Picea abies* and *Abies alba* saplings along a natural light gradient. *Tree Physiology*, **21**(12–13), 959–967.

Grassi, G. and Giannini, R. (2005). Influence of light and competition on crown and shoot morphological parameters of Norway spruce and silver fir saplings. *Annals of Forest Science*, **62**(3), 269–274.

Gratzer, G., Darabant, A., Chhetri, P.B., Rai, P.B., and Eckmüllner, O. (2004). Interspecific variation in the response of growth, crown morphology, and survivorship to light of six tree species in the conifer belt of the Bhutan Himalayas. *Canadian Journal of Forest Research*, **34**(5), 1093–1107.

Gray, A.N. and Spies, T.A. (1996). Gap size, within-gap position and canopy structure effects on conifer seedling establishment. *Journal of Ecology*, **84**(5), 635–645.

Greene, D.F., Zasada, J.C., Sirois, L., Kneeshaw, D., Morin, H., Charron, I., and Simard, M.J. (1999). A review of the regeneration dynamics of North American boreal forest tree species. *Canadian Journal of Forest Research*, **29**(6), 824–839.

Gronewold, C., D'Amato, A.W., and Palik, B.J. (2012). Relationships between growth, quality, and stocking within managed old-growth northern hardwoods. *Canadian Journal of Forest Research*, **42**(6), 1115–1125.

Groot, A., Gauthier, S., and Bergeron, Y. (2004). Stand dynamics modelling approaches for multicohort management of eastern Canadian boreal forests. *Silva Fennica*, **38**(4), 437–448.

Grulke, N.E. (2011). The nexus of host and pathogen phenology: understanding the disease triangle with climate change. *New Phytologist*, **189**(1), 8–11.

Guldin, J.M. (1991). Uneven-aged bdq regulation of Sierra Nevada USA mixed conifers. *Western Journal of Applied Forestry*, **6**(2), 27–32.

Guldin, J.M. (2011). Experience with the selection method in pine stands in the southern United States, with implications for future application. *Forestry*, **84**(5), 539–546.

Guldin, J.M. and Baker, J.B. (1988). Yield comparisons from even-aged and uneven-aged loblolly-shortleaf pine stands. *Southern Journal of Applied Forestry*, **12**(2), 107–114.

Guldin, J.M. and Fitzpatrick, M.W. (1991). Comparison of log quality from even-aged and uneven-aged loblolly pine stands in south Arkansas. *Southern Journal of Applied Forestry*, **15**(1), 10–17.

Guldin, J.M., Iffrig, G.F., and Flader, S.L. (2008). *Pioneer Forest: a half century of sustainable uneven-aged forest management in the Missouri Ozarks*. USDA Forest Service, Southern Research Station. SRS-GTR-108.

Gustafsson, L., Baker, S.C., Bauhus, J., Beese, W.J., Brodie, A., Kouki, J., Lindenmayer, D.B., Löhmus, A., Martínez Pastur, G., Messier, C., Neyland, M., Palik, B., Sverdrup-Thygeson, A., Volney, W.J.A., Wayne, A., and Franklin, J.F. (2012). Retention forestry to maintain multifunctional forests: a world perspective. *Bioscience*, **62**(7), 633–645.

Hägglund, B. (1981). Evaluation of forest site productivity. *Forestry Abstracts*, **42**(11), 515–527.

Haight, R.G. (1987). Evaluating the efficiency of even-aged and uneven-aged stand management. *Forest Science*, **33**(1), 116–134.

Haight, R.G. and Monserud, R.A. (1990). Optimizing any-aged management of mixed-species stands.1. Performance of a coordinate-search process. *Canadian Journal of Forest Research*, **20**(1), 15–25.

Haight, R.G., Brodie, J.D., and Adams, D.M. (1985). Optimizing the sequence of diameter distributions and selection harvests for uneven-aged stand management. *Forest Science*, **31**(2), 451–462.

Haila, Y., Hanski, I.K., Niemela, J., Punttila, P., Raivio, S., and Tukia, H. (1994). Forestry and the boreal fauna—matching management with natural forest dynamics. *Annales Zoologici Fennici*, **31**(1), 187–202.

Hale, S.E., Gardiner, B.A., Wellpott, A., Nicoll, B.C., and Achim, A. (2012). Wind loading of trees: influence of tree size and competition. *European Journal of Forest Research*, **131**(1), 203–217.

Hall, D.O. (1983). Financial maturity for even-aged and all-aged stands. *Forest Science*, **29**(4), 833–836.

Hall, D.O. and Bruna, J.A. (1983). A management decision framework for winnowing simulated all-aged stand prescriptions. USDA Forest Service, Intermountain Forest and Range Experiment Station. General Technical Report INT-GTR-147.

Hallin, W.E. (1951). Unit area control in California forests. USDA Forest Service, California Forest and Range Experiment Station. Research Note 77.

Hallin, W.E. (1959). The application of unit area control in the management of ponderosa-jeffrey pine at Black Mountains Experimental Forest. USDA Techical Bulletin 1191. US Government Printing Office, Washington, DC.

Hanewinkel, M. (2001). Economic aspects of the transformation from even-aged pure stands of Norway spruce to uneven-aged mixed stands of Norway spruce and beech. *Forest Ecology and Management*, **151**(1–3), 181–193.

Hanewinkel, M. (2002). Comparative economic investigations of even-aged and uneven-aged silvicultural systems: a critical analysis of different methods. *Forestry*, **75**(4), 473–481.

Hanewinkel, M. (2004). Spatial patterns in mixed coniferous even-aged, uneven-aged and conversion stands. *European Journal of Forest Research*, **123**(2), 139–155.

Hanewinkel, M. and Pretzsch, H. (2000). Modelling the conversion from even-aged to uneven-aged stands of Norway spruce (*Picea abies* L. Karst.) with a distance-dependent growth simulator. *Forest Ecology and Management*, **134**(1–3), 55–70.

Hann, D.W. and Bare, B.B. (1979). *Uneven-aged forest management: state of the art (or science?)*. USDA Forest Service, Intermountain Forest and Range Experiment Station. General Technical Report INT-GTR-50.

Hansen, G.D. and Nyland, R.D. (1987). Effects of diameter distributions on the growth of simulated uneven-aged sugar maple stands. *Canadian Journal of Forest Research*, **17**(1), 1–8.

Hanson, J.J. and Lorimer, C.G. (2007). Forest structure and light regimes following moderate wind storms: Implications for multi-cohort management. *Ecological Applications*, **17**(5), 1325–1340.

Hao, Q., Meng, F., Zhou, Y., and Wang, J. (2005). A transition matrix growth model for uneven-aged mixed-species forests in the Changbai Mountains, northeastern China. *New Forests*, **29**(3), 221–231.

Harcombe, P.A. (1987). Tree life tables. *Bioscience*, **37**(8), 557–568.

Harcombe, P.A. and Marks, P.L. (1978). Tree diameter distributions and replacement processes in southeast Texas forests. *Forest Science*, **24**(2), 153–166.

Harrington, T.B. (2006). Five-year growth responses of Douglas-fir, western hemlock, and western redcedar seedlings to manipulated levels of overstory and understory competition. *Canadian Journal of Forest Research*, **36**(10), 2439–2453.

Hartsfield, A. and Ostermeier, D. (2003). Certification: the view from FSC-certified land managers. *Journal of Forestry*, **101**(8), 32–36.

Harvey, B.D., Leduc, A., Gauthier, S., and Bergeron, Y. (2002). Stand-landscape integration in natural disturbance-based management of the southern boreal forest. *Forest Ecology and Management*, **155**(1–3), 369–385.

Hasse, W.D. and Ek, A.E. (1981). A simulated comparison of yield for even- versus uneven-aged management of northern hardwood stands. *Journal of Environmental Management*, **12**, 235–246.

Hawksworth, F.G. and Wiens, D. (1996). *Dwarf mistletoes: biology, pathology, and systematics*. USDA Forest Service. USDA Agriculture Handbook 709. US Government Printing Office, Washington, DC.

Hawley, R.C. (1921). *The Practice of Silviculture*. John Wiley & Sons, Inc., New York.

Heggenstaller, D.J., Zenner, E.K., Brose, P.H., and Peck, J.E. (2012). How much older are Appalachian oaks belowground than above-ground? *Northern Journal of Applied Forestry*, **29**(3), 155–157.

Heinrichs, S. and Schmidt, W. (2009). Short-term effects of selection and clear cutting on the shrub and herb layer vegetation during the conversion of even-aged Norway spruce stands into mixed stands. *Forest Ecology and Management*, **258**(5), 667–678.

Heitzman, E. and Nyland, R.D. (1991). Cleaning and early crop-tree release in northern hardwood stands: a review. *Northern Journal of Applied Forestry*, **8**(3), 111–115.

Helgerson, O.T., Newton, M., and McNabb, D.H. (1992). Site preparation. In S.D. Hobbs, S.D. Tesch, P.W. Owston, R.E. Stewart, J.C.I. Tappeineir, and G.E. Wells, eds. *Reforestation Practices in Southwestern Oregon and Northern California*, pp. 232–256. Oregon State University, Corvallis.

Helliwell, D.R. (1997). Dauerwald. *Forestry*, **70**(4), 375–379.

Helliwell, R. (2013). *Continuous Cover Management of Woodlands: A Brief Introduction*. Rodney Helliwell, York.

Helms, J.A., ed. (1998), *The Dictionary of Forestry*. Society of American Foresters, Bethesda.

Helms, J.A. and Porter, D.J. (2009). Natural forestry, unpublished manuscript.

Helms, J.A. and Standiford, R.B. (1985). Predicting release of advance reproduction of mixed conifer species in California following overstory removal. *Forest Science*, **31**(1), 3–15.

Helsinki Declaration (1993). *Resolution H1: General Guidelines for the Sustainable Management of Forests in Europe*.

Second Ministerial Conference on the Protection of Forests in Europe.

Herbert, D.A., Fownes, J.H., and Vitousek, P.M. (1999). Hurricane damage to a Hawaiian forest: nutrient supply rate affects resistance and resilience. *Ecology*, **80**(3), 908–920.

Heske, F. (1938). *German Forestry*, Yale University Press, New Haven.

Hessburg, P.F. and Agee, J.K. (2003). An environmental narrative of inland northwest United States forests, 1800–2000. *Forest Ecology and Management*, **178**(1–2), 23–59.

Hill, M.O., Roy, D.B., and Thompson, K. (2002). Hemeroby, urbanity and ruderality: bioindicators of disturbance and human impact. *Journal of Applied Ecology*, **39**(5), 708–720.

Hills, G.A. (1953). The use of site in forest management. *Forestry Chronicle*, **29**(2), 128–136.

Hitsuma, G., Ota, T., Kanazashi, T., and Masaki, T. (2006). Seven-year changes in growth and crown shape of *Thujopsis dolabrata* var. *hondai* saplings after release from suppression. *Journal of Forest Research*, **11**(4), 281–287.

Hobbs, S.D., Tesch, S.D., Owston, P.W., Stewart, R.E., Tappeineir, J.C.I., and Wells, G.E., eds. (1992). *Reforestation Practices in Southwestern Oregon and Northern California*, Oregon State University, Corvallis.

Hodges, J.D. and Gardiner, E.S. (1993). Ecology and physiology of oak regeneration. In D.L. Loftis and C.E. McGee, eds. *Oak Regeneration: Serious Problems, Practical Recommendations*, pp. 54–65. USDA Forest Service, Southeastern Forest Experiment Station. General Technical Report SE-GTR-84.

Hodgkins, E.J. (1956). Testing soil-site index tables in southwestern Alabama. *Journal of Forestry*, **54**(4), 261–266.

Hofgaard, A. (1993). 50 years of change in a Swedish boreal old-growth *Picea-Abies* forest. *Journal of Vegetation Science*, **4**(6), 773–782.

Hosius, B., Leinemann, L., Konnert, M., and Bergmann, F. (2006). Genetic aspects of forestry in the central Europe. *European Journal of Forest Research*, **125**(4), 407–417.

Howarth, R.B. (2009). Discounting, uncertainty, and revealed time preference. *Land Economics*, **85**(1), 24–40.

Howe, G.E. (1989). Genetic effects of even-aged and uneven-aged silviculture. In *Proceedings of the National Silviculture Workshop: Silvicultural Challenges and Opportunities in the 1990's*, pp. 84–91. USDA Forest Service. Timber Management, Washington, DC.

Hughes, A.R., Inouye, B.D., Johnson, M.T.J., Underwood, N., and Vellend, M. (2008). Ecological consequences of genetic diversity. *Ecology Letters*, **11**(6), 609–623.

Hull, R.B. and Buhyoff, G.J. (1986). The scenic beauty temporal distribution method: an attempt to make scenic beauty assessments compatible with forest planning efforts. *Forest Science*, **32**(2), 271–286.

Hull, R.B., Robertson, D.P., and Kendra, A. (2001). Public understandings of nature: a case study of local knowledge about "natural" forest conditions. *Society & Natural Resources*, **14**(4), 325–340.

Hummel, S. and Agee, J.K. (2003). Western spruce budworm defoliation effects on forest structure and potential fire behavior. *Northwest Science*, **77**(2), 159–169.

Hunter, M.L., Jr. (1990). *Wildlife, Forests and Forestry: Principles of Managing Forests for Biological Diversity*. Prentice-Hall, Englewood Cliffs.

Hunter, M.L., Jr. (1993). Natural fire regimes as spatial models for managing boreal forests. *Biological Conservation*, **65**(2), 115–120.

Hurst, J.M., Stewart, G.H., Perry, G.L.W., Wiser, S.K., and Norton, D.A. (2012). Determinants of tree mortality in mixed old-growth *Nothofagus* forest. *Forest Ecology and Management*, **270**, 189–199.

Husch, B., Beers, T.W., and Kershaw, J.A. (2003). *Forest Mensuration*, 4th edn. John Wiley & Sons, Hoboken.

Huss, J. (1990). The development of the "Dauerwald" (continuous forest) concept up to the Third Reich. *Forst und Holz*, **45**(7), 163–171.

Hussendörfer, E. and Konnert, M. (2000a). Impact of forest management on genetic variation of silver fir and European beech populations. *Forest Snow and Landscape Research*, **75**(1–2), 187–204.

Hussendörfer, E. and Konnert, M. (2000b). Investigations of genetic variation of silver fir (*Abies alba* Mill.) in uneven-aged forests ("Plenterwald") in comparison with even-aged forests ("Altersklassenwald"). *Forstwissenschaftliches Centralblatt*, **119**(4), 208–225.

Hyink, D. and Moser, J. (1983). A generalized framework for projecting forest yield and stand structure using diameter distributions. *Forest Science*, **29**(1), 85–95.

Iffrig, G.F., Trammel, C.E., and Cunningham, T. (2004). Pioneer Forest: a case study in sustainable forest management. In S.L. Flader, ed. *Toward Sustainability For Missouri Forests*, pp. 193–204. USDA Forest Service, North Central Research Station. General Technical Report NC-239.

International Model Forest Network (2012). *2012 Annual Report: Celebrating 20 Years of Innovation and Impact*. International Model Forest Network, Ottawa.

Irland, L.C. (2000). Ice storms and forest impacts. *Science of the Total Environment*, **262**(3), 231–242.

Isaac, L.A. (1930). Seed flight in the Douglas fir region. *Journal of Forestry*, **28**(4), 492–499.

Isaac, L.A. (1943). *Reproductive Habits of Douglas-fir*. Charles Lathrop Pack Forestry Foundation, Washington, DC.

Isaac, L.A. (1956). *Place of partial cutting in old-growth stands of the Douglas-fir region*. USDA Forest Service, Pacific

Northwest Forest and Range Experiment Station. Research Paper 16.

Isaac, L.A. (1967). *Leo A. Isaac: Douglas-Fir Research in the Pacific Northwest, 1920–1956. Interview by A.R. Fry.* University of California, Regional Oral History Office, Berkeley.

Ishii, H. and Ford, E.D. (2001). The role of epicormic shoot production in maintaining foliage in old *Pseudotsuga menziesii* (Douglas-fir) trees. *Canadian Journal of Botany*, **79**(3), 251–264.

Jack, S.B., Neel, W.L., and Mitchell, R.J. (2006a). The Stoddard-Neel approach: a conservation-oriented approach. In S. Jose, E.J. Jokela, and D.L. Miller, eds. *The Longleaf Pine Ecosystem: Ecology, Silviculture, and Restoration*, pp. 242–245. Springer, New York.

Jack, S.B., Mitchell, R.J., and Pecot, S.D. (2006b). Silvicultural alternatives in a longleaf pine/wiregrass woodland in Southwest Georgia: understory hardwood response to harvest-created gaps. In K. F. Connor, ed. *13th Biennial Southern Silvicultural Research Conference*, pp. 85–89. USDA Forest Service, Southern Research Station. General Technical Report SRS-GTR-92.

Jaeck, L.L., Oliver, C.D., and DeBell, D.S. (1984). Young stand development in coastal western hemlock as influenced by 3 harvesting regimes. *Forest Science*, **30**(1), 117–124.

Janowiak, M.K., Nagel, L.M., and Webster, C.R. (2008). Spatial scale and stand structure in northern hardwood forests: implications for quantifying diameter distributions. *Forest Science*, **54**(5), 497–506.

Janowiak, M.K., Swanston, C.W., Nagel, L.M., Webster, C.R., Palik, B.J., Twery, M.J., Bradford, J.B., Parker, L.R., Hille, A.T., and Johnson, S.M. (2011). *Silvicultural decision-making in an uncertain climate future: a workshop-based exploration of considerations, strategies, and approaches.* USDA Forest Service, Northern Research Station. General Technical Report NRS-GTR-81.

Johnson, P.S. (1977). *Predicting oak stump sprouting and sprout development in the Missouri Ozarks.* USDA Forest Service, North Central Forest Experiment Station. Research Paper NC-149.

Johnson, P.S., Shifley, S.R., and Rogers, R. (2009). *The Ecology and Silviculture of Oaks*, 2nd edn. CABI Publishing, Wallingford.

Johnson, R.L., Brunson, M.W., and Kimura, T. (1994). Using image-capture technology to assess scenic value at the urban forest interface: a case-study. *Journal of Environmental Management*, **40**(2), 183–195.

Jokela, E.J., Martin, T.A., and Vogel, J.G. (2010). Twenty-five years of intensive forest management with southern pines: important lessons learned. *Journal of Forestry*, **108**(7), 338–347.

Jones, D.A. and O'Hara, K.L. (2012). Carbon density in managed coast redwood stands: implications for forest carbon estimation. *Forestry*, **85**(1), 99–110.

Jones, J.R. (1969). *Review and comparison of site evaluation methods.* USDA Forest Service, Rocky Mountain Forest and Range Experiment Station. Research Paper RM-RP-51.

Karjalainen, E. and Komulainen, M. (1999). The visual effect of felling on small-and medium-scale landscapes in north-eastern Finland. *Journal of Environmental Management*, **55**(3), 167–181.

Keane, R.E., Hessburg, P.F., Landres, P.B., and Swanson, F.J. (2009). The use of historical range and variability (HRV) in landscape management. *Forest Ecology and Management*, **258**, 1025–1037.

Kearney, A.R. and Bradley, G.A. (2011). The effects of viewer attributes on preference for forest scenes: contributions of attitudes, knowledge, demographic factors, and stakeholder group membership. *Environment and Behavior*, **43**(2), 147–181.

Keeley, J.E. (2009). Fire intensity, fire severity, and burn severity: a brief review and suggested usage. *International Journal of Wildland Fire*, **18**(1), 116–126.

Kelty, M.J., Kittredge, D.B., Kyker-Snowman, T., and Leighton, A.D. (2003). The conversion of even-aged stands to uneven-aged structure in southern New England. *Northern Journal of Applied Forestry*, **20**(3), 109–116.

Kenderes, K., Aszalos, R., Ruff, J., Barton, Z., and Standovar, T. (2007). Effects of topography and tree stand characteristics on susceptibility of forests to natural disturbances (ice and wind) in the Borzsony Mountains (Hungary). *Community Ecology*, **8**(2), 209–220.

Kenefic, L.S. and Nyland, R.D., eds. (2005), *Proceedings of the conference on diameter-limit cutting in northeastern forests.* USDA Forest Service, Northeastern Research Station. General Technical Report GTR-NE-342.

Kenefic, L.S., Sendak, P.E., and Brissette, J.C. (2005). Comparison of fixed diameter-limit and selection cutting in northern conifers. *Northern Journal of Applied Forestry*, **22**(2), 77–84.

Kenk, G. and Guehne, S. (2001). Management of transformation in central Europe. *Forest Ecology and Management*, **151**(1–3), 107–119.

Kerr, G. (1999). The use of silvicultural systems to enhance the biological diversity of plantation forests in Britain. *Forestry*, **72**(3), 191–205.

Kerr, G. (2014). The management of silver fir forests: de Liocourt (1898) revisited. *Forestry*, 87(1), 29–38.

Kessler, W.B., Salwasser, H., Cartwright, C.W., and Caplan, J.A. (1992). New perspectives for sustainable natural resources management. *Ecological Applications*, **2**(3), 221–225.

Keyes, C.R., Maguire, D.A., and Tappeiner, J.C. (2009). Recruitment of ponderosa pine seedlings in the Cascade Range. *Forest Ecology and Management*, **257**(2), 495–501.

Kimmins, J.P. (1990). Modeling the sustainability of forest production and yield for a changing and uncertain future. *Forestry Chronicle*, **66**(3), 271–280.

Kimmins, J.P. (1993). Ecology, environmentalism and green religion. *Forestry Chronicle*, **69**(3), 285–289.

Kimmins, J.P. (1997). *Forest Ecology: A Foundation for Sustainable Management*, 2nd edn. Prentice-Hall, Inc., Upper Saddle River.

Kira, T. (1978). Community architecture and organic matter dynamics in tropical lowland rain forests of southeast Asia with special reference to Pasoh Forest, West Malaysia. In P.B. Tomlinson and M.H. Zimmermann, eds. *Tropical Trees as Living Systems*, pp. 561–590. Cambridge University Press, Cambridge.

Kira, T. and Shidei, T. (1967). Primary production and turnover of organic matter in different forest ecosystems of the western Pacific. *Journal of Japanese Ecology*, **17**(2), 70–87.

Kirkland, B.P. and Brandstrom, A.J.F. (1936). *Selective Timber Management in the Douglas-fir Region*. Charles Lathrop Pack Forestry Foundation, Washington, DC.

Kitajima, K. and Augspurger, C.K. (1989). Seed and seedling ecology of a monocarpic tropical tree, *Tachigalia versicolor*. *Ecology*, **70**(4), 1102–1114.

Klinka, K., Krajina, V.J., Ceska, A., and Scagel, A.M. (1989). *Indicator Plants of Coastal British Columbia*. UBC Press, Vancouver.

Klopcic, M. and Bončina, A. (2011). Stand dynamics of silver fir (*Abies alba* Mill.)-European beech (*Fagus sylvatica* L.) forests during the past century: a decline of silver fir? *Forestry*, **84**(3), 259–271.

Kluender, R., Lortz, D., McCoy, W., Stokes, B., and Klepac, J. (1998). Removal intensity and tree size effects on harvesting cost and profitability. *Forest Products Journal*, **48**(1), 54–59.

Kneeshaw, D., Williams, H., Nikinmaa, E., and Messier, C. (2002). Patterns of above- and below-ground response of understory conifer release 6 years after partial cutting. *Canadian Journal of Forest Research*, **32**(2), 255–265.

Kneeshaw, D.D., Kobe, R.K., Coates, K.D., and Messier, C. (2006). Sapling size influences shade tolerance ranking among southern boreal tree species. *Journal of Ecology*, **94**(2), 471–480.

Knoke, T. (2012). The economics of continuous cover forestry. In T. Pukkala and K. von Gadow, eds. *Continuous Cover Forestry*, 2nd edn., pp. 167–193. Springer, Dordrecht.

Knoke, T., Moog, M., and Plusczyk, N. (2001). On the effect of volatile stumpage prices on the economic attractiveness of a silvicultural transformation strategy. *Forest Policy and Economics*, **2**(3–4), 229–240.

Knoke, T. and Plusczyk, N. (2001). On economic consequences of transformation of a spruce (*Picea abies* (L.) Karst.) dominated stand from regular into irregular age structure. *Forest Ecology and Management*, **151**(1–3), 163–179.

Kobe, R.K. and Coates, K.D. (1997). Models of sapling mortality as a function of growth to characterize interspecific variation in shade tolerance of eight tree species of northwestern British Columbia. *Canadian Journal of Forest Research*, **27**(2), 227–236.

Koch, N.E. and Skovsgaard, J.P. (1999). Sustainable management of planted forests: some comparisons between Central Europe and the United States. *New Forests*, **17**(1–3), 11–22.

Köhl, M. and Baldauf, T. (2012). Resource assessment techniques for continuous cover forestry. In T. Pukkala and K. von Gadow, eds. *Continuous Cover Forestry*, 2nd edn., pp. 273–291. Springer, Dordrecht.

Kollenberg, C.L. and O'Hara, K.L. (1999). Leaf area and tree increment dynamics of even-aged and multiaged lodgepole pine stands in Montana. *Canadian Journal of Forest Research*, **29**(6), 687–695.

Kollmuss, A., Zink, H., and Polycarp, C. (2008). *Making Sense of Voluntary Carbon Market: A Comparision of Carbon Offset Standards*. World Wildlife Fund, Germany.

Kolström, M. (1998). Ecological simulation model for studying diversity of stand structure in boreal forests. *Ecological Modelling*, **111**(1), 17–36.

Konnert, M., Hosius, B., and Hussendorfer, E. (2007). Genetic impacts of forest management—results and research need. *Forst und Holz*, **62**(1), 8–14.

Korsgaard, S. (1989). The stand table projection simulation model. In H. E. Burkhart, M. Rauscher, and K. Johann. *Artificial Intelligence and Growth Models for Forest Management Decisions*, pp. 4343–4359. Virginia Polytechnic and State University, Blacksburk.

Koski, V. (1970). A study of pollen dispersal as a mechanism of gene flow in conifers. *Metsatieteellisen tutkimuslaitoksen julkaisuja*, **70**(4), 5–78.

Kozlowski, T.T. and Palardy, S.G. (1997). *Growth Control in Wood Plants*. Academic Press, San Diego.

Krajicek, J.E., Brinkman, K.A., and Gingrich, S.F. (1961). Crown competition: a measure of density. *Forest Science*, **7**(1), 35–42.

Krajina, V.J. (1965). Biogeoclimatic zones and biogeocoenoses of British Columbia. *Ecology of Western North America*, **1**, 1–17.

Krajina, V.J. (1969). Ecology of forest trees in British Columbia. *Ecology of Western North America*, **2**, 1–146.

Krakowski, J. and El-Kassaby, Y.A. (2003). Impacts of alternative silviculture systems on mating systems and genetic *diversity of forest tree species*. Canadian Forest Service, Laurentian Forestry Centre, Quebec Region. Information Report No. LAU-X-128.

Kunisaka, T. and Imada, M. (1996). Tree diameter distrubtion of stands each species in large-scale cool-temperate forest of Kyushu, southern Japan. *Journal of Faculty of Agriculture, Kyushu, University*, 41(1–2), 45–56.

Kusbach, A., Long, J.N., Van Miegroet, H., and Shultz, L.M. (2012). Fidelity and diagnostic species concepts in vegetation classification in the Rocky Mountains, northern Utah, USA. *Botany-Botanique*, 90(8), 678–693.

Kuuluvainen, T., Tahvonen, O., and Aakala, T. (2012). Even-aged and uneven-aged forest management in boreal Fennoscandia: a review. *Ambio*, 41(7), 720–737.

Lähde, E., Laiho, O., and Norokorpi, Y. (1999). Diversity-oriented silviculture in the boreal zone of Europe. *Forest Ecology and Management*, 118(1–3), 223–243.

Lähde, E., Laiho, O., Norokorpi, Y., and Saksa, T. (1994a). Structure and yield of all-sized and even-sized conifer-dominated stands on fertile sites. *Annales des Sciences Forestieres*, 51(2), 97–109.

Lähde, E., Laiho, O., Norokorpi, Y., and Saksa, T. (1994b). Structure and yield of all-sized and even-sized Scots pine-dominated stands. *Annales des Sciences Forestieres*, 51(2), 111–120.

Lähde, E., Laiho, O., Norokorpi, Y. and Saksa, T. (1999). Stand structure as the basis of diversity index. *Forest Ecology and Management*, 115(2–3), 213–220.

Laiho, O., Lähde, E., and Pukkala, T. (2011). Uneven- vs even-aged management in Finnish boreal forests. *Forestry*, 84(5), 547–556.

Lamprecht, H. (1989). *Silviculture in the Tropics*. Deutsche Gesellschaft fur Technische zussammenarbeit (GTZ) Gmbh., Eschborn.

Landres, P.B., Morgan, P., and Swanson, F.J. (1999). Overview of the use of natural variability concepts in managing ecological systems. *Ecological Applications*, 9(4), 1179–1188.

Landsberg, J.J. and Waring, R.H. (1997). A generalised model of forest productivity using simplified concepts of radiation-use efficiency, carbon balance and partitioning. *Forest Ecology and Management*, 95(3), 209–228.

Larsen, D.R., Metzger, M.A., and Johnson, P.S. (1997). Oak regeneration and overstory density in the Missouri Ozarks. *Canadian Journal of Forest Research*, 27(6), 869–875.

Larsen, J.B. (1995). Ecological stability of forests and sustainable silviculture. *Forest Ecology and Management*, 73(1–3), 85–96.

Larsen, J.B. (2005). Near-natural forestry. *DST—Dansk Skovbrugs Tidsskrift*, 90(1/2), 400.

Larsen, J.B. and Nielsen, A.B. (2007). Nature-based forest management—Where are we going?: Elaborating forest development types in and with practice. *Forest Ecology and Management*, 238(1–3), 107–117.

Larson, P.R. (1969). *Wood Formation and the Concept of Wood Quality*. Bulletin No. 74, Yale University, School of Forestry, New Haven.

Larson, A.J. and Churchill, D. (2012). Tree spatial patterns in fire-frequent forests of western North America, including mechanisms of pattern formation and implications for designing fuel reduction and restoration treatments. *Forest Ecology and Management*, 267, 74–92.

Larson, A.J., Stover, K.C., and Keyes, C.R. (2012). Effects of restoration thinning on spatial heterogeneity in mixed-conifer forest. *Canadian Journal of Forest Research*, 42(8), 1505–1517.

Lavender, D.P., Parish, R., Johnson, C.M., Montgomery, G., Vyse, A., Willis, R.A., and Winston, D., eds. (1990). *Regenerating British Columbia's Forests*. University of British Columbia Press, Vancouver.

Leak, W.B. (1964). An expression of diameter distribution of unbalanced, uneven-aged stands and forests. *Forest Science*, 10(1), 39–50.

Leak, W.B. (1996). Long-term structural change in uneven-aged northern hardwoods. *Forest Science*, 42(2), 160–165.

Leak, W.B. (2002). *Origin of sigmoid diameter distributions*. USDA Forest Service, Northeastern Research Station. Research Paper RP-NE-718.

Leak, W.B. and Filip, S.M. (1977). Thirty-eight years of group selection in New England northern hardwoods. *Journal of Forestry*, 75(10), 641–643.

Leak, W.B. and Sendak, P.E. (2002). Changes in species, grade, and structure over 48 years in a managed New England northern hardwood stand. *Northern Journal of Applied Forestry*, 19(1), 25–27.

Ledig, F.T. (1986). Conservation strategies for forest gene resources. *Forest Ecology and Management*, 14(2), 77–90.

Ledig, F.T. and Fryer, J.H. (1974). *Genetics of pitch pine*. USDA Forest Service. Research Paper WO-27.

Leibundgut, H. (1945). Waldbauliche untersuchungen uber den aufbau von Plenterwaldern. *Mitteil Schweiz Anst Forst Versuchswesen*, 24(1), 219–296.

Leopold, A. (1949). *Sand County Almanac*. Oxford University Press, New York.

Leverenz, J.W. (1981). Shoot structure and productivity in conifers. In S. Linder, ed. *Understanding and Predicting Tree Growth. Studia Forestalia Suecica*, volume 160, pp. 135–137. Swedish University of Agricultural Sciences, Uppsala.

Liang, J. and Picard, N. (2013). Matrix model of forest dynamics: an overview and outlook. *Forest Science*, 59(3), 359–378.

Lie, M.H., Josefsson, T., Storaunet, K.O., and Ohlson, M. (2012). A refined view on the "Green lie": Forest structure and composition succeeding early twentieth century selective logging in SE Norway. *Scandinavian Journal of Forest Research*, **27**(3), 270–284.

Liebhold, A.M. (2012). Forest pest management in a changing world. *International Journal of Pest Management*, **58**(3), 289–295.

Lieffers, V.J., Messier, C., Stadt, K.J., Gendron, F., and Comeau, P.G. (1999). Predicting and managing light in the understory of boreal forests. *Canadian Journal of Forest Research*, **29**(6), 796–811.

Lieffers, V.J., Mugasha, A.G., and Macdonald, S.E. (1993). Ecophysiology of shade needles of *Picea glauca* saplings in relation to removal of competing hardwoods and degree of prior shading. *Tree Physiology*, **12**(3), 271–280.

Lindenmayer, D.B., Franklin, J.F., Lõhmus, A., Baker, S.C., Bauhus, J., Beese, W., Brodie, A., Kiehl, B., Kouki, J., Martínez Pastur, G., Messier, C., Neyland, M., Palik, B., Sverdrup-Thygeson, A., Volney, J., Wayne, A., and Gustafsson, L. (2012). A major shift to the retention approach for forestry can help resolve some global forest sustainability issues. *Conservation Letters*, **5**(6), 421–431.

Lindner, M., Maroschek, M., Netherer, S., Kremer, A., Barbati, A., Garcia-Gonzalo, J., Seidl, R., Delzon, S., Corona, P., Kolstrom, M., Lexer, M.J., and Marchetti, M. (2010). Climate change impacts, adaptive capacity, and vulnerability of European forest ecosystems. *Forest Ecology and Management*, **259**(4), 698–709.

Liu, Q. and Hytteborn, H. (1991). Gap structure, disturbance and regeneration in a primeval *Picea abies* forest. *Journal of Vegetation Science*, **2**(3), 391–402.

Loewenstein, E.F., Johnson, P.S., and Garrett, H.E. (2000). Age and diameter structure of a managed uneven-aged oak forest. *Canadian Journal of Forest Research*, **30**, 1060–1070.

Loewenstein, E. (2005). Conversion of uniform broad-leaved stands to an uneven-aged structure. *Forest Ecology and Management*, **215**(1–3), 103–112.

Loftis, D.L. (1990a). Predicting postharvest performance of advance red oak reproduction in the southern Appalachians. *Forest Science*, **36**(4), 908–916.

Loftis, D.L. (1990b). A shelterwood method for regenerating red oak in the southern Appalachians. *Forest Science*, **36**(4), 917–929.

Lohmander, P. and Helles, F. (1987). Windthrow probability as a function of stand characteristics and shelter. *Scandinavian Journal of Forest Research*, **2**(1–4), 227–238.

Long, J.N. (1985). A practical approach to density management. *Forestry Chronicle*, **61**(1), 23–27.

Long, J.N. (2009). Emulating natural disturbance regimes as a basis for forest management: a North American view. *Forest Ecology and Management*, **257**(9), 1868–1873.

Long, J.N. and Daniel, T.W. (1990). Assessment of growing stock in uneven-aged stands. *Western Journal of Applied Forestry*, **5**(3), 93–96.

Long, J.N. and Shaw, J.D. (2010). The influence of compositional and structural diversity on forest productivity. *Forestry*, **83**(2), 121–128.

Long, J.N. and Smith, F.W. (1990). Determinants of stemwood production in *Pinus contorta* var *latifolia* forests: the influence of site quality and stand structure. *Journal of Applied Ecology*, **27**(3), 847–856.

López-Torres, I., Ortuño-Pérez, S., García-Robredo, F., and Fullana-Belda, C. (2013). Are the economically optimal harvesting strategies of uneven-aged *Pinus nigra* stands always sustainable and stabilizing? *Forests*, **4**(4), 830–848.

Lorimer, C.G. (1989). Relative effects of small and large disturbances on temperate hardwood forest structure. *Ecology*, **70**(3), 565–567.

Lorimer, C.G. and Frelich, L.E. (1984). A simulation of equilibrium diameter distributions of sugar maple (*Acer saccharum*). *Bulletin of the Torrey Botanical Club*, **111**(2), 193–199.

Lugo, A.E. (2008). Visible and invisible effects of hurricanes on forest ecosystems: an international review. *Austral Ecology*, **33**(4), 368–398.

Lundquist, J.E. and Hamelin, R.C., eds. (2005). *Forest Pathology: From Genes to Landscapes*. American Phytopathological Society, Minneapolis.

Lutz, J. and Halpern, C. (2006). Tree mortality during early forest development: a long-term study of rates, causes, and consequences. *Ecological Monographs*, **76**(2), 257–275.

MacKinney, A.L., Schumacher, F.X., and Chaiken, L.E. (1937). Construction of yield tables for nonnormal loblolly pine stands. *Journal of Agricultural Research*, **54**, 531–545.

Madgwick, H.A.I., Jackson, D.S., and Knight, P.J. (1977). Above-ground dry matter, energy, and nutrient concents of trees in an age series of *Pinus radiata* plantations. *New Zealand Journal of Forestry Science*, **7**(3), 445–468.

Magill, A.W. (1992). *Managed and natural landscapes: what do people like?* USDA Forest Service, Pacific Southwest Research Station. PSW-RP-213.

Magnuson, J.J. (1990). Long-term ecological research and the invisible present. *Bioscience*, **40**(7), 495–501.

Magnussen, S. and Peschl, A. (1981). The influence of shading on the photosynthesis and transpiration of silver and grand fir seedlings. *Allgemeine Forst und Jagdzeitung*, **152**(5), 82–93.

Maguire, D. (2005). Uneven-aged management: Panacea, viable alternative, or component of a grander strategy? *Journal of Forestry*, **103**(2), 73–74.

Magurran, A.E. (1988). *Ecological Diversity and its Measurement*. Princeton University Press, Princeton.

Mailly, D. and Kimmins, J. (1997). Growth of *Pseudotsuga menziesii* and *Tsuga heterophylla* seedlings along a light gradient: resource allocation and morphological acclimation. *Canadian Journal of Botany*, **75**(9), 1424–1435.

Makkonen, O. (1967). Ancient forestry. An historical study. Part I. Facts and information on trees. *Acta Forestalia Fennica*, **82**(3), 1–84.

Makkonen, O. (1975). On the multiple use of forests in ancient times. *Silva Fennica*, **9**(2), 116–129.

Malcolm, D.C., Mason, W.L., and Clarke, G.C. (2001). The transformation of conifer forests in Britain—regeneration, gap size and silvicultural systems. *Forest Ecology and Management*, **151**(1–3), 7–23.

Malcolm, J.R. and Harvey, B.D. (2013). The need for multicohort management in boreal forests. *Forestry Chronicle*, **89**(3), 271–274.

Manabe, T., Shimatani, K., Kawarasaki, S., Aikawa, S., and Yamamoto, S. (2009). The patch mosaic of an old-growth warm-temperate forest: patch-level descriptions of 40-year gap-forming processes and community structures. *Ecological Research*, **24**(3), 575–586.

Mani, S. and Parthasarathy, N. (2006). Tree diversity and stand structure in inland and coastal tropical dry evergreen forests of peninsular India. *Current Science*, **90**(9), 1238–1246.

Margolis, H., Oren, R., Whitehead, D., and Kaufmann, M.R. (1995). Leaf area dynamics of conifer forests. In W.K. Smith and T.M. Hinckley, eds. *Ecophysiology of Coniferous Forests*, pp. 181–224. Academic Press, San Diego.

Marquardt, P.E., Echt, C.S., Epperson, B.K., and Pubanz, D.M. (2007). Genetic structure, diversity, and inbreeding of eastern white pine under different management conditions. *Canadian Journal of Forest Research*, **37**(12), 2652–2662.

Marquis, D.A. (1965). Controlling light in small clearcuttings. USDA Forest Service, Northeastern Forest Experiment Station. Research Paper RP-NE-39.

Marris, E. (2011). *Rambunctious Garden: Saving Nature in a Post-wild World*. Bloomsbury.

Martin, P., Bancroft, B., Day, K., and Peel, K. (2005). A new basis for understory stocking standards for partially harvested stands in the British Columbia interior. *Western Journal of Applied Forestry*, **20**(1), 5–12.

Martin Vicente, A. and Fernández Alés, R. (2006). Long term persistence of dehesas. Evidences from history. *Agriforestry Systems*, **67**(1), 19–28.

Maser, C. (1994). *Sustainable Forestry: Philosophy, Science, and Economics*. St. Lucie Press, Delray Beach.

Mason, B. and Kerr, G. (2004). *Transforming even-aged conifer stands to continuous cover management*. Information Note 40. UK Forestry Commission, Edinburgh.

Mason, B., Kerr, G., and Simpson, J. (1999). *What is continuous cover forestry?* Information Note 29. UK Forestry Commission, Edinburgh.

Mason, W.L. (2002). Are irregular stands more windfirm? *Forestry*, **75**(4), 347–355.

Mason, W.L., Edwards, C., and Hale, S.E. (2004). Survival and early seedling growth of conifers with different shade tolerance in a Sitka spruce spacing trial and relationship to understorey light climate. *Silva Fennica*, **38**(4), 357–370.

Mathiasen, R. (1996). Dwarf mistletoes in forest canopies. *Northwest Science*, **70**(1), 61–71.

Matthews, J.D. (1989). *Silvicultural Systems*. Oxford University Press, Oxford.

McCaughey, W.W. and Ferguson, D.E. (1988). Response of advance regeneration to release in the Inland Mountain West: a summary. In W.C. Schmidt, ed. *Future Forests of the Mountain West: A Stand Culture Symposium*, pp. 255–266. USDA Forest Service, Intermountain Research Station. General Technical Report, INT-GTR-243.

McDonald, P.M. (1976). *Forest regeneration and seedling growth from five cutting methods in north central California*. USDA Forest Service, Pacific Southwest Forest and Range Experiment Station. Research Paper PSW-RP-115.

McDonald, P.M. and Reynolds, P.E. (1999). *Plant community development after 28 years in small group-selection openings*. USDA Forest Service, Pacific Southwest Research Station. PSW-RP-241.

McEvoy, T.J. (2004). *Positive Impact Forestry: A Sustainable Approach to Managing Woodlands*. Island Press, Washington, DC.

McGinley, K., Alvarado, R., Cubbage, F., Diaz, D., Donoso, P.J., Goncalves Jacovine, L.A., de Silva, F.L., MacIntyre, C., and Monges Zalazar, E. (2012). Regulating the sustainability of forest management in the Americas: Cross–country comparisons of forest legislation. *Forests*, **3**(3), 467–505.

McLintock, T.F. and Bickford, C.A. (1957). *A proposed site index for red spruce in the northeast*. USDA Forest Service, Northeastern Experiment Station. Station Paper NE-93.

McMahon, J.P. (1999). International expectations for sustainable forestry: a view from the US forest industry. *New Forests*, **17**(1–3), 329–338.

Mendoza, G. and Setyarso, A. (1986). A transition matrix forest growth model for evaluating alternative harvesting schemes in Indonesia. *Forest Ecology and Management*, **15**(3), 219–228.

Messier, C., Puettmann, K.J. and Coates, K.D., eds. (2013). *Managing Forests as Complex Adaptive Systems*. Earthscan from Routledge, Abingdon.

Messier, C., Doucet, R., Ruel, J., Claveau, Y., Kelly, C., and Lechowicz, M. (1999). Functional ecology of advance regeneration in relation to light in boreal forests. *Canadian Journal of Forest Research*, **29**(6), 812–823.

Metslaid, M., Jogiste, K., Nikinmaa, E., Moser, W.K., and Porcar-Castell, A. (2007). Tree variables related to growth response and acclimation of advance regeneration of Norway spruce and other coniferous species after release. *Forest Ecology and Management*, **250**(1–2), 56–63.

Meyer, H.A. (1943). Management without rotation. *Journal of Forestry*, **41**(2), 126–132.

Meyer, H.A. (1952). Structure, growth, and drain in balanced uneven-aged forests. *Journal of Forestry*, **50**(2), 85–92.

Meyer, H.A. and Stevenson, D.D. (1943). The structure and growth of virgin beech-birch-maple-hemlock forests in northern Pennsylvania. *Journal of Agricultural Research*, **67**(2), 465–484.

Millar, C.I., Stephenson, N.L., and Stephens, S.L. (2007). Climate change and forests of the future: managing in the face of uncertainty. *Ecological Applications*, **17**(8), 2145–2151.

Milner, K., Running, S., and Coble, D. (1996). A biophysical soil-site model for estimating potential productivity of forested landscapes. *Canadian Journal of Forest Research*, **26**(7), 1174–1186.

Mitchell, A.K. (2001). Growth limitations for conifer regeneration under alternative silvicultural systems in a coastal montane forest in British Columbia, Canada. *Forest Ecology and Management*, **145**(1–2), 129–136.

Mitchell, A.K. and Arnott, J.T. (1995). Effects of shade on the morphology and physiology of Amabilis fir and western hemlock. *New Forests*, **10**(1), 79–98.

Mitchell, A.K., Koppenaal, R., Goodmanson, G., Benton, R., and Bown, T. (2007). Regenerating montane conifers with variable retention systems in a coastal British Columbia forest: 10-year results. *Forest Ecology and Management*, **246**(2–3), 240–250.

Mitchell, R.J., Hiers, J.K., O'Brien, J.J., Jack, S.B., and Engstrom, R.T. (2006). Silviculture that sustains: the nexus between silviculture, frequent prescribed fire, and conservation of biodiversity in longleaf pine forests of the southeastern United States. *Canadian Journal of Forest Research*, **36**(11), 2724–2736.

Mitchell, S.J. (2000). Stem growth responses in Douglas–fir and Sitka spruce following thinning: implications for assessing wind-firmness. *Forest Ecology and Management*, **135**(1–3), 105–114.

Mitchell, S.J. (2013). Wind as a natural disturbance agent in forests: a synthesis. *Forestry*, **86**(2), 147–157.

Mitchell, S.J. and Beese, W.J. (2002). The retention system: reconciling variable retention with the principles of silvicultural systems. *Forestry Chronicle*, **78**(3), 397–403.

Mitscherlich, G. (1952). Der Tannen-Fichten-(Buchen)-Plenterwald: eine ertragskundliche Studie. *Schriftenreihe der Badischen Forstlichen Versuchanstalt*, **8**, 1–42.

Mitton, J.B. (1992). The dynamic mating systems of conifers. *New Forests*, **6**(1–4), 197–216.

Mlinšek, D. (1996). From clear-cutting to a close-to-nature silvicultural system. *IUFRO News*, **25**(4), 6–8.

Mohren, G.M.J. and Burkhart, H.E. (1994). Contrasts between biologically-based process models and management-oriented growth and yield models. *Forest Ecology and Management*, **69**(1–3), 1–5.

Moller, A. (1922). *Der Dauerwaldgedanke: Sein Sinn und seine Bedeutung*. Springer Verlag, Heidelberg.

Möller, C.M., Muller, D., and Nielson, J. (1954). Graphic presentation of dry matter production in European beech. *Det Forstlige Forsogsvaesen i Danmark*, **21**, 327–335.

Monserud, R.A. and Sterba, H. (1996). A basal area increment model for individual trees growing in even- and uneven-aged forest stands in Austria. *Forest Ecology and Management*, **80**(1–3), 57–80.

Morgan, P., Aplet, G.H., Haufler, J.B., Humphries, H.C., Moore, M.M., and Wilson, W.D. (1994). Historical range of variability: a useful tool for evaluating ecosystem change. *Journal of Sustainable Forestry*, **2**(1/2), 87–111.

Morsbach, H.W. (2002). *Common Sense Forestry*. Chelsea Green Publishing Company, White River Junction.

Moser, K.W. (2006). The Stoddard–Neel system: case studies. In S. Jose, E.J. Jokela, and D.L. Miller, eds. *The Longleaf Pine Ecosystem: Ecology, Silviculture, and Restoration.*, pp. 246–249. Springer New York, New York.

Moser, J. (1972). Dynamics of an uneven-aged forest stand. *Forest Science*, **18**(3), 184–191.

Moser, J. and Hall, O. (1969). Deriving growth and yield functions for uneven-aged forest stands. *Forest Science*, **15**(2), 183–186.

Mount, J.R. (2010). *Torching Conventional Forestry: The Artful Application of Science*. Auberry Press, Fresno.

Müller, F. (1990). Naturverjüngung und genetische Vielfalt. *Österreichische Forstzeitung*, **12**(2), 58–60.

Munger, T.T. (1950). A look at selective cutting in Douglas–fir. *Journal of Forestry*, **48**(2), 97–99.

Munger, T.T., Brandstrom, A.J.F., and Kolbe, E.L. (1936). Maturity selection system applied to ponderosa pine. *West Coast Lumberman*, **63**(11), 33–44.

Munro, D.D. (1974). Forest growth models: a prognosis. In J. Fries, ed. *Growth Models for Tree and Stand Simulation*. IUFRO Working Party S4.01.4. Proceedings of Meetings in 1973, pp. 7–21. Royal College of Forestry, Stockholm.

Murawski, D.A., Gunatilleke, I.A.U.N., and Bawa, K.S. (1994). The effects of selective logging on inbreeding in *Shorea megistophylla* (Dipterocarpaceae) from Sri Lanka. *Conservation Biology*, **8**(4), 997–1002.

Murphy, T.E.L., Adams, D.L., and Ferguson, D.E. (1999). Response of advance lodgepole pine regeneration to overstory removal in eastern Idaho. *Forest Ecology and Management*, **120**(1–3), 235–244.

Muscolo, A., Sidari, M., Bagnato, S., Mallamaci, C., and Mercurio, R. (2010). Gap size effects on above- and below-ground processes in a silver fir stand. *European Journal of Forest Research*, **129**(3), 355–365.

Muscolo, A., Sidari, M., and Mercurio, R. (2007). Influence of gap size on organic matter decomposition, microbial biomass and nutrient cycle in Calabrian pine (*Pinus laricio*, Poiret) stands. *Forest Ecology and Management*, **242**(2–3), 412–418.

Mustain, A.P. (1978). History and philosophy of silviculture management systems in use today. In *Uneven–aged Silviculture & Management in the United States*, pp. 11–17. USDA Forest Service. General Technical Report WO-24.

Nason, J.D., Herre, E.A., and Hamrick, J.L. (1998). The breeding structure of a tropical keystone plant resource. *Nature*, **391**(6668), 685–687.

National Academy of Sciences (2000). *Environmental Issues in Pacific Northwest Forest Management*. National Academy Press, Washington, DC.

National Research Council (1991). *Managing Global Genetic Resources: Forest Trees*. National Academy Press, Washington, DC.

Neale, D.B. (1985). Genetic-implications of shelterwood regeneration of Douglas-fir in southwest Oregon. *Forest Science*, **31**(4), 995–1005.

Neale, D.B., Devey, M.E., Jermstad, K.D., Ahuga, M.R., Alosi, M.C., and Marshall, K.A. (1992). Use of DNA markers in forest tree improvement research. *New Forests*, **6**(1–4), 391–407.

Neel, L., Sutter, P.S., and Way, A.G. (2010). *The Art of Managing Longleaf: A Personal History of the Stoddard–Neel Approach*. University of Georgia Press, Athens.

Neumann, M. and Starlinger, F. (2001). The significance of different indices for stand structure and diversity in forests. *Forest Ecology and Management*, **145**(1–2), 91–106.

North, M., Chen, J.Q., Smith, G., Krakowiak, L., and Franklin, J. (1996). Initial response of understory plant diversity and overstory tree diameter growth to a green tree retention harvest. *Northwest Science*, **70**(1), 24–35.

North, M.P. and Keeton, W.S. (2008). Emulating natural disturbance regimes: an emerging approach for sustainable forest management. In R. Lafortezza, J. Chen, G. Sanesi, and T.R. Crow, eds. *Patterns and Processes in Forest Landscapes*, pp. 341–372. Springer, Dordrecht.

Nyland, R.D. (2002). *Silviculture: Concepts and Applications*, 2nd edn, Waveland Press, Long Grove.

Nyland, R.D. (2003). Even- to uneven-aged: the challenges of conversion. *Forest Ecology and Management*, **172**(2–3), 291–300.

Nyland, R.D. (2010). The shelterwood method: adapting to diverse management objectives. *Journal of Forestry*, **108**(8), 419–420.

O'Hara, K.L. (1988). Stand structure and growing space efficiency following thinning in an even-aged Douglas-fir stand. *Canadian Journal of Forest Research*, **18**(7), 859–866.

O'Hara, K.L. (1989). Stand growth efficiency in a Douglas-fir thinning trial. *Forestry*, **62**(4), 409–418.

O'Hara, K.L. (1991). A biological justification for pruning in coastal Douglas-fir stands. *Western Journal of Applied Forestry*, **6**(3), 59–63.

O'Hara, K.L. (1996). Dynamics and stocking-level relationships of multi-aged ponderosa pine stands. *Forest Science*, **42**(4), 1–34.

O'Hara, K.L. (1998). Silviculture for structural diversity: a new look at multiaged system. *Journal of Forestry*, **96**(7), 4–10.

O'Hara, K.L. (2001). The silviculture of transformation—a commentary. *Forest Ecology and Management*, **151**(1–3), 81–86.

O'Hara, K.L. (2002). The historical development of uneven-aged silviculture in North America. *Forestry*, **75**(4), 339–346.

O'Hara, K.L. (2006). Multiaged forest stands for protection forests: concepts and applications. *Forest Snow and Landscape Research*, **80**(1), 45–55.

O'Hara, K.L. (2007). Pruning wounds and occlusion: a long-standing conundrum in forestry. *Journal of Forestry*, **105**(3), 131–138.

O'Hara, K.L. and Gersonde, R.F. (2004). Stocking control concepts in uneven-aged silviculture. *Forestry*, **77**(2), 131–143.

O'Hara, K.L., Hasenauer, H., and Kindermann, G. (2007a). Sustainability in multi-aged stands: an analysis of long-term plenter systems. *Forestry*, **80**(2), 163–181.

O'Hara, K.L., Lähde, E., Laiho, O., Norokorpi, Y., and Saksa, T. (1999). Leaf area and tree increment dynamics on a fertile mixed-conifer site in southern Finland. *Annals of Forest Science*, **56**(3), 237–247.

O'Hara, K.L., Lähde, E., Laiho, O., Norokorpi, Y., and Saksa, T. (2001). Leaf area allocation as a guide to stocking control in multi-aged, mixed-conifer forests in southern Finland. *Forestry*, **74**(2), 171–185.

O'Hara, K.L., Latham, P.A., Hessburg, P., and Smith, B.G. (1996). A structural classification for inland northwest forest vegetation. *Western Journal of Applied Forestry*, **11**(3), 97–102.

O'Hara, K.L., Leonard, L.P., and Keyes, C.R. (2012). Variable-density thinning and a marking paradox: comparing prescription protocols to attain stand variability in coast redwood. *Western Journal of Applied Forestry*, **27**(3), 143–149.

O'Hara, K.L. and Nagel, L.M. (2006). A functional comparison of productivity in even-aged and multiaged stands: a synthesis for *Pinus Ponderosa*. *Forest Science*, **52**(3), 290–303.

O'Hara, K.L. and Nagel, L.M. (2013). The stand: revisiting a central concept in forestry. *Journal of Forestry*, **111**(5), 335–340.

O'Hara, K.L. and Oliver, C.D. (1999). A decision system for assessing stand differentiation potential and prioritizing precommercial thinning treatments. *Western Journal of Applied Forestry*, **14**(1), 7–13.

O'Hara, K.L. and Ramage, B.S. (2013). Silviculture in an uncertain world: utilizing multi-aged management systems to integrate disturbance. *Forestry*, **86**(4), 401–410.

O'Hara, K.L., Seymour, R.S., Tesch, S.D., and Guldin, J.M. (1994). Silviculture and our changing profession: leadership for shifting paradigms. *Journal of Forestry*, **92**(1), 8–13.

O'Hara, K.L., Stancioiu, P.T., and Spencer, M.A. (2007b). Understory stump sprout development under variable canopy density and leaf area in coast redwood. *Forest Ecology and Management*, **244**(1–3), 76–85.

O'Hara, K.L. and Valappil, N.I. (1999). Masam—a flexible stand density management model for meeting diverse structural objectives in multiaged stands. *Forest Ecology and Management*, **118**(1–3), 57–71.

O'Hara, K.L., Valappil, N.I., and Nagel, L.M. (2003). Stocking control procedures for multiaged ponderosa pine stands in the inland northwest. *Western Journal of Applied Forestry*, **18**(1), 5–14.

O'Hara, K.L., Youngblood, A., and Waring, K.M. (2010). Maturity selection versus improvement selection: lessons from a mid-20th century controversy in the silviculture of ponderosa pine. *Journal of Forestry*, **108**(8), 397–407.

O'Keefe, T. (1990). Holistic (new) forestry: significant difference or just another gimmick? *Journal of Forestry*, **88**(4), 23–24.

Oldeman, R.A.A. (1978). Architecture and energy exchange of dicotyledonous trees in the forest. In P.B. Tomlinson and M.H. Zimmermann, eds. *Tropical Trees as Living Systems*, pp. 535–560. Cambridge University Press, Cambridge.

Oldeman, R.A.A. (1990). *Forests, Elements of Silvology*. Springer, Berlin.

Oliver, C.D. (1992). Similarities of stand structures and stand development processes throughout the world—some evidence and applications to silviculture through adaptive management. In M.J. Kelty, B.C. Larson, and C.D. Oliver, eds. *The Ecology and Silviculture of Mixed-Species Forests: A Festchrift for David M. Smith*, pp. 11–26. Kluwer Academic Publishers, Dordrecht.

Oliver, C.D. (1976). Growth response of suppressed hemlocks after release. In W. Atkinson and R.J. Zasoski, eds. *Western Hemlock Management Conference*, pp. 266–272. University of Washington, College of Forest Resources, Seattle.

Oliver, C.D. (1981). Forest development in North America following major disturbances. *Forest Ecology and Management*, **3** (1980–81), 153–168.

Oliver, C.D. and Larson, B.C. (1990). *Forest Stand Dynamics*. McGraw–Hill, Inc, New York.

Oliver, C.D. and Larson, B.C. (1996). *Forest Stand Dynamics*, update edn. John Wiley & Sons, New York.

Oliver, C.D. and O'Hara, K.L. (2005). Effects of restoration at the stand level. In J.A. Stanturf and P. Madsen, eds. *Restoration of Boreal and Temperate Forests*, pp. 31–59. CRC Press, Boca Raton.

Oliver, W.W. (1986). *Growth of California red fir advance regeneration after overstory removal and thinning*. USDA Forest Service, Pacific Southwest Research Station Research Paper. Research Paper PSW-RP-180.

Oliver, W.W. and Dolph, K.L. (1992). Mixed-conifer seedling growth varies in response to overstory release. *Forest Ecology and Management*, **48**(1–2), 179–183.

Olson, D.F. and Della-Bianca, L. (1959). *Site index comparisons for several tree species in the Virginia-Carolina Piedmont*. USDA Forest Service, Southeastern Forest Experiment Station. Station Paper 104.

Omule, A.Y., Mitchell, A.K., and Wagner, W.L. (2011). *Fertilization and thinning effects on a Douglas-fir ecosystem at Shawnigan Lake: 32-year growth response*. Information Report FI-X-005. Canadian Forest Service, Canadian Wood Fibre Centre, Victoria.

Örlander, G. and Karlsson, C. (2000). Influence of shelterwood density on survival and height increment of *Picea abies* advance growth. *Scandinavian Journal of Forest Research*, **15**(1), 20–29.

Orois, S.S., Chang, S.J. and, Gadow, K. (2004). Optimal residual growing stock and cutting cycle in mixed uneven–aged maritime pine stands in Northwestern Spain. *Forest Policy and Economics*, **6**(2), 145–152.

Osunkjoya, O., Ash, J., Hopkins, M., and Graham, A. (1992). Factors affecting survival of tree seedlings in north Queensland rain-forests. *Oecologia*, **91**(4), 569–578.

Page, L., Cameron, A., and Clarke, G. (2001). Influence of overstorey basal area on density and growth of advance regeneration of Sitka spruce in variably thinned stands. *Forest Ecology and Management*, **151**(1–3), 25–35.

Pâquet, J. and Bélanger, L. (1997). Public acceptability thresholds of clearcutting to maintain visual quality of boreal balsam fir landscapes. *Forest Science*, **43**(1), 46–55.

Paquette, A., Bouchard, A., and Cogliastro, A. (2006). Successful under-planting of red oak and black cherry in early-successional deciduous shelterwoods of North America. *Annals of Forest Science*, **63**(8), 823–831.

Parent, S. and Messier, C. (1995). Effects of light gradient on height growth and crown architecture of a naturally regenerated balsam fir. *Canadian Journal of Forest Research*, **25**(6), 878–885.

Parish, R., Antos, J.A., and Fortin, M.J. (1999). Stand development in an old-growth subalpine forest in southern interior British Columbia. *Canadian Journal of Forest Research*, **29**(9), 1347–1356.

Parish, R. and Antos, J.A. (2006). Slow growth, long-lived trees, and minimal disturbance characterize the dynamics of an ancient, montane forest in coastal British Columbia. *Canadian Journal of Forest Research*, **36**(11), 2826–2838.

Park, A. and Talbot, C. (2012). Assisted migration: uncertainty, risk and opportunity. *Forestry Chronicle*, **88**(4), 412–419.

Parker, G.R., Leopold, D.J., and Eichenberger, J.K. (1985). Tree dynamics in an old-growth, deciduous forest. *Forest Ecology and Management*, **11**(1–2), 31–57.

Pautasso, M. (2009). Geographical genetics and the conservation of forest trees. *Perspectives in Plant Ecology Evolution and Systematics*, **11**(3), 157–189.

Pearson, G.A. (1942). Improvement selection cutting in ponderosa pine. *Journal of Forestry*, **40**(10), 753–766.

Peck, J.E., Zenner, E.K., and Palik, B. (2012). Variation in microclimate and early growth of planted pines under dispersed and aggregated overstory retention in mature managed red pine in Minnesota. *Canadian Journal of Forest Research*, **42**(2), 279–290.

Pejchar, L. and Mooney, H.A. (2009). Invasive species, ecosystem services and human well-being. *Trends in Ecology & Evolution*, **24**(9), 497–504.

Peltola, H., Nykanen, M.L., and Kellomaki, S. (1997). Model computations on the critical combination of snow loading and windspeed for snow damage of Scots pine, Norway spruce and Birch sp. at stand edge. *Forest Ecology and Management*, **95**(3), 229–241.

Peng, C.H. (2000). Growth and yield models for uneven-aged stands: past, present and future. *Forest Ecology and Management*, **132**(2–3), 259–279.

Perera, A.H., Buse, L.J., and Weber, M.G., eds. (2004), *Emulating Natural Forest Landscape Disturbances: Concepts and Applications*, Columbia University Press, New York.

Petritan, A.M., Von Luepke, B. and Petritan, I.C. (2007). Effects of shade on growth and mortality of maple (*Acer pseudoplatanus*), ash (*Fraxinus excelsior*) and beech (*Fagus sylvatica*) saplings. *Forestry*, **80**(4), 397–412.

Pfister, R.D., Kovalchik, B.L., Arno, S.F., and Presby, R.C. (1977). *Forest habitat types of Montana*. USDA Forest Service, Intermountain Research Station. General Technical Report INT-GTR-34.

Phillips, D. and Shure, D. (1990). Patch-size effects on early succession in southern Appalachian forests. *Ecology*, **71**(1), 204–212.

Pickett, S.T.A. and White, P.S., eds. (1985). *The Ecology of Natural Disturbance and Patch Dynamics*. Academic Press Inc., San Diego.

Pilarski, M., ed. (1994). *Restoration Forestry: An International Guide to Sustainable Forestry Practices*. Kivaki Press, Durango.

Pinard, M.A. and Cropper, W.P. (2000). Simulated effects of logging on carbon storage in dipterocarp forest. *Journal of Applied Ecology*, **37**(2), 267–283.

Pollan, M. (1998). Preserving a view: Should people "garden" a nature area in order to retain a farm look? *Chicago Tribune*. April 25, 1998.

Pommerening, A. (2002). Approaches to quantifying forest structures. *Forestry*, **75**(3), 305–324.

Pommerening, A. (2006). Transformation to continuous cover forestry in a changing environment. *Forest Ecology and Management*, **224**(3), 227–228.

Pommerening, A. and Murphy, S. (2004). A review of the history, definitions and methods of continuous cover forestry with special attention to afforestation and restocking. *Forestry*, **77**(1), 27–44.

Poorter, H. and Nagel, O. (2000). The role of biomass allocation in the growth response of plants to different levels of light, CO(2), nutrients and water: a quantitative review. *Australian Journal of Plant Physiology*, **27**(12), 1191–1191.

Porter, K.B., Hemens, B., and Maclean, D.A. (2004). Using insect-caused patterns of disturbance in northern New Brunswick to inform forest management. In A.H. Perrera, L.J. Buse, and M.G. Weber, eds. *Emulating Natural Forest Landscaped Disturbances: Concepts and Applications*, pp. 135–145. Columbia University Press, New York.

Powers, R.F. and Wiant, H.V. (1970). Sprouting of old-growth coastal redwood stumps on slopes. *Forest Science*, **16**(3), 339–341.

Prabhu, R., Colfer, C.J.P., and Dudley, R. (1999). *Guidelines for Developing, Testing and Selecting Criteria and Indicators for Sustainable Forest Management*. Center for International Forestry Research (CIFOR), Jakarta.

Pretzsch, H. (2005). Stand density and growth of Norway spruce (*Picea abies* (L.) Karst.) and European beech (*Fagus sylvatica* L.): evidence from long-term experimental plots. *European Journal of Forest Research*, **124**(3), 193–205.

Pretzsch, H. (2009). *Forest Dynamics, Growth and Yield*. Springer, Berlin.

Pretzsch, H., Biber, P., and Dursky, J. (2002). The single tree-based stand simulator SILVA: construction, application and evaluation. *Forest Ecology and Management*, **162**(1), 3–21.

Pro Silva. (2012). *Pro Silva Principles*. Pro Silva—Association of European Foresters Practicing Management Which Follows Natural Processes, Truttenhausen.

Puettmann, K.J., Coates, K.D., and Messier, C. (2009). *A Critique of Silviculture: Managing for Complexity*. Island Press, Washington, DC.

Puettmann, K.J. (2011). Silvicultural challenges and options in the context of global change: "simple" fixes and opportunities for new management approaches. *Journal of Forestry*, **109**(6), 321–331.

Pukkala, T., Lähde, E., and Laiho, O. (2010). Optimizing the structure and management of uneven-sized stands of Finland. *Forestry*, **83**(2), 129–142.

Pukkala, T., Lähde, E., and Laiho, O. (2011). Variable-density thinning in uneven-aged forest management–a case for Norway spruce in Finland. *Forestry*, **84**(5), 557–565.

Pukkala, T., Lähde, E., and Laiho, O. (2012). Continuous cover forestry in Finland—recent research results. In T. Pukkala and K. von Gadow, eds. *Continuous Cover Forestry*, 2nd edn., pp. 85–128. Springer, Dordrecht.

Putz, F.E. (1983). Treefall pits and mounds, buried seeds, and the importance of soil disturbance to pioneer trees on Barro-Colorado Island, Panama. *Ecology*, **64**(5), 1069–1074.

Putz, F.E. (1984). The natural-history of lianas on Barro-Colorado Island, Panama. *Ecology*, **65**(6), 1713–1724.

Putz, F.E., Sist, P., Fredericksen, T., and Dykstra, D. (2008). Reduced-impact logging: challenges and opportunities. *Forest Ecology and Management*, **256**(7), 1427–1433.

Putz, F.E., Zuidema, P.A., Pinard, M.A., Boot, R.G.A., Sayer, J.A., Sheil, D., Sist, P., and Vanclay, J.K. (2012). Improved tropical forest management for carbon retention. *Plos Biology*, **6**(7), 1368–1369.

Raffa, K.F., Aukema, B.H., Bentz, B.J., Carroll, A.L., Hicke, J.A., Turner, M.G., and Romme, W.H. (2008). Cross-scale drivers of natural disturbances prone to anthropogenic amplification: the dynamics of bark beetle eruptions. *Bioscience*, **58**(6), 501–517.

Raison, R.J., Brown, A.G., and Flinn, D.W., eds. (2001). *Criteria and Indicators for Sustainable Forest Management*. CABI Publishing, Wallingford.

Raymond, P., Bedard, S., Roy, V., Larouche, C., and Tremblay, S. (2009). The irregular shelterwood system: review, classification, and potential application to forests affected by partial disturbances. *Journal of Forestry*, **107**(8), 405–413.

Rayner, M.E. (1992). Evaluation of six site classifications for modelling timber yield of regrowth karri (*Eucalyptus diversicolor* F.Muell.). *Forest Ecology and Management*, **54**(1–4), 315–336.

Reid, D.E.B., Silins, U., and Lieffers, V.J. (2003). Stem sapwood permeability in relation to crown dominance and site quality in self-thinning fire-origin lodgepole pine stands. *Tree Physiology*, **23**(12), 833–840.

Reif, A. and Walentowski, H. (2008). The assessment of naturalness and its role for nature conservation and forestry inEurope. *Waldökologie, Landschaftsforschung und Naturschutz*, **6**, 63–76.

Reineke, L.H. (1933). Perfecting a stand-density index for even-aged forests. *Journal of Agricultural Research*, **46**, 627–638.

Reinikainen, M., D'Amato, A.W., and Fraver, S. (2012). Repeated insect outbreaks promote multi-cohort aspen mixedwood forests in northern Minnesota, USA. *Forest Ecology and Management*, **266**, 148–159.

Reyer, C.P.O., Leuzinger, S., Rammig, A., Wolf, A., Bartholomeus, R.P., Bonfante, A., de Lorenzi, F., Dury, M., Gloning, P., Abou Jaoude, R., Klein, T., Kuster, T.M., Martins, M., Niedrist, G., Riccardi, M., Wohlfahrt, G., de Angelis, P., de Dato, G., Francois, L., Menzel, A., and Pereira, M. (2013). A plant's perspective of extremes: terrestrial plant responses to changing climatic variability. *Global Change Biology*, **19**(1), 75–89.

Reyer, C., Lasch, P., Mohren, G.M.J., and Sterck, F.J. (2010). Inter-specific competition in mixed forests of Douglas-fir (*Pseudotsuga menziesii*) and common beech (*Fagus sylvatica*) under climate change—a model-based analysis. *Annals of Forest Science*, **67**(8), 805.

Reynolds, R.R., Baker, J.B., and Ku, T.T. (1984). *Four decades of selection management of the Crossett Farm Forestry forties*. Bulletin 872. University of Arkansas, Division of Agriculture, Agriculture Experiment Station, Fayetteville.

Ribe, R.G. (1989). The aesthetics of forestry: what has empirical preference research taught us. *Environmental Management*, **13**(1), 55–74.

Ribe, R.G. (1999). Regeneration harvests versus clearcuts: public views of the acceptability and aesthetics of Northwest Forest Plan harvests. *Northwest Science*, **73**, 102–117.

Ribe, R.G. (2006). Perceptions of forestry alternatives in the US Pacific Northwest: information effects and acceptability distribution analysis. *Journal of Environmental Psychology*, **26**(2), 100–115.

Ribe, R.G. (2009). In-stand scenic beauty of variable retention harvests and mature forests in the US Pacific Northwest: the effects of basal area, density, retention pattern and down wood. *Journal of Environmental Management*, **91**(1), 245–260.

Ribe, R.G., Ford, R.M., and Williams, K.J.H. (2013). Clearfell controversies and alternative timber harvest designs: how acceptability perceptions vary between Tasmania and the U.S. Pacific Northwest. *Journal of Environmental Management*, **114**, 46–62.

Ricklefs, R.E. (1977). Environmental heterogeneity and plant species diversity: a hypothesis. *American Naturalist*, **111**(978), 376–381.

Ritter, E. (2005). Litter decomposition and nitrogen mineralization in newly formed gaps in a Danish beech (*Fagus sylvatica*) forest. *Soil Biology & Biochemistry*, **37**(7), 1237–1247.

Roberds, J.H. and Conkle, M.T. (1984). Genetic-structure in loblolly pine stands: allozyme variation in parents and progeny. *Forest Science*, **30**(2), 319–329.

Roberge, J.M., Mönkkönen, M., Toivanen, T., and Kotiaho, J.S. (2013). Retention forestry and biodiversity conservation: a parallel with agroforestry. *Nature Conservation*, **4**, 23–29.

Roberts, M.R. (2007). A conceptual model to characterize disturbance severity in forest harvests. *Forest Ecology and Management*, **242**(1), 58–64.

Robinson, G. (1994). *The Forest and the Trees: A Guide to Excellent Forestry*. Island Press, Washington, DC.

Roe, A.L., Alexander, R.R., and Andrews, M.D. (1970). *Engelmann spruce regeneration practices in the Rocky Mountains*. USDA Forest Service. Production Research Report 115.

Rojo, J. and Orois, S. (2005). A decision support system for optimizing the conversion of rotation forest stands to continuous cover forest stands. *Forest Ecology and Management*, **207**(1–2), 109–120.

Romm, J. (1994). Sustainable forests and sustainable forestry. *Journal of Forestry*, **92**(7), 35–39.

Royama, T. (1984). Population dynamics of the spruce budworm *Choristoneura fumiferana*. *Ecological Monographs*, **54**(4), 429–462.

Royama, T., MacKinnon, W.E., Kettela, E.G., Carter, N.E., and Hartling, L.K. (2005). Analysis of spruce budworm outbreak cycles in New Brunswick, Canada, since 1952. *Ecology*, **86**(5), 1212–1224.

Rubin, B.D., Manion, P.D., and Faber-Langendoen, D. (2006). Diameter distributions and structural sustainability in forests. *Forest Ecology and Management*, **222**(1–3), 427–438.

Ruel, J., Messier, C., Doucet, R., Claveau, Y., and Comeau, P. (2000). Morphological indicators of growth response of coniferous advance regeneration to overstorey removal in the boreal forest. *Forestry Chronicle*, **76**(4), 633–642.

Runkle, J.R. (1982). Patterns of disturbance in some old-growth mesic forests of eastern North America. *Ecology*, **63**(5), 1533–1546.

Runkle, J.R. (1985). Disturbance regimes in temperate forests. In S.T.A. Pickett and P.S. White, eds. *The Ecology of Natural Disturbance and Patch Dynamics*, pp. 17–33. Academic Press, Inc., San Diego.

Runkle, J.R., Stewart, G.H. and, Veblen, T.T. (1995). Sapling diameter growth in gaps for 2 *Nothofagus* species in New-Zealand. *Ecology*, **76**(7), 2107–2117.

Running, S.W. and Coughlan, J.C. (1988). A general model of forest ecosystem processes for regional applications. I. Hydrologic balance, canopy gas exchange and primary production processes. *Ecological Modelling*, **42**(2), 125–154.

Ruth, R.H. and Yoder, R.A. (1953). *Reducing wind damage in the forests of the Oregon coast range*. USDA Forest Service, Pacific Northwest Forest and Range Experiment Station. Research Paper 7.

Saksa, T. (2004). Regeneration process, from seed crop to saplings—a case study in uneven-aged Norway spruce-dominated stands in southern Finland. *Silva Fennica*, **38**(4), 371–381.

Salwasser, H. (1992). From new perspectives to ecosystem management. *Conservation Biology*, **6**(3), 469–472.

Salwasser, H. (1994). Ecosystem management: can it sustain diversity and productivity. *Journal of Forestry*, **92**(8), 6–10.

Sampson, R.N. and Adams, D.L., eds. (1994). *Assessing Forest Ecosystem Health in the Inland West. Proceedings of the American Forests Workshop*. Haworth Press, Binghamton.

Sander, I.L., Johnson, P.S., and Watt, R.F. (1976). *A guide for evaluating the adequacy of oak advance reproduction*. USDA Forest Service, North Central Forest Experiment Station. NC-23.

Sander, I.L. (1971). Height growth of new oak sprouts depends on size of advance reproduction. *Journal of Forestry*, **69**(11), 809–811.

Sato, Y. (2009). *Criteria and Indicators for the Conservation and Sustainable Management of Temperate and Boreal Forests: the Montreal Process*. Montreal Process, Tokyo.

Savill, P.S. and Evans, J. (1986). *Plantation Silviculture in Temperate Regions*. Oxford University Press, Oxford.

Savolainen, O. and Kärkkäinen, K. (1992). Effect of forest management on gene pools. *New Forests*, **6**(1–4), 329–345.

Schabel, H. and Palmer, S. (1999). The Dauerwald: its role in the restoration of natural forests. *Journal of Forestry*, **97**(11), 20–25.

Schaberg, P.G., DeHayes, D.H., Hawley, G.J., and Nijensohn, S.E. (2008). Anthropogenic alterations of genetic diversity within tree populations: implications for forest ecosystem resilience. *Forest Ecology and Management*, **256**(5), 855–862.

Schindler, D., Bauhus, J., and Mayer, H. (2012). Wind effects on trees. *European Journal of Forest Research*, **131**(1), 159–163.

Schliemann, S.A. and Bockheim, J.G. (2011). Methods for studying treefall gaps: a review. *Forest Ecology and Management*, **261**(7), 1143–1151.

Schoppa, F.N. and Gregorius, H.R. (2001). Is autochthony an operational concept? In G. Müller-Starck and R. Schubert, eds. *Genetic Response of Forest Systems to Changing Environmental Conditions*, pp. 185. Springer, Dordrecht.

Schuler, T.M. (2004). Fifty years of partial harvesting in a mixed mesophytic forest: composition and productivity. *Canadian Journal of Forest Research*, **34**(5), 985–997.

Schütz, J.-P. (1975). Dynamique et conditions d'équilibre de peuplements jardinés sur les staions de la hêtraie à sapin. *Schweizerische Zeitschrift fur Forestwesen Journal Forestier Suisse*, **126**(9), 637–670.

Schütz, J.-P. (1994). History and current importance of uneven-aged silviculture in Europe. *Allgemeine Forst und Jagdzeitung*, **165**(5–6), 106–114.

Schütz, J.-P. (1997). Conditions of equilibrium in fully irregular, uneven-aged forests: the state-of-the-art in European Plenter forests. In W.H. Emmingham, ed. *Proceedings of the IUFRO Interdisciplinary Uneven-aged Management Symposium*, pp. 455–467. Oregon State University, Corvallis.

Schütz, J.-P. (1999). Close-to-nature silviculture: is this concept compatible with species diversity? *Forestry*, **72**(4), 359–366.

Schütz, J.-P. (2001a). *Der Plenterwald*. Parey Buchverlag, Berlin.

Schütz, J.-P. (2001b). Opportunities and strategies of transforming regular forests to irregular forests. *Forest Ecology and Management*, **151**(1–3), 87–94.

Schütz, J.-P. (2002a). Uneven-aged silviculture: tradition and practices. *Forestry*, **75**(1), 327–328.

Schütz, J.-P. (2002b). Silvicultural tools to develop irregular and diverse forest structures. *Forestry*, **75**(4), 329–337.

Schütz, J.-P. (2006). Modelling the demographic sustainability of pure beech plenter forests in eastern Germany. *Annals of Forest Science*, **63**(1), 93–100.

Schütz, J.-P. and Pommerening, A. (2013). Can Douglas fir (*Pseudotsuga menziesii* (Mirb.) Franco) sustainably grow in complex forest structures? *Forest Ecology and Management*, **303**, 175–183.

Schütz, J.-P., Pukkala, T., Donoso, P., and Gadow, K. (2012). Historical emergence and current application of CCF. In T. Pukkala and K. von Gadow, eds. *Continuous Cover Forestry*, 2nd edn., pp. 1–28. Springer, Dordrecht.

Seastedt, T.R., Hobbs, R.J., and Suding, K.N. (2008). Management of novel ecosystems: are novel approaches required? *Frontiers in Ecology and the Environment*, **6**(10), 547–553.

Seavey, S.R. and Bawa, K.S. (1986). Late-acting self-incompatibility in angiosperms. Botanical Review, **52**(2), 195–219.

Segura, G., Brubaker, L.B., Franklin, J.F., Hinckley, T.M., Maguire, D.A., and Wright, G. (1994). Recent mortality and decline in mature *Abies amabilis*: the interaction between site factors and tephra deposition from Mount St. Helens. *Canadian Journal of Forest Research*, **24**(6), 1112–1122.

Seidel, K.W. (1980). *A guide for comparing height growth of advance reproduction and planted seedlings*. USDA Forest Service, Pacific Northwest Forest and Range Experiment Station. Research Note PNW-RN-360.

Seidel, K.W. (1983). *Growth of supressed grand fir and Shasta red fir in central Oregon after release and thinning: 10-year results*. USDA Forest Service, Pacific Northwest Forest and Range Experiment Station. Research Note PNW-RN-404.

Seidel, K.W. (1985). *Growth response of suppressed true fir and mountain hemlock after release*. USDA Forest Service, Pacific Northwest Forest and Range Experiment Station. Research Paper PNW-RP-344.

Seidl, R. and Lexer, M.J. (2013). Forest management under climatic and social uncertainty: trade-offs between reducing climate change impacts and fostering adaptive capacity. *Journal of Environmental Management*, **114**, 461–469.

Seidl, R., Rammer, W., Jaeger, D., Currie, W.S., and Lexer, M.J. (2007). Assessing trade-offs between carbon sequestration and timber production within a framework of multi-purpose forestry in Austria. *Forest Ecology and Management*, **248**(1–2), 64–79.

Seidl, R., Schelhaas, M.-J., Lindner, M., and Lexer, M.J. (2009). Modelling bark beetle disturbances in a large scale forest scenario model to assess climate change impacts and evaluate adaptive management strategies. *Regional Environmental Change*, **9**(2), 101–119.

Seidl, R., Schelhaas, M.-J., and Lexer, M.J. (2011). Unraveling the drivers of intensifying forest disturbance regimes in Europe. *Global Change Biology*, **17**(9), 2842–2852.

Sendak, P.E., Brissette, J.C., and Frank, R.M. (2003). Silviculture affects composition, growth, and yield in mixed northern conifers: 40-year results from the Penobscot Experimental Forest. *Canadian Journal of Forest Research*, **33**(11), 2116–2128.

Seydack, A.H.W. (1995). An unconventional approach to timber yield regulation for multi-aged, multispecies forests. 1. Fundamental considerations. *Forest Ecology and Management*, **77**(1–3), 139–153.

Seydack A.H.W. (2012). Regulation of timber yield sustainability for tropical and subtropical moist forests: ecosilvicultural paradigms and economic constraints. In T. Pukkala and K. von Gadow, eds. *Continuous Cover Forestry*, 2nd edn, pp. 129–165. Springer, Dordrecht.

Seymour, R.S. (2005). Integrating natural disturbance parameters into conventional silvicultural systems: experience from the Acadian Forest of northeastern North America. In C. Peterson and D.A. Maguire, eds. *Balancing ecosystem values: innovative experiments for sustainable forestry*, pp. 41–48. USDA Forest Service, *Pacific Northwest Research Station*. General Technical Report PNW-GTR-625.

Seymour, R.S. and Hunter, M.L., Jr. (1992). *New forestry in eastern spruce-fir forests: principles and applications to Maine*. Maine Agricultural Experimental Station. Miscellaneous Publication 716.

Seymour, R.S. and Hunter, M.L., Jr. (1999). Principles of ecological forestry. In M.L. Hunter Jr., ed. *Managing Biodiversity in Forest Ecosystems*, pp. 22–61. Cambridge University Press, Cambridge.

Seymour, R.S. and Kenefic, L.S. (2002). Influence of age on growth efficiency of *Tsuga canadensis* and *Picea rubens* trees in mixed-species, multiaged northern conifer stands. *Canadian Journal of Forest Research*, **32**(11), 2032–2042.

Seymour, R.S., White, A.S., and deMaynadier, P.G. (2002). Natural disturbance regimes in northeastern North America—evaluating silvicultural systems using natural scales and frequencies. *Forest Ecology and Management*, **155**(1–3), 357–367.

Shaw, C.G., III and Kile, G.A. (1991). *Armillaria root disease*. USDA Forest Service. Agriculture Handbook 691. US Government Printing Office, Washington, DC.

Shelby, B., Thompson, J., Brunson, M., and Johnson, R. (2003). Changes in scenic quality after harvest:a decade of ratings for six silviculture treatments. *Journal of Forestry*, **101**(2), 30–35.

Shepperd, W.D. (2001). Manipulations to regenerate aspen ecosystems. In W.D. Shepperd, D. Binkley, D.L. Bartos, T.J. Stohlgren, and L.G. Eskew, eds. *Symposium on sustaining aspen in western landscapes*, pp. 355–365. USDA Forest Service, Rocky Mountain Forest and Range Experiment Station. RMRS-P-18.

Shifley, S., Thompson, F., Dijak, W., Larson, M., and Millspaugh, J. (2006). Simulated effects of forest management alternatives on landscape structure and habitat suitability in the Midwestern United States. *Forest Ecology and Management*, **229**(1–3), 361–377.

Shindler, B. and Mallon, A.L. (2009). *Public acceptance of disturbance-based forest management: a study of the Blue River Landscape Strategy in the Central Cascades Adaptive Management Area*. USDA Forest Service, Pacific Northwest Research Station. Research Paper PNW-RP-581.

Shindler, B.A., Brunson, M., and Stankey, G.H. (2002). *Social acceptability of forest conditions and management practices: a problem analysis*. USDA Forest Service, Pacific Northwest Research Station. General Technical Report PNW-GTR-537.

Shinneman, D.J. and Baker, W.L. (2009). Historical fire and multidecadal drought as context for pinon-juniper woodland restoration in western Colorado. *Ecological Applications*, **19**(5), 1231–1245.

Shoulders, E. and Tiarks, A.E. (1980). Predicting height and relative performance of major southern pines from rainfall, slope, and available soil moisture. *Forest Science*, **26**(3), 437–447.

Shugart, H. (1984). *A Theory of Forest Dynamics: The Ecological Implications of Forest Succession Models*. Springer-Verlag, New York.

Shugart, H. and West, D. (1977). Development of an Appalachian deciduous forest succession model and its application to assessment of impact of chestnut blight. *Journal of Environmental Management*, **5**(2), 161–179.

Sipe, T.W. and Bazzaz, F.A. (1994). Gap partitioning among maples (*Acer*) in central New-England: shoot architecture and photosynthesis. *Ecology*, **75**(8), 2318–2332.

Sipe, T. and Bazzaz, F. (1995). Gap partitioning among maples (*Acer*) in central New-England: survival and growth. *Ecology*, **76**(5), 1587–1602.

Sist, P., Fimbel, R., Sheil, D., Nasi, R., and Chevallier, M. (2003). Towards sustainable management of mixed dipterocarp forests of South-east Asia: moving beyond minimum diameter cutting limits. *Environmental Conservation*, **30**(4), 364–374.

Skovsgaard, J.P. and Vanclay, J.K. (2008). Forest site productivity: a review of the evolution of dendrometric concepts for even-aged stands. *Forestry*, **81**(1), 13–31.

Skovsgaard, J.P. and Vanclay, J.K. (2013). Forest site productivity: a review of spatial and temporal variability in natural site conditions. *Forestry*, **86**(3), 305–315.

Slodicak, M. and Novak, J. (2006). Silvicultural measures to increase the mechanical stability of pure secondary Norway spruce stands before conversion. *Forest Ecology and Management*, **224**(3), 252–257.

Smith, D.M. (1962). *The Practice of Silviculture*, 7th edn., John Wiley & Sons, New York.

Smith, D.M. (1972). The continuing evolution of silvicultural practice. *Journal of Forestry*, **70**(2), 89–92.

Smith, D.M., Larson, B.C., Kelty, M.J., and Ashton, P.M.S. (1997). *The Practice of Silviculture: Applied Forest Ecology*, 9th edn., John Wiley & Sons, New York.

Smith, H.C. and Debald, P.S. (1978). Economics of even-aged and uneven-aged silviculture and management in eastern hardwoods. In *Uneven-aged silviculture & management in the United States*, pp. 125–141. USDA Forest Service. General Technical Report WO-24.

Solomon, D.S. and Blum, B.M. (1967). *Stump sprouting of four northern hardwoods*. USDA Forest Service, Northeastern Forest Experiment Station. Research Paper RP-NE-59.

Sousa, W.P. (1984). The role of disturbance in natural communities. *Annual Review of Ecology and Systematics*, **15**, 353–391.

Speicker, H., Hansen, J., Klimo, E., Skovsgaard, J.P., Sterba, H., and Teuffel, K., eds. (2004). *Norway Spruce Conversion—Options and Consequences*. European Forest Institute Research Reports, Vol. 18. Brill, Leiden.

Spies, T.A. (1998). Forest structure: a key to the ecosystem. *Northwest Science*, **72** (Special Issue 2), 34–39.

Spies, T., Franklin, J., and Klopsch, M. (1990). Canopy gaps in Douglas-fir forests of the Cascade Mountains. *Canadian Journal of Forest Research*, **20**(5), 649–658.

Spies, T.A. and Turner, M.G. (1999). Dynamic forest mosaics. In M.L. Hunter Jr., ed. *Maintaining Biodiversity in Forest Ecosystems*, pp. 95–160. Cambridge University Press, Cambridge.

Spilsbury, R.H. and Smith, D.S. (1947). *Forest site types of the Pacific Northwest: a preliminary report*. British Columbia Forest Service, Department of Lands and Forests. Forest Technical Publication T. 30.

Sprugel, D.G. (1976). Dynamic structure of wave-regenerated *Abies balsamea* forests in the north-eastern United States. *Journal of Ecology*, **64**(3), 889–912.

Sprugel, D.G. (1991). Disturbance, equilibrium, and environmental variability: what is natural vegetation in a changing environment. *Biological Conservation*, **58**(1), 1–18.

Sprugel, D.G., Brooks, J.R., and Hinckley, T.M. (1996). Effects of light on shoot geometry and needle morphology in *Abies amabilis*. *Tree physiology*, **16**(1–2), 91–98.

St. Clair, J.B. and Sniezko, R.A. (1999). Genetic variation in response to shade in coastal Douglas-fir. *Canadian Journal of Forest Research*, **29**(11), 1751–1763.

Stadt, K.J., Lieffers, V.J., Hall, R.J., and Messier, C. (2005). Spatially explicit modeling of PAR transmission and growth of *Picea glauca* and *Abies balsamea* in the boreal forests of Alberta and Quebec. *Canadian Journal of Forest Research*, **35**(1), 1–12.

Stage, A.R. (1973). *Prognosis model for stand development*. USDA Forest Service, Intermountain Research Station. Research Paper INT-RP-137.

Stancioiu, P.T. and O'Hara, K.L. (2006a). Leaf area and growth efficiency of regeneration in mixed species, multiaged forests of the Romanian Carpathians. *Forest Ecology and Management*, **222**(1–3), 55–66.

Stancioiu, P.T. and O'Hara, K.L. (2006b). Morphological plasticity of regeneration subject to different levels of canopy cover in mixed-species, multiaged forests of the Romanian Carpathians. *Trees*, **20**(2), 196–209.

Stancioiu, P.T. and O'Hara, K.L. (2006c). Regeneration growth in different light environments of mixed species, multiaged, mountainous forests of Romania. *European Journal of Forest Research*, **125**(2), 151–162.

Stanturf, J.A. and Madsen, P. (2002). Restoration concepts for temperate and boreal forests of North America and Western Europe. *Plant Biosystems*, **136**(2), 143–158.

Starke, R., Ziehe, M. and Muller-Starck, G. (1996). Viability selection in juvenile populations of European beech (*Fagus sylvatica* L.). *Forest Genetics*, **3**(4), 217–225.

Steen, H.K. (1990). *David M. Smith and the History of Silviculture*. Forest History Society, Durham.

Steinbrenner, E.C. (1979). Forest soil productivity relationships. In P.E. Heilman, H.W. Anderson, and D.M. Baumgartner, eds. *Forest Soils of the Douglas-fir Region*, pp. 199–229. Cooperative Extension Service, Washington State University, Pullman.

Ste-Marie, C., Nelson, E.A., Dabros, A., and Bonneau, M. (2011). Assisted migration: introduction to a multifaceted concept. *Forestry Chronicle*, **87**(6), 724–730.

Stephens, S.L., Millar, C.I., and Collins, B.M. (2010). Operational approaches to managing forests of the future in Mediterranean regions within a context of changing climates. *Environmental Research Letters*, **5**(2), 024003.

Stephens, S.L., Moghaddas, J.J., Hartsough, B.R., Moghaddas, E.E.Y., and Clinton, N.E. (2009). Fuel treatment effects on stand-level carbon pools, treatment-related emissions, and fire risk in a Sierra Nevada mixed-conifer forest. *Canadian Journal of Forest Research*, **39**(8), 1538–1547.

Sterba, H. (2004). Equilibrium curves and growth models to deal with forests in transition to uneven-aged structure—Application in two sample stands. *Silva Fennica*, **38**(4), 413–423.

Sterba, H. and Zingg, A. (2001). Target diameter harvesting—a strategy to convert even-aged forests. *Forest Ecology and Management*, **151**(1–3), 95–105.

Sterba, H. and Zingg, A. (2006). Distance dependent and distance independent description of stand structure. *Allgemeine Forst und Jagdzeitung*, **177**(8–9), 169–176.

Stewart, G.H., Rose, A.B., and Veblen, T.T. (1991). Forest development in canopy gaps in old-growth beech (*Nothofagus*) forests, New Zealand. *Journal of Vegetation Science*, **2**(5), 679–690.

Stine, P., Spies, T., Hessburg, P., Kramer, M., Fettig, C., Hanson, A., Lehmkuhl, J., O'Hara, K., Polivka, K., Singleton, P., Charnley, S., and Mershel, A. (2014). *The ecology and management of moist mixed-conifer forests in eastern Oregon and Washington; a synthesis of the relevant biophysical science and implications for future land management*. USDA Forest Service, Pacific Northwest Research Station PNW-GTR-897.

Styles, B.T. (1972). The flower biology of the Meliaceae and its bearing on tree breeding. *Silvae Genetica*, **21**(5), 175–182.

Suffling, R. and Perera, A.H. (2004). Characterizing natural forest disturbance regimes. In A.H. Perera, L.J. Buse, and M.G. Weber, eds. *Emulating Natural Forest Landscape Disturbances: Concepts and Applications*, pp. 43–54. Columbia University Press, New York.

Susse, R., Allegrini, C., Bruciamacchie, M., and Burrus, R. (2011). *Management of Irregular Forests: Developing the Full Potential of the Forest*. Association Futaie Irrebuliere, Besancon.

Swaine, M. and Hall, J. (1983). Early succession on cleared forest land in Ghana. *Journal of Ecology*, **71**(2), 601–627.

Swanson, F.H. (2011). *The Bitterroot & Mr. Brandborg: Clearcutting and the Struggle for Sustainable Forestry in the Northern Rockies*. University of Utah Press, Salt Lake City.

Tahvonen, O. (2009). Optimal choice between even- and uneven-aged forestry. *Natural Resource Modeling*, **22**(2), 289–321.

Tahvonen, O. (2011). Optimal structure and development of uneven-aged Norway spruce forests. *Canadian Journal of Forest Research*, **41**(12), 2389–2402.

Tahvonen, O., Pukkala, T., Laiho, O., Lähde, E., and Niinimäki, S. (2010). Optimal management of uneven-aged Norway spruce stands. *Forest Ecology and Management*, **260**(1), 106–115.

Takahashi, M., Mukouda, M., and Koono, K. (2000). Differences in genetic structure between two Japanese beech (*Fagus crenata* Blume) stands. *Heredity*, **84**(1), 103–115.

Tappeineir, J.C.I., Maguire, D.A., and Harrington, T.B. (2007). *Silviculture and Ecology of Western U.S. Forests*. Oregon State University Press, Corvallis.

Tarp, P., Buongiorno, J., Helles, F., Larsen, J.B., Meilby, H., and Strange, N. (2005). Economics of converting an even-aged *Fagus sylvatica* stand to an uneven-aged stand using target diameter harvesting. *Scandinavian Journal of Forest Research*, 20(1), 63–74.

Teck, R., Moeur, M., and Eav, B. (1996). Forecasting ecosystems with the forest vegetation simulator. *Journal of Forestry*, 94(12), 7–10.

Tesch, S.D., Baker-Katz, K., Korpela, E.J., and Mann, J.W. (1993). Recovery of Douglas-fir seedlings and saplings wounded during overstory removal. *Canadian Journal of Forest Research*, 23(8), 1684–1694.

Tesch, S.D. and Korpela, E.J. (1993). Douglas-fir and white fir advance regeneration for renewal of mixed-conifer forests. *Canadian Journal of Forest Research*, 23(7), 1427–1437.

Thirgood, J.V. (1981). *Man and the Mediterranean Forest: A History of Resource Depletion*. Academic Press, London.

Thomas, J.W., ed. (1979). *Wildlife habitats in managed forests: the Blue Mountains of Oregon and Washington*. USDA Agriculture Handbook No. 553. US Government Printing Office, Washington, DC.

Thompson, J.R., Duncan, S.L., and Johnson, K.N. (2009). Is there potential for the historical range of variability to guide conservation given the social range of variability? *Ecology and Society*, 14(1), 1–14.

Thorpe, H.C. and Daniels, L.D. (2012). Long-term trends in tree mortality rates in the Alberta foothills are driven by stand development. *Canadian Journal of Forest Research*, 42(9), 1687–1696.

Thorpe, H.C., Thomas, S.C., and Caspersen, J.P. (2008). Tree mortality following partial harvests is determined by skidding proximity. *Ecological Applications*, 18(7), 1652–1663.

Tigerstedt, P.M.A., Rudin, D., Niemela, T., and Tammisola, J. (1982). Competition and neighboring effect in a naturally regenerating population of Scots pine. *Silva Fennica*, 16(2), 122–129.

Tiscar Oliver, P.A., Lucas-Borja, M.E., and Candel Pérez, D. (2011). Changes in the structure and composition of two *Pinus nigra* subsp *salzmannii* forests over a century of different silvicultural treatments. *Forest Systems*, 20(3), 525–535.

Tonnes, S., Karjalainen, E., Lofstrom, I., and Neuvonen, M. (2004). Scenic impacts of retention trees in clear-cutting areas. *Scandinavian Journal of Forest Research*, 19(4), 348–357.

Torimaru, T., Itaya, A., and Yamamoto, S.-I. (2012). Quantification of repeated gap formation events and their spatial patterns in three types of old-growth forests: analysis of long-term canopy dynamics using aerial photographs and digital surface models. *Forest Ecology and Management*, 284, 1–11.

Totman, C. (1989). *The Green Archipelago: Forestry in Preindustrial Japan*. University of California Press, Berkeley.

Troup, R.S. (1928). *Silvicultural Systems*. Oxford University Press, London.

Tucker, G.F. and Emmingham, W.H. (1977). Morphological changes in leaves of residual western hemlock after clear and shelterwood cutting. *Forest Science*, 23(2), 195–203.

Tucker, G.F., Hinckley, T.M., Leverenz, J.W., and Jiang, S.M. (1987). Adjustments of foliar morphology in the acclimation of understory Pacific silver fir following clearcutting. *Forest Ecology and Management*, 21(3–4), 249–268.

Turner, M.G., Baker, W.L., Peterson, C.J., and Peet, R.K. (1998). Factors influencing succession: lessons from large, infrequent natural disturbances. *Ecosystems*, 1(6), 511–523.

Turner, M.G. (2010). Disturbance and landscape dynamics in a changing world. *Ecology*, 91(10), 2833–2849.

Turner, M.G., O'Neill, R.V., Gardner, R.H., and Milne, B.T. (1989). Effects of changing spatial scale on the analysis of landscape pattern. *Landscape Ecology*, 3(3–4), 153–162.

Tyler, C.M., Kuhn, B., and Davis, F.W. (2006). Demography and recruitment limitations of three oak species in California. *Quarterly Review of Biology*, 81(2), 127–152.

Vaartaja, O. (1952). On the recovery of released pine advance growth and its silvicultural importance. *Acta Forest Fennica*, 59(1), 1–133.

Mantgem, P.J., Nesmith, J.C.B., Keifer, M., Knapp, E.E., Flint, A., and Flint, L. (2013). Climatic stress increases forest fire severity across the western United States. *Ecology Letters*, 16(9), 1151–1156.

Vanclay, J.K. (1992). Assessing site productivity in tropical moist forests: a review. *Forest Ecology and Management*, 54(1–4), 257–287.

Vanclay, J.K. (1994). *Modelling Forest Growth and Yield: Applications to Mixed Tropical Forests*. CAB International, Wallingford.

Vanclay, J.K. (1995). Synthesis: growth models for tropical forests: a synthesis of models and methods. *Forest Science*, 41(1), 7–42.

Vanclay, J.K. and Henry, N.B. (1988). Assessing site productivity of indigenous cypress pine forest in southern Queensland. *Commonwealth Forestry Review*, 67(1), 53–64.

Vicente, J.R., Pinto, A.T., Araújo, M.B., Verburg, P.H., Lomba, A., Randin, C.F., Guisan, A., and Honrado, J.P. (2013). Using life strategies to explore the vulnerability of ecosystem services to invasion by alien plants. *Ecosystems*, 16(4), 678–693.

Vitousek, P.M. and Denslow, J.S. (1986). Nitrogen and phosphorus availability in treefall gaps of a lowland tropical rainforest. *Journal of Ecology*, **74**(4), 1167–1178.

Vose, J.M., Dougherty, P.M., Long, J.N., Smith, F.W., Gholz, H.L., and Curran, P.J. (1994). Factors influencing the amount and distribution of leaf area of pine stands. *Ecological Bulletins*, **43**, 102–114.

Vose, J.M., Peterson, D.L., and Patel-Weynand, T. (2012). *Effects of climatic variability and change on forest ecosystems: a comprehensive science synthesis for the U.S. forest sector*. USDA Forest Service, Pacific Northwest Research Station, PNW-GTR-870.

Vospernik, S., Monserud, R.A., and Sterba, H. (2010). Do individual-tree growth models correctly represent height:diameter ratios of Norway spruce and Scots pine? *Forest Ecology and Management*, **260**(10), 1735–1753.

Vyse, A., Ferguson, C., Simard, S.W., Kano, T., and Puttonen, P. (2006). Growth of Douglas-fir, lodgepole pine, and ponderosa pine seedlings underplanted in a partially-cut, dry Douglas-fir stand in south-central British Columbia. *Forestry Chronicle*, **82**(5), 723–732.

Wadsworth, F.H. (1992). Temperate zone roots of silviculture in the tropics. In M.J. Kelty, B.C. Larson, and C.D. Oliver, eds. *The Ecology and Silviculture of Mixed-Species Forests: A Festchrift for David M. Smith*, pp. 245–255. Kluwer Academic Publishers, Dordrecht.

Waring, K.M. and O'Hara, K.L. (2005). Silvicultural strategies in forest ecosystems affected by introduced pests. *Forest Ecology and Management*, **209**(1–2), 27–41.

Waring, R.H. (1983). Estimating forest growth and efficiency in relation to canopy leaf area. *Advances in Ecological Research*, **13**, 327–354.

Waring, R.H. and Running, S.W. (1998). *Forest Ecosystems: Analysis at Multiple Scales*. Academic Press, San Diego.

Watt, A.S. (1947). Pattern and process in the plant community. *Journal of Ecology*, **35**(1–2), 1–22.

Wayne, P. and Bazzaz, F. (1993). Birch seedling responses to daily time courses of light in experimental forest gaps and shadehouses. *Ecology*, **74**(5), 1500–1515.

Weaver, H. (1943). Fire as an ecological and silvicultural factor in the ponderosa pine region of the Pacific slope. *Journal of Forestry*, **41**(1), 7–15.

Webb, W.L. (1981). Relation of starch content to conifer mortality and growth loss after defoliation by the Douglas-fir tussock moth. *Forest Science*, **27**(2), 224–232.

Webster, C.R. and Lorimer, C.G. (2002). Single-tree versus group selection in hemlock-hardwood forests: are smaller openings less productive? *Canadian Journal of Forest Research*, **32**(4), 591–604.

Weeks, B.C., Hamburg, S.P. and, Vadeboncoeur, M.A. (2009). Ice storm effects on the canopy structure of a northern hardwood forest after 8 years. *Canadian Journal of Forest Research*, **39**(8), 1475–1483.

Weigel, D.R. and Peng, C.Y.J. (2002). Predicting stump sprouting and competitive success of five oak species in southern Indiana. *Canadian Journal of Forest Research*, **32**(4), 703–712.

Weiskittel, A.R., Hann, D.W., Kershaw, J.A., and Vanclay, J.K. (2011). *Forest Growth and Yield Modeling*. John Wiley & Sons, Ltd., West Sussex.

West, P.W. (1983). Comparison of stand density measures in even-aged regrowth eucalypt forests of southern Tasmania. *Canadian Journal of Forest Research*, **13**(1), 22–31.

West, P.W. (2006). *Growing Plantation Forests*. Springer, Berlin, Germany.

Westphal, C., Tremer, N., Oheimb, G.V., Hansen, J., Gadow, K., and Hardtle, W. (2006). Is the reverse J-shaped diameter distribution universally applicable in European virgin beech forests? *Forest Ecology and Management*, **223**(1/3), 75–83.

Wetzel, S. and Burgess, D. (2001). Understorey environment and vegetation response after partial cutting and site preparation in *Pinus strobus* L. stands. *Forest Ecology and Management*, **151**(1–3), 43–59.

White, P.S. and Pickett, S.T.A. (1985). Natural disturbance and patch dynamics: an introduction. In S.T.A. Pickett and P.S. White, eds. *The Ecology of Natural Disturbance and Patch Dynamics*, pp. 3–13. Academic Press, Inc., San Diego.

White, T.L., Adams, W.T., and Neale, D.B. (2007). *Forest Genetics*. CABI Publishing, Wallingford.

Whitmore, T.C. (1975). *Tropical Rain Forests of the Far East*. Clarendon Press, Oxford.

Whittaker, R.J., Willis, K.J., and Field, R. (2001). Scale and species richness: towards a general, hierarchical theory of species diversity. *Journal of Biogeography*, **28**(4), 453–470.

Williams, H., Messier, C., and Kneeshaw, D. (1999). Effects of light availability and sapling size on the growth and crown morphology of understory Douglas-fir and lodgepole pine. *Canadian Journal of Forest Research*, **29**(2), 222–231.

Wilson, F.G. (1979). Thinning as an orderly discipline: a graphic spacing schedule for red pine. *Journal of Forestry*, **77**(8), 483–486.

Wilson, J.S. and Oliver, C.D. (2000). Stability and density management in Douglas-fir plantations. *Canadian Journal of Forest Research*, **30**(6), 910–920.

Wiser, S.K., Allen, R.B., Benecke, U., Baker, G., and Peltzer, D. (2005). Tree growth and mortality after small-group harvesting in New Zealand old-growth *Nothofagus* forests. *Canadian Journal of Forest Research*, **35**(10), 2323–2331.

Wonn, H.T. and O'Hara, K.L. (2001). Height:diameter ratios and stability relationships for four northern Rocky Mountain tree species. *Western Journal of Applied Forestry*, **16**(2), 87–94.

Wood, N. (1971). *Clearcut: The Deforestation of America*. Sierra Club, San Francisco.

Woodall, C.W., Fiedler, C.E., and Milner, K.S. (2003). Stand density index in uneven-aged ponderosa pine stands. *Canadian Journal of Forest Research*, **33**(1), 96–100.

Wright, E.F., Canham, C.D. and Coates, K.D. (2000). Effects of suppression and release on sapling growth for 11 tree species of northern, interior British Columbia. *Canadian Journal of Forest Research*, **30**(10), 1571–1580.

Wycoff, W.R., Crookston, N.L. and Stage, A.R. (1982). *User's guide to the stand prognosis model*. USDA Forest Service, Intermountain Research Station. General Technical Report INT-GTR-133.

Wylie, F.R. and Speight, M.R. (2012). *Insect Pests in Tropical Forestry*. CABI Publishers, Wallingford.

Yamamoto, S. (1996). Gap regeneration of major tree species in different forest types of Japan. *Vegetatio*, **127**(2), 203–213.

Yamamoto, S.-I. (1998). Gap-disturbance regimes in different forest types of Japan. *Journal of Sustainable Forestry*, **6**(3/4), 223–235.

Yamamoto, S.-I. (2000). Forest gap dynamics and tree regeneration. *Journal of Forest Research*, **5**, 223–229.

Yamashita, K., Mizoue, N., Ito, S., Inoue, A., and Kaga, H. (2006). Effects of residual trees on tree height of 18- and 19-year-old *Cryptomeria japonica* planted in group selection openings. *Journal of Forest Research*, **11**(4), 227–234.

Yazdani, R., Muona, O., Rudin, D., and Szmidt, A.E. (1985). Genetic structure of a *Pinus sylvestris* L. seed-tree stand and naturally regenerated understory. *Forest Science*, **31**(2), 430–436.

York, R.A., Battles, J.J., and Heald, R.C. (2003). Edge effects in mixed conifer group selection openings: tree height response to resource gradients. *Forest Ecology and Management*, **179**(1/3), 107–121.

York, R.A., Heald, R.C., Battles, J.J., and York, J.D. (2004). Group selection management in conifer forests: relationships between opening size and tree growth. *Canadian Journal of Forest Research*, **34**(3), 630–641.

York, R. and Battles, J. (2008). Growth response of mature trees versus seedlings to gaps associated with group selection management in the Sierra Nevada, California. *Western Journal of Applied Forestry*, **23**(2), 94–98.

Yorke, D.M.B. (1992). *The Management of Continuous Cover Conifer Forests: An Alternative to Clear Felling*. Continuous Cover Forestry Group, Melksham.

Youngblood, A. and Ferguson, D. (2003). Changes in needle morphology of shade-tolerant seedlings after partial overstory canopy removal. *Canadian Journal of Forest Research*, **33**(7), 1315–1322.

Yousefpour, R., Jacobsen, J.B., Thorsen, B.J., Meilby, H., Hanewinkel, M., and Oehler, K. (2012). A review of decision-making approaches to handle uncertainty and risk in adaptive forest management under climate change. *Annals of Forest Science*, **69**(1), 1–15.

Zeide, B. (2005). How to measure stand density. *Trees-Structure and Function*, **19**(1), 1–14.

Zenner, E.K. (2005). Development of tree size distributions in Douglas-fir forests under differing disturbance regimes. *Ecological Applications*, **15**(2), 701–714.

Zenner, E.K., Acker, S.A., and Emmingham, W.H. (1998). Growth reduction in harvest-age, coniferous forests with residual trees in the western central Cascade Range of Oregon. *Forest Ecology and Management*, **102**(1), 75–88.

Zenner, E.K., Lähde, E., and Laiho, O. (2011). Contrasting the temporal dynamics of stand structure in even- and uneven-sized *Picea abies* dominated stands. *Canadian Journal of Forest Research*, **41**(2), 289–299.

Zenner, E.K., Peck, J.E., Lähde, E., and Laiho, O. (2012). Decomposing small-scale structural complexity in even- and uneven-sized Norway spruce-dominated forests in southern Finland. *Forestry*, **85**(1), 41–49.

Zhou, M., Buongiorno, J., and Liang, J. (2012). Bootstrap simulation, Markov decision process models, and role of discounting in the valuation of ecological criteria in uneven-aged forest management. In T. Pukkala and K. von Gadow, eds. *Continuous Cover Forestry*, 2nd edn., pp. 243–271. Springer, Dordrecht.

Zingg, A. (1999). English and German terminologies in forestry research on growth and yield: a few examples. *Forest Snow and Landscape Research*, **74**(2), 179–187.

Zobel, B.J. and Talbert, J. (1984). *Applied Forest Tree Improvement*. John Wiley & Sons, Inc., New York.

Index

A
Advance regeneration 12, 87–93
  definition of 12
  guidelines 92–93
  physiology of 92
  seedling origin 89–91
  size, effects of 89–91
  sprouts 87–89
Advanced growth effect 136
Aesthetics 124, 169–171
Agroforestry 49–50
All-aged 41, 68–69
Alternative silvicultural approaches 1–4
  names of 2, 4
Animal damage 21
Area control forest regulation 52
  "balanced stand" and 65–67
Artificial regeneration 84, 93–95, 97, 155–156
  genetics and 152–154
  seedling types 94–95
Artificial selection 145
Ash deposition and tree damage 21
Assisted colonization 120
Assisted migration 120
Average annual yield (AAY) 139–140

B
"Balanced stand" 65–67
BDq method 72–73, 110
Biodiversity 1, 49, 82–83, 118–120, 123
Biogeocoenosis 126
Biological legacies 12
Boreal forests 8–9, 13, 35, 93, 104, 119, 155

C
Carbon allocation 158
Carbon storage 120–121
Chablis 40
Chemical treatments 95–97
Climate change 120, 166–168, 178
Climax 41–42, 126
Close-to-nature forestry 4, 173
Coarse filter approach 118
Coarse woody debris 1, 12, 120, 123, 170
Competing vegetation 95, 102
Competition control 95, 102
Components, see Stand components
Continuous cover forestry 4

Conversion 107–116
  definition of 107
Coupe 37
Creaming, see High grading
Crop tree release 118
Crown morphology 27–32
Cutting cycle 32, 35, 48, 65–66
  cutting severity and 65–66
  definition of 65
  density management zone and 65–66
  genetic implications of 151–154
Cyclic patterns in forestry 6–10, 116, 161–162, 174

D
Dauerwald 2
Decurrent growth form 29
Deforestation 169
Density management zone 64–67
  definition of 64
  leaf area index and 64
Developmental pathways 2, 41, 108, 110, 111, 134, 137, 168, 170
Diameter distributions 67–68, 112–114, 113, 128–130
  negative exponential 67–68
  q-factor and 68
  reverse-J 67–68
  rotated-sigmoid 68, 113
Diameter-limit cutting 69–70, 81, 112–113, 156
  lower diameter-limit 69–70
  upper diameter-limit 70
Differentiation 34–35, 114
  definition of 34
Diminution quotient, see q-factor
Dioecioius species 86, 148–149
Direct seeding 95
Disturbance 11–20, 158–166
  definition of 11
  genetics and 151–152
  stand structure and 159
Disturbance emulation 11–12, 21, 55–57, 151, 173, 176–177
Disturbance integration 167–168
Disturbance regime 12–16
  duration 14
  frequency 15
  intensity 15

  invisible effects of 12
  magnitude 14
  rotation period 15
  seasonality 15
  severity 14
  shape 15
  size 15
  synergism 16
  variation 15–16
Disturbance types 16–21
Diversity indices 63
DMZ, see Density management zone
Dysgenic selection 112–113, 145, 151, 156

E
Ecological forestry 4
Economic productivity 142–143
Ecosystem management 4, 146
Ecosystem services 124
Edge effects 43–44, 85–86
Elasticity, see Resilience
Epinastic control 29
Equilibrium curve 67, 70–72, 113, 123
"Equilibrium stand", see "Balanced stand"
Eugenic selection 151
Even-aged 2–3, 104, 107–115, 121, 140
  definition of 3
  economic production of 142–143
  relative production of 140
  social acceptability 174
  thinning methods 98, 100
  transformation of 107–115
Excurrent growth form 29

F
Femelschlag 54–55, 110
Fertilization 104–105
  leaf area index and 104–105
Fine filter approach 118
Fire 18–19, 96–97, 105–106, 147, 164–166
Freestyle silviculture 57
"Free" silviculture approaches 57, 79, 102
  Pioneer Forest and 80–81
  Stoddard-Neel and 82–83
Freeing silviculture 178
Fuel ladders 164
Future range of variability 146–147, 171

G

Gaps 37–47, 109
  definition of 37
  development 39–42
  extended gap 38
  latitude, effects 42–47
  light regimes 42–43
  partitioning 43–44
  replacement models 47
  size 38–39, 44–47
Gap light index 42–43
Gap phase 41
Gap replacement models 132
Genetic conservation 145–147
Genetic diversity 145–157
Genetic drift 145
Genetics of multiaged silviculture 151–157
"Green lie" 8–9
Green-tree retention 2
Gross primary production 137–138
Group selection 2, 37, 85
  definition of 48
  stocking control and 77
Groups 37–38
Growing space 11–12, 16–21, 22–23, 27–32, 39, 59–81, 93, 103–104, 110–114, 133, 139–143, 151
  definition of 11
  occupancy 23, 63–66, 71, 73–75, 104, 141–143
Growing space efficiency 61–62, 113, 141–142
Growth efficiency, see Growing space efficiency
Growth and yield equations 130
Growth and yield models 129–132
  distance dependent 131–132
  distance independent 131–132
Growth projection 128–132

H

Height:diameter ratios 100, 114–115, 159
Hemeroby 173
Herbivory 21, 35
High grading 69, 145, 161–162
  definition of 69
High shade 24–25
Historic range of variability 146–147, 151, 171, 177–178

I

Ice storms 20
Improvement cutting 99
Improvement planting 94
Improvement selection 79
Inbreeding 148
Individual tree models 131–132
Insects 19–20, 100, 160–163
Intensive management 98, 100, 101
Invasives 160–164
Irregular shelterwood 54–55, 109
Irregular structures 3–4, 13, 54–55, 159, 178

L

LAI, see Leaf area index
Landscape patterns 118–119
Leaf area allocation, see Multiaged stocking assessment model
Leaf area index 16, 27, 31, 75–78, 104, 137–139, 141–142
  definition of 31, 75
  development of 31–32, 136–139
Lianas 38
Light-limited systems 22–27
Light regimes 22–25, 42–43
Low shade 24–25

M

MASAM, see Multiaged stocking assessment model
Matrix models 131
Maturity selection 79
Maximum size-density 33–34
Mean annual increment (MAI) 139–140
Mechanistic models 132
Moisture-limited systems 27–29
Monoecious 86, 150
Mortality 14, 20, 26, 33–35
  climate change and 166
  disturbance-caused 14
  modeling 141–142
  stocking control and 59, 65
Multiaged
  definition of 1
  economic production of 142–143
  relative production of 140
  social acceptability 174
Multiaged stocking assessment model 75–77
Multicohort 2, 13

N

Natural forest management 147
Natural regeneration 84–93
  seed dispersal 85–86
  seedlings 89–91
Natural selection 145
Naturalness 107, 171–174, 179
  definition of 171
  "hands-off" approach and 106
Net primary production 137–138
New forestry 4
Niche 43–44, 118, 163
  growth niche 44
  regeneration niche 44
Niche partitioning 43–44
Nutrition, see Fertilization

O

Oak regeneration 24, 27–28, 80–81, 89
Old forest 42, 46, 52, 90, 119, 138, 161–162
Old growth, see Old forest
Operations research 132–133
Optimization tools 132–133

P

Patches
  definition of 37
Pathogens 20, 160–164
Periodic increment 61, 140
Photosynthetically active radiation (PAR) 22
Physiology of even-aged and multiaged stands 136–139, 141–142
Pioneer Forest 80–81
Planted forest 3, 93–95, 154–156
  definition of 3
Plenterwald 2
Plenter system 6, 60, 70–72, 103, 154, 156, 172
  naturalness of 172
Pollan, Michael 174
Pollen dispersal 147–150
Polycyclic 2
Preparatory treatments 109
Prescribed burning 96–97, 106, 164–166
Pro Silva 4
Productivity
  economic 142–143
  modeling relative 142
  relative productivity 136–143
  site quality 125–128
Protective functions 124
Pruning 101, 102–104, 117–118

Q

q-factor 72–73

R

Regeneration methods 2, 5–6, 12, 48–49, 59, 87, 145, 152–156, 171, 173
  genetics and 152–154
Rehabilitation 110–111
Reinforcement planting 94
Release 30, 32–33, 44, 55–56, 83, 90–93, 99, 102, 151
  definition of 32
  size, effects of 89–91
  shade tolerance 91
  vigor, effects of 90
Release treatments 102
  definition of 102
  vegetation control 102
Resilience 120, 158–159
Resistance 120, 158–159, 162
Respiration 138
Restoration 121–122
Reverse-J diameter distribution 67–68
Rotated sigmoid distribution, see Diameter distributions

S

Salvage 167
Savannah woodlands 27, 161–162

Scandinavia 8–9
Scenic beauty estimation 170–171
Schlägl Forest 113–114
SDI, *see* Stand density index
Seed bank 85, 112, 150
Seed dispersal 85–86, 147–150
Seed transfer guidelines 120
Seed wall 85
Selection methods 2–4, 44–47, 48–57
 group 2, 52–53
 single tree 2, 48–51
Selection ratio 152
"Selective" cutting 7–8, 154, 156
Selfing 148
Self-thinning 33–35, 41, 153
Shade tolerance 25–27
 classification 25
Shifting mosaic 41
Shelterwood 54–55, 109, 116, 154
 preparatory treatments 109
 transformation, relation to 109, 116
Silvicultural pathways, *see* Developmental pathways
Silvicultural systems 2, 48–49, 57
Silvigenesis 41
Single tree selection 2, 48–51
 definition of 48
Site index 125–128
 definition of 126
Site preparation 95–97
Site quality assessment 125–128
 complex stand approaches 127–128
 biogeocoenosis 126
 soil-site approaches 126
 vegetation approaches 126
Smith, David M. 10, 58, 177, 179
Snags 1, 12, 120, 123, 135
Social acceptability 174
Specific leaf area 30, 92
Sprouting 87–89
 functional equilibrium 88–89
 overstory density 89
 stump size 88

Stand components 1, 59, 61, 64, 74–81, 100, 102, 113, 133, 142
Stand density 61–80, 98–100, 117–118
 absolute 63–64
 relative 63–64, 73–75, 100
Stand density index (SDI) 64, 73–75, 76, 78
Stand health 100, 158–168
Stand structure 1–4, 9–10, 11–12, 13, 37, 50–57, 59–62, 63–64, 67–81, 107, 108–111, 118–119
 definition of 1
 diversity and 118–119
Stand table projection 131
Stocking control 59–83
 definition of 59
 first principles approach to 75
 group selection and 77
 multiple objectives 117–124
 regeneration and 62–63
 reverse-J and 77
 stand density index and 73–75, 76, 78
Stocking regulation, *see* Stocking control
Stoddard-Neel 57, 82–83
Stratified mixed-species stands 109–110, 161–162
Stump sprouting, *see* Sprouting
Sunflecks 24
Sustainability 5–9, 59, 62, 65–68, 70, 72–73, 80–83, 84, 111–112, 120–121, 147, 151, 154
Sustainable forest management 123–124
 definition of 123

T
Target stand structure 108
Temperate forests 2, 7–8, 12, 45, 47, 55–56, 80–83, 161–162
Thinning 98–102, 104, 153–154
 free 99, 102
 genetics and 152–154
 species composition 100

Timber production 117–118
Transformation 100, 107–115, 134
 definition of 107
 gap formation 109–111
 optimization of 133
 silviculture of 108–111
 thinning and 110–115
 tree longevity and 115
Tree architecture 29–32, 92
Tree seedling bank 12
Tree stability 100, 101, 114–115
Tropical forests 2, 32, 38–39, 43–45, 49–50, 93, 99, 119, 121, 127, 154
Two-aged 1, 2, 23, 49, 51, 53–61, 72, 85, 96, 99–100, 108–109

U
Uneven-aged 1–3
 definition of 1
Unit area control 55–56, 79

V
Variable-density thinning 55, 57, 100
Variable retention 2–3, 53–54, 109, 153–154
Vigor 50–51, 88–93, 138
Volume/guiding diameter limit (VGDL) 79
Volume production 117–118, 139–142

W
Water 124
Wildlife habitat 118–119, 124
Wind damage 16–18, 158–160
Windthrow, *see* Wind damage
Whole stand growth models 129–132
Wood quality 102–103
Woodlands 27–28

Y
Yield tables 129

Printed and bound by CPI Group (UK) Ltd, Croydon, CR0 4YY

Intermountain Research Station. Research Paper INT-RP-348.

Bergeron, Y., Cyr, D., Drever, C.R., Flannigan, M., Gauthier, S., Kneeshaw, D., Lauzon, E., Leduc, A., Le Goff, H., Lesieur, D., and Logan, K. (2006). Past, current, and future fire frequencies in Quebec's commercial forests: implications for the cumulative effects of harvesting and fire on age-class structure and natural disturbance-based management. *Canadian Journal of Forest Research*, **36**(11), 2737–2744.

Bergeron, Y. and Harvey, B. (1997). Basing silviculture on natural ecosystem dynamics: An approach applied to the southern boreal mixedwood forest of Quebec. *Forest Ecology and Management*, **92**(1–3), 235–242.

Bergeron, Y., Harvey, B., Leduc, A., and Gauthier, S. (1999). Forest management guidelines based on natural disturbance dynamics: stand- and forest-level considerations. *Forestry Chronicle*, **75**(1), 49–54.

Berrill, J.-P. and O'Hara, K.L. (2014). Estimating site productivity in irregular stand structures by indexing basal area or volume increment of the dominant species. *Canadian Journal of Forest Research*, **44**(1), 92–100.

Bhagwat, S.A., Willis, K.J., Birks, H.J.B., and Whittaker, R.J. (2008). Agroforestry: a refuge for tropical biodiversity? *Trends in Ecology & Evolution*, **23**(5), 261–267.

Bickford, C.A., Baker, F.S., and Wilson, F.G. (1957). Stocking, normality, and measurement of stand density. *Journal of Forestry*, **55**(2), 99–104.

Biging, G.S. and Dobbertin, M. (1992). A comparison of distance-dependent competition measures for height and basal area growth of individual conifer trees. *Forest Science*, **38**(3), 695–720.

Biging, G.S. and Dobbertin, M. (1995). Evaluation of competition indexes in individual tree growth models. *Forest Science*, **41**(2), 360–377.

Binkley, D. (1986). *Forest Nutrition Management*. John Wiley & Sons, New York.

Bliss, J.C. (1990). Missing the target. *Journal of Forestry*, **88**(11), 64.

Bliss, J.C. (2000). Public perceptions of clearcutting. *Journal of Forestry*, **98**(12), 4–9.

Blondel, J. (2006). The "design" of Mediterranean landscapes: a millennial story of humans and ecological systems during the historic period. *Human Ecology*, **34**(5), 713–729.

Bohrerova, Z., Bohrer, G., Cho, K.D., Bolch, M.A., and Linden, K.G. (2009). Determining the viability response of pine pollen to atmospheric conditions during long-distance dispersal. *Ecological Applications*, **19**(3), 656–667.

Bolte, A., Ammer, C., Lof, M., Madsen, P., Nabuurs, G., Schall, P., Spathelf, P., and Rock, J. (2009). Adaptive forest management in central Europe: climate change impacts, strategies and integrative concept. *Scandinavian Journal of Forest Research*, **24**(6), 473–482.

Bolte, A. and Degen, B. (2010). Forest adaptation to climate change—options and limitations. *Landbauforschung*, **60**(3), 111–117.

Bolton, N.W. and D'Amato, A.W. (2011). Regeneration responses to gap size and coarse woody debris within natural disturbance-based silvicultural systems in northeastern Minnesota, USA. *Forest Ecology and Management*, **262**(7), 1215–1222.

Bonan, G.B. (2008). Forests and climate change: forcings, feedbacks, and the climate benefits of forests. *Science*, **320**(5882), 1444–1449.

Bončina, A. (2011). History, current status and future prospects of uneven-aged forest management in the Dinaric region: an overview. *Forestry*, **84**(5), 467–478.

Bond, W.J. and Midgley, J.J. (2001). Ecology of sproutin in wood plants: the persistence niche. *Trends in Ecology and Evolution*, **16**(1), 45–51.

Bontemps, J.D. and Bouriaud, O. (2014). Predictive approaches to forest site productivity: recent trends, challenges, and future perspectives. *Forestry*, 87(1), 109–128.

Bormann, D.B. and Likens, G.E. (1979). *Pattern and Process in a Forested Ecosystem*. Springer-Verlag, New York.

Bossel, H. (1991). Modeling forest dynamics: moving from description to explanation. *Forest Ecology and Management*, **42**(1–2), 129–142.

Botkin, D.B. (1990). *Discordant Harmonies: A New Ecology for the Twenty-First Century*. Oxford University Press, New York.

Botkin, D.B., Wallis, J.R., and Janak, J.F. (1972). Some ecological consequences of a computer model of forest growth. *Journal of Ecology*, **60**(3), 849–872.

Bourne, R. (1951). A fallacy in the theory of growing stock. *Forestry*, **24**(1–2), 6–18.

Bowman, D.M.J.S. and Kirkpatrick, J.B. (1986). Establishment, suppression and growth of *Eucalyptus delagatensis* R.T. Baker in multiaged forests. I. The effect of fire on mortality and seedling establishment. *Australian Journal of Botany*, **34**(1), 63–72.

Bradley, G.A. and Kearney, A.R. (2007). Public and professional responses to the visual effects of timber harvesting: different ways of seeing. *Western Journal of Applied Forestry*, **22**(1), 42–54.

Bradshaw, F.J. (1992). Quantifying edge effect and patch size for multiple-use silvicuture—a discussion paper. *Forest Ecology and Management*, **48**(3–4), 249–264.

Bragg, D.C., Shelton, M.G., and Zeide, B. (2003). Impacts and management implications of ice storms on forests in the southern United States. *Forest Ecology and Management*, **186**(1–3), 99–123.

Brandeis, T.J., Newton, M., and Cole, E.C. (2001). Underplanted conifer seedling survival and growth in thinned